The
European Communities'
Health and Safety
Legislation

The European Communities' Health and Safety Legislation

VOLUME 1

The Social Dimension
1957–1991

EDITED BY

Alan C. Neal

Professor of Law and Director,
International Centre for Management,
Law and Industrial Relations,
University of Leicester

AND

Frank B. Wright

Lecturer in Law and Director,
European Occupational Health and Safety Law Unit,
University of Salford

Routledge
Taylor & Francis Group

LONDON AND NEW YORK

First published 1992 by Spon Press

2 Park Square, Milton Park, Abingdon, Oxon OX14 4RN
711 Third Avenue, New York, NY 10017, USA

Routledge is an imprint of the Taylor & Francis Group, an informa business

First issued in hardback 2017

Transferred to Digital Printing 2004

ISBN 978-0-412-46690-8 (pbk)
ISBN 978-1-138-43095-2 (hbk)

A catalogue record for this book is available from the British Library
Library of Congress Cataloging-in-Publication data available

CONTENTS

PART THREE
Directives

PREFACE

This book brings together, for the first time within one convenient volume, the wide range of legislation relating to occupational health and safety which has been developed at the level of the European Communities. It also sets that legislation in the economic and industrial relations context which has largely influenced that development throughout the past thirty years, and provides a survey of the gradual expansion of Community activities in this important field.

Inspiration to embark upon the task of producing this volume stems from experience derived from work done by the authors both within the United Kingdom and for the European Commission. Too often, important studies and surveys relating to health and safety at work have been embarked upon without recourse to basic materials and fundamental data. All too often, in consequence, serious shortcomings are to be found in the eventual results produced. In particular, as lawyers, the authors have been concerned at the apparent unfamiliarity with basic European legislative instruments which occasionally displays itself in policy-making documentation and elsewhere.

As one response to such problems, therefore, the authors decided, early in 1990, to embark upon a number of "mapping" initiatives, seeking to compile and organise basic legislative and other regulatory material from all of the Member States, as well as from the European Community itself. This activity has now resulted in the production of a complete collection of national legislation in the field of occupational health and safety from all of the Member States of the European Communities. At the same time, the associated Community materials have been compiled and organised. This volume contains the first fruits of that latter initiative.

During the course of undertaking the formidable task of legislative compilation and organisation, the authors have been assisted by a large number of experts throughout the Member States. To all of these we express our sincere thanks. In particular, however, we should like to single out Mrs. Sheila Pantry, Head of Information and Library Services in the British Health and Safety Executive, whose support and encouragement (often in the face of an absence of enthusiasm, or, even, hostility, towards the ambitions of the authors) has helped to overcome the many hurdles in the way of this ambitious project.

The material presented in this volume reflects the legislative position as at 1 July 1991.

Alan C. Neal
Frank B. Wright

The field of occupational health and safety has often been regarded as occupying something of a "Cinderella" position, involving a wide range of important regulation, inspection, and policing for the safeguarding of workers' health and hygiene at work, but rarely accorded the recognition which could be said to be its due. However, as the completion of the "1992" process towards the Single Market reaches its zenith, two particular factors have combined to catapult this area of activity to the forefront — albeit, some would argue, rather belatedly.

First, a dramatic programme of legislation at the level of the European Communities, following the introduction of Article 118a into the Treaty of Rome (by virtue of the Single European Act 1986), has confirmed how important this field can be in the broader context of Labour Law and industrial relations regulation. Second, the designation of 1992 as the European Year of Health and Safety at the Workplace, with its attendant high profile accorded to the work of DG/V/E of the Commission of the European Communities, promises to establish a basis from which more appropriate recognition can be accorded to matters of health, safety, and hygiene at work.

In this Preface, there is an initial presentation of the historical foundations for the policies developed by the European Communities in relation to health and safety at work. This indicates the limited approach adopted in the early years of the "Community of Six", before the impetus provided by the later periods of the Commission's social action programmes. The impact of the Single European Act is then considered, before an indication is given of the key role of occupational health and safety regulation within the context of the Community Charter of Fundamental Social Rights of Workers, together with the associated Commission Action Programme to implement the ambitions of that Social Charter. Finally, reference is made to the steadily increasing pace of implementation for measures in this field, in the context of the broadening "social dimension" of the European Communities.

The Context for Health and Safty Regulation at the European Level

The regulation of industrial health and safety has constituted a vital part of the work of the European Communities ever since the foundation of the European Coal and Steel Community forty years ago.

Various reasons have been put forward to explain the importance of this area, of which four may be considered to carry particular force. First, it is said that common health and safety standards assist economic integration, since products cannot circulate freely within the Community if prices for similar items differ in various Member States on account of variable health and safety costs. Second, it is argued that the reduction of the human, social and economic costs of accidents and ill-health borne by a workforce of some 138 million, together with their families, brings about an increase in the quality of life for the whole Community. Third, the introduction of more efficient work practices is said to bring with it increased productivity and better industrial relations. Finally, it is argued that the regulation of certain risks, such as those arising from massive explosions, should be harmonized at a supra-national level because of the scale of resource costs and (an echo of the first reason canvassed above) because any disparity in the substance and application of such provisions produces distortions of competition and affects product prices.

It has also to be admitted, however, that regulatory intervention in this area tends to receive a higher degree of political support in the aftermath of serious incidents, such as the major disaster which occurred on board the Piper Alpha drilling rig in July 1988. A timely harnessing of such political will can, therefore, it is argued, facilitate the passage not only of specific measures in the aftermath of the particular incident, but also serve to provide an impetus for more general aspects of the health and safety programme, including the promotion of social change — moves which might not normally enjoy the same degree of political consensus.

Health and Safety and the "Social Dimension"

From the outset, the European Economic Community has taken the view that, while its principal aim was one of economic expansion, this could not be seen as an end in itself. At the same time, an improvement in the quality of life, in the environment, and in the workplace, should be an important subsidiary aim. Specific provisions concerned with development of the social policy — or what has often come to be referred to as the "social dimension" — of the Community are contained in Articles 117 — 128 of the Treaty of Rome, and, in recent years, the promotion of this "social" alongside the "economic" dimension of the single market envisaged in the 1986 Single European Act has provided a focal point for major political controversies between the Member States of the gradually enlarging Community.

A particular impetus to a broad commitment to health and safety at work has resulted from the fact that, during the last four decades, new technology, new substances and processes, and new dimensions of scale in the industrial process have brought in their train new and increased risks. Indeed, it is noteworthy that one of the most recent Joint Opinions delivered by the Community's "social partners", exercising their functions within the "social dialogue" provisions of Article 118b of the revised Treaty of Rome, has been directly concerned with new technologies, work organization and adaptability of the labour market, including the health and safety aspects to which these give rise.

Moreover (and, arguably, most significantly), the continuing very high number of workplace accidents throughout Europe gives rise to tragic and costly consequences, most of which must, eventually, be met by the Community and all of its citizens.

Developing a Community-Level Policy for Occupational Health and Safety

In the context of such deep and intractable problems, the scope of the terms of Article 118 of the 1957 Treaty of Rome were perceived early on as being especially important. This included, in particular, a statement to the effect that "the Commission shall have the task of promoting close co-operation between Member States in the social field, particularly in matters relating to ... prevention of occupational accidents and diseases; ... occupational hygiene...".

Given the terms of Article 118, therefore, it was readily accepted that it should be the duty of the European Commission to initiate, promote, and develop a European preventive policy with regard to occupational risks. Eventually, it was decided that the

Commission should approach this first by gathering information, second by encouraging co-operation and co-ordination, and third by legislating at Community level.

At the outset, it was argued that such actions must be complementary to those taken by Member States, and that the purpose would be to *persuade* the responsible authorities in the Member States, together with both sides of industry, to join forces in combatting risks of all kinds. In recent years, however, the stance adopted by the Commission has appeared to be increasingly interventionist, and a significant body of European-level legislation has now been produced (and continues to emerge) in relation to occupational health and safety.

The gradual development of a Community-level policy in this field over the 35 years since the inception of the Common Market has taken place in a number of phases. During the initial period, as the Member States went through the first throes of common collaboration and regulation, some of the foundations were laid for an integrated policy to improve working conditions in terms of the hygiene and workplace safety of workers in the European labour force.

The "First Steps"

Thus, the early years between 1957 and 1974 can be seen as the "first steps". In this phase, the newly-formed Community focussed upon the area of occupational illness and disease, concentrating upon the listing of various risks, harmonizing the arrangements for compensating workers who might fall victims to such risks, and, later, laying down requirements that medical check-ups be provided for workers who were identified as being exposed to various occupational risks. Such an initially cautious approach of "mapping" the area, identifying the problems, and seeking to promote improvements in the welfare support available for victims, has been continued throughout the last three decades. Indeed, one of the most recent initiatives taken at the outset of the 1990s has been for the Commission to publish a Recommendation to the Member States concerning the adoption of a European schedule of occupational diseases.[1]

At first, therefore, the Communities confined themselves to the encouragement of research, the promotion of exchanges of experience, and the development of common legislative guidelines. However, it soon became evident that effective solutions to workplace health and safety problems were still proving elusive. Indeed, by 1974 (shortly following the first enlargement of the Communities from their original six founder members to nine Member States), the Communities were recording, each year, nearly 100,000 deaths and 12 million injuries arising from accidents of all types, within a workforce of some 104 million. Industrial accidents and occupational diseases clearly merited special attention.

The response came in the commencement of a series of "Action Programmes", each of which set out broad goals for developing further the Community's occupational health and safety protective framework, and each of which has then been followed by a variety of legislative initiatives — mostly in the guise of directives to be implemented in appropriate manner by each of the Member States within their own national systems of occupational health and safety protection.

---

1. Commission Recommendation of 22nd May 1990 to the Member States concerning the adoption of a European schedule of occupational diseases (90/326/EEC).

The 1974 Social Action Programme

A first, highly significant, move towards implementing a Community-wide health and safety policy of prevention was taken in 1974, with the first step in a new phase of "programmatization".

On 21st January 1974, the Council of the European Communities provided by Resolution for the adoption of a Social Action Programme.[2] The 1974 Resolution — which was not confined only to matters touching upon health and safety at the workplace — adopted the views of the Summit Conference of Heads of State or Government held in Paris in 1972, that economic expansion should result in an improvement in the quality of life as well as in the standard of living. Furthermore, it underlined the need for increased participation by both sides of industry in the economic, employment, and social policy decisions of the Community. The 1974 Resolution included a number of elements directly concerned with the issues of occupational health and safety, and, by this means, the Council of the European Communities gave recognition and support to what had already been a growing awareness throughout Europe of the crucial importance of health and safety at work.

Amongst the matters set out in the 1974 Social Action Programme was the general ambition "to establish an action programme for workers aimed at the humanization of their living and working conditions, with particular reference to: ... improvement in safety and health conditions at work; ... the gradual elimination of physical and psychological stress which exists in the place of work and on the job, especially through improving the environment and seeking ways of increasing job satisfaction ...". More specifically, the Council resolved upon "The establishment of an initial action programme, relating in particular to health and safety at work, the health of workers and improved organization of tasks, beginning in those economic sectors where working conditions appear to be the most difficult". Amongst the earliest steps mentioned in the 1974 Resolution to underpin the goals of the Social Action Programme were "the setting up of a European General Industrial Safety Committee and the extension of the competence of the Mines Safety and Health Commission".

It should also be noted that this move in the direction of a specialized Community-level policy on occupational health and safety was not something undertaken in isolation. In particular, many of the measures proposed in the 1974 Social Action Programme and subsequently were a response to policy initiatives in other, closely allied, fields. Thus, for example, there was close concern for the direction being taken by the Community in the area of industrial policy, as set out in the Council Resolution on that matter of 17th December 1973.[3] Furthermore, co-ordination was necessary with the goals set out in the Community's Action Programme on the Environment, which had been approved by the Council on 22nd November 1973.[4]

2. Council Resolution of 21st January 1974 concerning a social action programme of the European Communities (74/c 13/1).

3. Council Resolution of 17th December 1973 on industrial policy (OJ C 117, 31.12.73, p.1).

4. Action Programme on the Environment, approved by Council Resolution on 22nd November 1973 (OJ 1973 C 112/1).

The Advisory Committee

In addition to the arrangements put in place to bring about the implementation of specific measures, closer collaboration at the European level was to be sought — as had been envisaged in the Social Action Programme — through a newly constituted body: the Advisory Committee for Safety, Hygiene and Health Protection at Work. Established by Council Decision of 27th June 1974, that body has now had almost two decades of experience operating in the field.[5]

The Advisory Committee is chaired by the Commissioner responsible for the Directorate-General for Employment, Industrial Relations and Social Affairs (currently Madame Vasso Papandreou), and consists of 72 full members; there being for each Member State two government representatives, two trade union representatives, and two representatives of employers organizations. An alternate is appointed for each full member.

The role of the Advisory Committee, which has responsibility for all sectors of the economy except those falling within the ambit of the ECSC and EAEC Treaties, is to "assist the Commission in the preparation and implementation of activities in the fields of safety, hygiene and health protection at work". Because of its constitution and membership, the Advisory Committee is much more important and pro-active than its title suggests, so that, over the years, it has had a significant influence on strategic policy development, acting alongside the European Parliament and the Economic and Social Committee. More specifically, the Committee is responsible for the following matters within its general frame of reference:

(a) conducting, on the basis of information available to it, exchanges of views and experience regarding existing or planned regulations;

(b) contributing towards the development of a common approach to problems existing in the fields of safety, hygiene and health protection at work and towards the choice of Community priorities as well as measures necessary for implementing them;

(c) drawing the Commission's attention to areas in which there is an apparent need for the acquisition of new knowledge and for the implementation of appropriate educational and research projects;

(d) defining, within the framework of Community action programmes, and in co-operation with the Mines Safety and Health Commission:

   (i) the criteria and aims of the campaign against the risk of accidents at work and health hazards within the undertaking;

   (ii) methods enabling undertakings and their employees to evaluate and to improve the level of protection;

(e) contributing towards keeping national administrations, trade unions and employers' organizations informed of Community measures in order to facilitate their co-operation and to encourage initiatives promoted by them aiming at exchanges of experience and at laying down codes of practice;

---

5. Council Decision of 27th June 1974 on the setting up of an advisory committee on safety, hygiene and health protection at work (74/325/EEC).

(*f*)   submitting opinions on proposals for directives and on all measures proposed by the Commission which are of relevance to health and safety at work.

In addition to these functions, the Committee prepares an annual report, which the Commission then forwards to the Council, the Parliament, the Economic and Social Committee, and the Consultative Committee of the European Coal and Steel Community.

The Dublin Foundation

Mention should also be made of the fact that, on 26th May 1975, the Council established a specialized autonomous Community body with the aim of planning for and establishing better living and working conditions in Europe.[6] This body, the European Foundation for the Improvement of Living and Working Conditions, is located just outside Dublin, in the Republic of Ireland. Its administrative structure comprises a tripartite Administrative Board, a Director and Deputy Director, and a Committee of Experts. The Foundation is primarily engaged in applied research in areas of social policy, the application of new technologies, and the improvement and protection of the environment, in an effort to identify, cope with, and forestall problems in the working environment.

The role and activities of the Dublin Foundation have been reinforced by the anticipation of the completed "Internal Market", and its objective now is to provide the Community institutions and other policy makers with a flow of information which would be useful to the Community in its aim of ensuring that economic and social advances are promoted equally within the Single Market. The current programme is operationally divided into six linked areas: *ie.* social dialogue, the structuring of working life, health and safety, the environment, living conditions, and future technologies.

Legislative Initiatives 1974—1978

A variety of initiatives were undertaken in the period following the adoption of the 1974 Social Action Programme. Thus, during 1977 and 1978, there was significant activity on the legislative front, with two important directives being enacted in this period.

The political initiative for the first of these — a Council Directive on the approximation of the laws, regulations and administrative provisions of the Member States relating to the provision of safety signs at places of work[7] — pre-dates the Resolution of 21st January 1974 establishing the Social Action Programme. The safety signs Directive, which takes as its legal base Article 100, came about in the wake of persistent pressure from the European Parliament and the Economic and Social Committee. These bodies had been concerned that the number and types of signs were proliferating, and that, if this were allowed to continue, not only would there be

6.   The European Foundation for the Improvement of Living and Working Conditions is described as an autonomous Community body. It was established by a regulation of the EC Council of Ministers of 26th May 1975, following joint deliberations between the social partners, national governments and Community institutions on the ways and means of solving the ever-growing problems associated with improving living and working conditions.

7.   Council Directive of 25th July 1977 on the approximation of the laws, regulations and administrative provisions of the Member States relating to the provision of safety signs at places of work (77/576/EEC). The original Directive has since been updated by a further Commission Directive of 21st June 1979 amending the Annexes to Council Directive 77/576/EEC on the approximation of the laws, regulations and administrative provisions of the Member States relating to the provision of safety signs at places of work (79/640/EEC).

confusion, but, far from preventing accidents, this might turn out to be a contributory cause of them. Moreover, neither Member States nor International Standards Bodies had up to date and comprehensive legislation in this area.

The Directive which was eventually adopted was not applicable to signs used in the rail, road, inland waterway, marine or air transport sectors, or in relation to signs laid down for the placing of dangerous substances and preparations on the market. The operation of the Directive was also excluded in relation to coal mines. However, where it did apply, the Directive required all Member States to use four categories of safety signs without text in order to indicate certain hazards. The categories laid down were *(i)* prohibition signs, *(ii)* warning signs, *(iii)* mandatory signs and *(iv)* emergency signs, each of which categories has to have its own distinctive form and colour.

The second piece of Community legislation in this period — a Council Directive on the approximation of the laws, regulations and administrative provisions of the Member States on the protection of the health of workers exposed to vinyl chloride monomer[8] — also emerged as a result of political initiatives which pre-dated the Social Action Programme. It was adopted in June 1978 as a response to world-wide concern over the appearance of a rare liver cancer in workers in the plastics industry, the cause of which had been identified by a number of investigations and studies as lying in the substance vinyl chloride monomer.

This 1978 Directive on vinyl chloride monomer was the first piece of Community legislation in the field of health and safety at work to deal specifically with control of worker exposure to a chemical carcinogen. It not only marked an important step towards the eventual development by the European Communities of an overall policy relating to chemical agents, but can also be seen to have set a pattern for this kind of legislative initiative at the Community level, which has been continued steadily throughout the 20 years since its enactment.

The First Action Programme on Safety and Health at Work 1978

Although the 1974 Social Action Programme incorporated a number of elements touching upon the regulation of occupational health and safety within the European Community, these represented only one part of an instrument whose scope and ambition was much more generally concerned with developing the "social dimension" of the Community.

However, in 1978, a specific initiative was effected for this field, in the shape of a First Action Programme for the European Communities on safety and health at work.[9] This Action Programme had been drawn up in an attempt to deal directly with some of the occupational health and safety problems outlined above. The Action Programme was adopted by the Council of Ministers in a Resolution of 29th June 1978, in which the Council agreed that the following actions could be undertaken during the period up until the end of 1982:

8. Council Directive of 29th June 1978 on the approximation of the laws, regulations and administrative provisions of the Member States on the protection of the health of workers exposed to vinyl chloride monomer (78/610/EEC).

9. Council Resolution of 29th June 1978 on an action programme of the European Communities on Safety and Health at Work (78/c 165/1).

(i)   action to focus effort on measures leading to improvements in accident and disease aetiology;

(ii)   action to provide protection against dangerous substances;

(iii)   action serving towards the prevention of the dangers and harmful effects of machines; and

(iv)   action to bring about an improvement of human attitudes in the workplace.

Accident and Disease Aetiology

The first of these areas of activity — accident and disease aetiology connected with work research — involved two major elements. First there was to be established, in collaboration with the Statistical Office of the European Communities, a common statistical methodology in order to assess with sufficient accuracy the frequency, gravity and causes of accidents at work, as well as the mortality, sickness and absenteeism rates in the case of diseases connected with work. Second, there was to be promotion of exchanges of knowledge, establishment of the conditions for close co-operation between research institutes, and identification of the subjects for research to be worked on jointly.

Protection Against Dangerous Substances

The second area of activity — protection against dangerous substances — involved a number of general strands. In the first place, action was to be taken to standardize the terminology and concepts relating to exposure limits for toxic substances. There was then to be harmonization of the exposure limits for a certain number of substances, taking into account exposure limits already in existence. At the same time, steps were to be taken to develop a preventive and protective action for substances recognized as being carcinogenic; in particular, by fixing exposure limits, sampling requirements and measuring methods, and satisfactory conditions of hygiene at the workplace, and by specifying prohibitions where necessary. In addition, action would be taken to establish, for certain specific toxic substances such as asbestos, arsenic, cadmium, lead and chlorinated solvents, exposure limits, limit values for human biological indicators, sampling requirements and measuring methods, and satisfactory conditions of hygiene at the workplace.

These specific steps were to be underpinned by a range of more general initiatives. In particular, the ambition was to establish a common methodology for the assessment of the health risks connected with physical, chemical and biological agents present at the workplace — notably, by research into criteria of harmfulness and by determining the reference values from which to obtain exposure limits. It was also made a priority to establish information notices on the risks relating to, and handbooks on the handling of, a certain number of dangerous substances; *eg.* pesticides, herbicides, carcinogenic substances, asbestos, lead, mercury, cadmium and chlorinated solvents.

Prevention of the Dangers and Harmful Effects of Machines

As regards the prevention of the dangers and harmful effects of machines, the 1978 Action Programme proposed steps to establish limit levels for noise and vibrations at the workplace, and to determine practical ways and means of protecting workers and reducing sound levels at places of work. It was also intended to establish permissible sound levels for building site equipment and other machines. In addition, a joint study was to be undertaken into the application of the principles of accident prevention, and of accident prevention and ergonomics in the design, construction and utilization of plant and machinery, as well as to promote this application in certain pilot sectors, including agriculture. Finally, there was to be an analysis of the provisions and measures governing the monitoring of the effectiveness of safety and protection arrangements, while an exchange of experience in this field was also to be organized.

Improvement of Human Attitudes in the Workplace

Monitoring and inspection, along with measures aimed at an improvement of human attitudes at the workplace, formed the substance of the final area for action under the 1978 Action Programme. Thus, there was to be developed a common methodology for monitoring both pollutant concentrations and the measurement of environmental conditions at places of work. There was also to be a carrying out of comparative programmes and establishment of reference methods for the determination of the most important pollutants. Promotion of new monitoring and measuring methods for the assessment of individual exposure — in particular through the application of sensitive biological indicators — also formed a major plank in this area. Meanwhile, special attention was to be given to the monitoring of exposure in the case of women — especially of expectant mothers and adolescents. The Action Programme also undertook to initiate a joint study into the principles and methods of application of industrial medicine, with a view to promoting better protection of workers' health.

Amongst other aspects dealt with in this phase of the 1978 Action Programme, steps were to be taken to establish the principles and criteria applicable to special monitoring in relation to assistance or rescue teams in the event of an accident or disaster, maintenance and repair teams and the isolated worker. There was also to be an exchange of experience concerning the principles and methods of organization of inspection by public authorities in the fields of safety, hygiene at work and occupational medicine. Finally, outline schemes were to be drawn up at Community level for introducing and providing information on safety and hygiene matters at the workplace to particular categories of workers (*eg.* migrant workers, newly recruited workers, and workers who have changed jobs).

Parallel Community Developments in Relation to Health and Safety

It should also be mentioned here that, alongside the steps set out in the 1978 Action Programme, the European Community, in the course of its continuing task of removing technical barriers to trade, has, over the last two decades, achieved considerable improvements in terms of product health and safety. Notable examples of this are to be found in the field of product performance standards, which establish limits for the noise produced by certain new machines sold for use at work. Furthermore, similar impressive progress has also been achieved in relation to the labelling and packaging of dangerous substances. In this latter case, for example, Community-level regulations provide for common symbols, a common description of risks, and for the dissemination of safety advice for both existing and new substances.

Although this aspect of the Community's work is concerned essentially with the establishment of the Internal Market, it is, of course, necessary for the Commission to ensure that, if working conditions are affected when such Directives are made, these are either maintained or improved. Indeed, it is important to note that, notwithstanding admitted differences in the stages of development reached by various Member States, the introduction of initiatives right across the field of health and safety at work and in related fields under no circumstances entitles any form of "trading down" for national standards.

The First Framework Directive 1980

Probably the single most important measure taken under the first Action Programme was the enactment of a so-called "framework Directive", agreed in 1980, on the harmonization of measures for the protection of workers with respect to chemical, physical and biological agents at work.[10]

This framework Directive, 80/1107/EEC, had two principal objectives: (1) the elimination or limitation of exposure to chemical and physical agents in the workplace, and (2) the protection of workers who are likely to be exposed to such agents. All workers are covered, except for those employed in sea and air transport, or those who are exposed to radiation covered by the Treaty establishing the European Atomic Energy Community.

The Directive sought to achieve these objectives by controlling the exposure of the worker to physical, chemical and biological agents, through *(i)* limiting the use of these agents in the workplace, *(ii)* limiting the numbers of workers exposed or likely to be exposed by establishing limit values, *(iii)* measuring procedures for evaluating results, and *(iv)* by individual and collective protection measures. Health surveillance and monitoring must be provided. Furthermore, workers must be kept informed both of the potential risks connected with their exposure and of necessary protective measures which need to be taken. Emergency measures need to be put in place to provide for abnormal exposures, and the workers have to be kept informed of these.

Daughter Directives and the Establishment of Protective "Trigger Mechanisms"

Four individual or "daughter" Directives based on the 1980 framework Directive have also subsequently been adopted. These Directives, which concern the protection of workers from the risks related to exposure to metallic lead and its ionic compounds at work;[11] the protection of workers from the risks related to exposure to asbestos at work;[12] the protection of workers from the risks related to exposure to noise at work;[13] and the banning of certain specified agents and/or certain work activities,[14] regulate every area of commercial activity throughout the twelve Member States, with the exception of air and sea transport.

10. Council Directive of 27th November 1980 on the protection of workers from harmful exposure to chemical, physical and biological agents at work (80/1107/EEC). This has subsequently been updated by a Council Directive of 16th December 1988 amending Directive 80/1107/EEC on the protection of workers from the risks related to exposure to chemical, physical and biological agents at work (88/642/EEC).

11. Council Directive of 28th July 1982 on the protection of workers from the risks related to exposure to metallic lead and its ionic compounds at work (82/605/EEC).

12. Council Directive of 19th September 1983 on the protection of workers from the risks related to exposure to asbestos at work (83/477/EEC).

13. Council Directive of 12th May 1986 on the protection of workers from the risks related to exposure to noise at work (86/188/EEC).

14. Council Directive of 9th June 1988 on the protection of workers by the banning of certain specified agents and/or certain work activities (88/364/EEC).

In order to achieve their objectives the framework Directive and its "daughter" Directives use three types of provisions which are interlinked. First, there are general rules and protections which apply in all circumstances. Second, there are protective provisions which are to be followed if a special "trigger mechanism" is activated. Third, provisions are included which deal with maximum permitted exposure levels (limit values) and counter-measures which must be taken when these are exceeded. It is provided that the content of this Directive is to be steadily revised, taking into account, in particular, progress made in scientific knowledge and technology, and in the light of experience gained in applying the Directive.

Of these provisions, the most interesting is the so-called "trigger mechanism". Thus, for example, it is provided in the Asbestos Directive that where any activity is likely to involve a risk of exposure to dust arising from asbestos or materials containing asbestos, "this risk must be assessed in such a way as to determine the nature and degree of exposure". Where this assessment shows asbestos fibre concentration below the maximum limit values, but at or above another specified level, a whole range of special protective provisions come into play. The specified trigger level is 0.25 fibre per cm^3, calculated over an eight hour reference period, and/or a cumulative dose of fifteen fibre days per cm^3 over three months, also calculated or measured with reference to an eight hour period. In addition, it is provided that there are to be maximum limits. If these limits are exceeded, work in that area must cease until adequate measures have been taken for the protection of the workers concerned.

The Lead, Noise, and Asbestos Directives all provide for *an assessment* to be made in order to identify the workplaces affected by their provisions and to determine the conditions under which these provisions are to apply. An employer must be competent to carry out the assessment himself, or must arrange for suitable competent persons to carry it out on his behalf. A formal systematic assessment of health risks is important in that it provides the basis for decisions on the level of controls monitoring and health surveillance necessary to comply with the Directives. Concise but detailed records must be made and retained for reference by the employer, his employees, and governmental inspection services.

The Second Action Programme 1984 and Enactment of the Single European Act 1986

On 27th February 1984 the Council agreed upon a second programme of action, which was to be put into effect during the period up to the end of 1988.[15] This new programme, which was presented in seven sections, extended the work commenced under the first Action Programme, and placed particular emphasis upon (i) protection against dangerous agents, (ii) protection against accidents and dangerous occurrences, (iii) organizational aspects of health and safety monitoring, (iv) training and information, (v) statistics, (vi) research and (vii) co-operation.

The introduction of the 1984 Action Programme came about during a period of preparation for completion of the internal market. This process included the eventual 1986 amendments to the Treaty of Rome, and an associated package of measures designed to complete the Internal Market by 1992.

---

15. Council Resolution of 27th February 1984 on a second programme of action of the European Communities on Safety and Health at Work (84/c 67/02).

As a first stage in this process, on 1st July 1987, the Single European Act came into effect, making a number of changes to the original Rome Treaty of 1957. These changes included matters relating to occupational health and safety: in particular, the fact that Article 118 of the Treaty of Rome was augmented by Article 21 of the Single European Act to include new health and safety provisions. The objectives of these new provisions were to be achieved by means of directives providing for minimum requirements to be achieved by gradual implementation. Most notably, a new Article 118a provided that "Member States shall pay particular attention to encouraging improvements, especially in the working environment, as regards the health and safety of workers, and shall set as their objective the harmonization of conditions in this area, while maintaining the improvements made". The new Article went on to provide that "In order to help achieve the objective laid down in the first paragraph, the Council, acting by a qualified majority on a proposal from the Commission, in co-operation with the European Parliament and after consulting the Economic and Social Committee, shall adopt, by means of directives, minimum requirements for gradual implementation, having regard to the conditions and technical rules obtaining in each of the Member States".

Under this new legal base, therefore, the Member States are instructed to pay particular attention to encouraging improvements in worker health and safety. The objective is to be the harmonization of conditions, whilst at the same time maintaining existing standards. A matter of particular importance in respect of the new Article 118a is the fact that legislation concerning health and safety can now be agreed in the Council by a process of "qualified majority", allowing the larger Member States greater voting weight than their smaller partners, and, it is said, enabling the Community to move more swiftly in placing health and safety measures on the European statute book. There is, however, one particular constraining factor which should be mentioned; namely that Article 118a insists that Directives "shall avoid imposing administrative, financial and legal constraints in a way which would hold back the creation and development of small and medium-sized undertakings".

The Period Following the Single European Act 1986

The "Programmatization" phase has seen two clearly divided elements, as indicated above. First, came the period 1974 — 1988, in which the broad approach principles, contained in the 1974 Social Action Programme, were established in a more detailed and concrete manner through the first and second Health and Safety Action Programmes of 1978 and 1984.

However, following the enactment of the Single European Act, the Commission took an early decision to act more forcefully on the basis of the newly inserted Article 118a, and (to a lesser extent) to promote a more wide-ranging "social dialogue" under the terms of Article 118b. Even before the 1984 Action Programme had been completed, therefore, a new Programme on Safety, Hygiene and Health at Work was issued by the Commission in 1988, making it clear that the new legal base provided for in the revised Treaty was to be taken as contributing a significant boost for action in these fields.[16]

16. Council Resolution of 21st December 1987 on safety, hygiene and health at work (88/c 28/03). This had been preceded by a Commission Communication on its programme concerning safety, hygiene and health at work (88/c 28/02).

The 1988 Programme also served as a foundation for the most recent phase of activity, which derives much of its inspiration from the 1989 Community Charter of Fundamental Social Rights of Workers,[17] and from Part X of the Commission's associated Action Programme of 29th November 1989.[18]

The Social Charter 1989

Much debate and controversy surrounds this instrument, which emerged after a protracted period of political wrangling between the President of the European Commission and, in particular, the British Prime Minister at the time.

In December 1987, the Commission agreed its broad plan of social policy initiatives which it intended to propose as part of the implementation package based upon the new Article 118a of the Treaty of Rome. There had been increasing concern that the stimulation of business should be accompanied by what was commonly being described as a "social dimension". It was argued that, if ordinary people are to have a sense of participating in 1992, the gains in efficiency and competition should be balanced by tangible benefits which can be felt by citizens throughout the Community. In the light of these pressures, therefore, the Commission decided to pursue the social dimension through a Community Charter of Fundamental Social Rights (the "Social Charter").

A draft for such a Charter was discussed at the European Council at Madrid on 26th—27th June 1989, following which a communiqué was issued confirming that:

(i) The European Council considered that in the course of the construction of the single European market social aspects should be given the same importance as economic aspects and should accordingly be developed in a balanced fashion;

(ii) The European Council re-affirmed its earlier conclusions on the achievement of the Internal Market as the most efficient method of creating jobs and ensuring maximum well-being for all Community citizens. Job development and creation would be given top priority in the achievement of the Internal Market; and

(iii) The Council agreed to continue its discussions with a view to adopting the measures necessary to achieve the social dimension of the Single Market, taking account of fundamental social rights. For this purpose the role to be played by Community standards, national legislation and contractual relations had to be clearly established.

The Madrid communiqué also welcomed progress in the fields of health and safety legislation, vocational training, and the social dialogue between employers and employees.

The eventual text of the Community Social Charter, which was adopted by eleven of the twelve Member States at a meeting of Heads of State or Government in Strasbourg on 9th December 1989, begins with a Preamble explaining the principles upon which it is based. The reasons for the Charter are given as the Treaty's commitment to improved living and working conditions, and the European Council's call for a balanced development of social and economic aspects of the Single Market. Its aim is to "declare solemnly"

17. Community Charter of Fundamental Social Rights (COM(89) 471 final).

18. Communication from the Commission concerning its Action Programme relating to the implementation of the Community Charter of Basic Social Rights for Workers (COM(89) 568 final).

that implementation of the Single Market must take full account of the social dimension and ensure the development of the social rights of workers. The Preamble then goes on to summarize the social objectives of the Community; job creation, social consensus, improvements in living and working conditions, greater freedom of movement, social protection, more education and training, equal opportunities, and respect for the rights of third country workers. It also touches upon implementation of the Charter. Having listed the relevant provisions of the EEC Treaty, it says that there must be a clear division of responsibility between the Community, national legislation, and collective agreements, which must be based on the principle of subsidiarity.

The refusal of the British Government to sign the Community Social Charter has served to infuse the policies for implementation of many of the initiatives contained in that instrument with some political ill-feeling and bitterness. Nevertheless, it has been clear that, despite the non-binding nature of the "solomn declaration" of eleven Member States, most of the matters contained in the Social Charter have, in fact, received support from all twelve of the Community Member States. In relation to those areas where disagreements are discernible, however, a number of seemingly intractable problems remain to be addressed — not least amongst which is the issue of what practical limits the "principle of subsidiarity" might place upon the regulatory activities of the Community's legislative organs.

Scope of the Community Social Charter

The Social Charter, as agreed in December 1989, covers twelve categories of "fundamental social rights":

(i)   *Freedom of movement*

The right to freedom of movement would be restricted only on grounds of public order, public safety or public health. It would give workers from one Member State, working in another Member State, the freedom to work in any occupation or profession on equal terms and with the same working conditions as nationals of the host country.

(ii)   *Employment and remuneration*

In each Member State, workers would receive fair remuneration, sufficient for a "decent" standard of living. Protection would be given to part-time workers. There would be limits on the withholding of wages, and workers would have free access to job placement services.

(iii)   *Improvement of living and working conditions*

There should be an improvement in working conditions, particularly in terms of limits on working time such as minimum annual paid leave and a weekly break from work. Particular mention is made of the need for improved conditions for those not on open-ended contracts (for example part-time or seasonal workers).

(iv)   *Social protection*

Workers, including the unemployed, should receive adequate social protection and social security benefits.

(v)   *Freedom of association and collective bargaining*

This involves the right to organize trade unions and to choose whether or not to join them, to conclude collective agreements, and to take collective action. Such action includes the right to strike, except where existing legislation stipulates exceptions (including the armed forces, the police and government service).

(vi)   Vocational training

It is noteworthy that this includes most university courses and some secondary education). Workers should be able to train, and re-train, throughout their working lives.

(vii)   Equal treatment for men and women

Action should be intensified to remove discrimination and provide support to prevent a clash between responsibilities in the home and at work.

(viii)   Information, consultation and participation for workers

This should apply especially in multinational companies and in particular at times of restructuring, redundancies or the introduction of new technology. Worker participation would be developed "in such a way as to take account" of existing rules and traditions.

(ix)   Health protection and safety at the workplace

(x)   Protection of children and adolescents

The minimum employment age should be no lower than the minimum school-leaving age, and in any case not lower than 15 years. The hours which those aged under 18 can work should be limited, and they should not generally work at night. Once compulsory education is over, young people must be entitled to receive adequate vocational training. If they are already working, this should take place during working hours.

(xi)   Elderly persons

Workers should be assured of resources providing a decent standard of living in retirement. Others should have sufficient resources and appropriate medical and social assistance.

(xii)   Disabled people

All disabled people should have additional help towards social and professional integration.

The final provisions cover implementation of the Charter. Member States are given responsibility in accordance with national practices for guaranteeing the rights in the Charter and implementing the necessary measures. The Commission is asked to submit proposals on areas within its competence.

Since that date there has been general acceptance of regulation in fields such as health and safety at work and it is clear that within the Community as a whole there is a groundswell of support for the Social Charter. Undoubtedly, Member States are anxious to show that workers, children and old people should benefit from the Community as well as shareholders and managers.

Post-1988 Policies and the 1989 Framework Directive

Central to the Commission's post-1988 approach to regulating health and safety at work has been the enactment of the so-called "framework Directive" 89/391/EEC on the introduction of measures to encourage improvements in the safety and health of workers at work.[19] This makes a significant step forward from the approach witnessed in the earlier "framework Directive" 80/1107/EEC on the protection of workers from harmful exposure to chemical, physical and biological agents at work. In particular, the 1989 Directive, while endorsing and adopting the approach of "self-assessment" already manifested in earlier instruments, also sought to establish a variety of *basic duties*, especially for the employer.

Furthermore, the promotion of "social dialogue" in the field of health and safety at work was explicitly incorporated into detailed provisions in the 1989 Directive, introducing significant requirements for information, consultation and participation for workers and their representatives at the workplace. The social dialogue in this area has since been underpinned by a recent Joint Opinion issued by the social partners, following the boost given by a meeting at the Palais d'Egmont on 12th January 1989.[20]

This 1989 framework Directive, which is addressed to Member States, requires compliance by 31st December 1992, and was the first Directive to be proposed as part of the implementation of Article 118a of the EEC Treaty. The overall object of the framework Directive is stated as being, "to introduce measures to encourage improvements in the safety and health of workers at work", and in its Preamble declares that:

(i)   Member States have a responsibility to encourage improvements in the safety and health of workers on their territory and that

(ii)   "taking measures to protect the health and safety of workers at work also helps, in certain cases, to preserve the health and possibly the safety of persons residing with them".

It is further noted "that the incidence of accidents at work and occupational diseases is still too high" and that "preventive measures must be introduced or imported without delay in order to safeguard the safety and health of workers and ensure a higher degree of protection".

The Directive contains re-stated general principles concerning, in particular, the prevention of occupational risks, the protection of safety and health and the informing, consultation and training of workers and their representatives, as well as principles concerning the implementation of such measures. This measure constitutes a first attempt to provide an overall complement to the technical harmonization directives

---

19. Council Directive of 12th June 1989 on the introduction of measures to encourage improvements in the safety and health of workers at work (89/391/EEC). For consideration of the impact of the 1989 framework Directive, see *M.Biagi*, "From Conflict to Participation in Safety: Industrial Relations and the Working Environment in Europe 1992", (1990) 6 *International Journal of Comparative Labour Law and Industrial Relations* 67; *A.Neal*, "The European Framework Directive on the Health and Safety of Workers: Challenges for the United Kingdom?",(1990) 6 *International Journal of Comparative Labour Law and* Industrial Relations 80; *M.Weiss*, "The Industrial Relations of Occupational Health: The Impact of the Framework Directive on the Federal Republic of Germany", (1990) 6 *International Journal of Comparative Labour Law and Industrial Relations* 119; *M-F.Mialon*, "Safety at Work in French Firms and the Effect of the European Directives of 1989", (1990) 6 *International Journal of Comparative Labour Law and Industrial Relations* 129; *L.Montuschi*, "Health and Safety Provision in Italy: The Impact of the EEC Framework Directive", (1990) 6 *International Journal of Comparative Labour Law and Industrial Relations* 146; and *E. Gonzales-Posada Martinez*, "Health and Safety at Work in the European Economic Community and the Impact on Spain", (1990) 6 *International Journal of Comparative Labour Law and Industrial Relations* 159. The practice of participation in relation to occupational health and safety is considered in *H.Krieger*, "Participation of Employees' Representatives in the Protection of the Health and Safety of Workers in Europe", (1990) 6 *International Journal of Comparative Labour Law and Industrial Relations* 217.

20. Joint Opinion of 10th January 1991 on new technologies, work organization and adaptability of the labour market.

designed to complete the internal market. The 1989 Directive also brings within its scope the provisions of the framework Directive on risks arising from use at work of chemical, physical and biological agents.

A recent comparative analysis of existing occupational health law within the Member States indicates that all will need to amend their legislation to some extent in order properly to implement the 1989 Directive. The extent of such amendments at national level varies dramatically between Member States, although even the most developed national systems of occupational health and safety regulation will face some need for change, at the very least in order to accommodate certain administrative and organizational aspects.

Scope of the 1989 Framework Directive

The scope of the 1989 Directive's provisions is widely drawn. Thus, *an employer* is defined as "any natural or legal person who has an employment relationship with the worker and has responsibility for the undertaking and/or establishment", while the term *worker* means "any person employed by an employer, including trainees and apprentices but excluding domestic servants". Some obligations are also owed by employers to "workers from any outside undertakings and/or establishments engaged in work in his undertaking and/or establishment". This includes the duty to provide information and appropriate instruction. In situations where several undertakings share a workplace there are specific provisions requiring co-operation between these in relation to measures of protection, prevention and information for their respective workforces.

The overall objectives indicated in Article 1 of the framework Directive may be summarized as being:

(i) Humanization of the working environment;

(ii) Accident prevention and health protection at the workplace;

(iii) To encourage information, dialogue and balanced participation on safety and health by means of procedures and instruments;

(iv) To promote throughout the Community, the harmonious development of economic activities, a continuous and balanced expansion and an accelerated rise in the standard of living;

(v) To encourage the increasing participation of management and labour in decisions and initiatives;

(vi) To establish the same level of health protection for workers in all undertakings, including small and medium-sized enterprises and to fulfil the single market requirements of the Single European Act 1986; and

(vii) The gradual replacement of national legislation by Community legislation.

As a general approach, the Directive (which applies to all sectors of activity, both in the public and the private sectors — although Member States are permitted to exclude certain public services, such as the Police and the Army) lays down minimum standards and requirements for health and safety matters. Although it is not concerned with the protection of the public from work activities, non-employees may, subject to certain provisions, be protected by virtue of the Directive.

General Duties Placed upon the Employer

An approach adopted by the 1989 framework Directive which has given rise to much comment has been the establishment across the European Community of a number of "general duties". This particular method of introducing occupational health and safety protection draws upon successful experience in some of the more developed national systems for regulation in this field.

Various kinds of duty for employers emerge from the framework Directive, which may be enunciated as follows:[21]

1.   *Duties of Awareness.*

These are threefold, consisting, first, in a broad duty which follows from the Preamble to the Directive — for the employer to keep himself informed of the latest advances in technology and scientific findings concerning workplace design; second, in a duty to identify and to evaluate risks to the safety and health of workers in the undertaking or establishment; and third, a duty — closely linked to duties of instruction below — to be aware of the capabilities of workers at the undertaking or establishments as regards health and safety.

2.   *Duties to Take Direct Action to Ensure Safety and Health.*

These firm duties constitute the "heart" of the Directive. They comprise *(a)* a duty to eliminate avoidable risks; and *(b)* a duty to reduce the dangers posed by unavoidable risks.

3.   *Duties of Strategic Planning to Avoid Risks to Safety and Health.*

These two duties reflect the need for both a general overview of safety and health needs and the implementation of a specific programme at the undertaking or establishment. The former is covered by the duty in Article 6(2)(g) to develop a "coherent overall prevention policy". The latter is detailed in Article 7 duties to develop and implement a system for the protection and prevention of occupational risks.

4.   *Duties to Train and Direct the Workforce.*

The general duties of training for the workforce in matters of safety and health are set out in Article 12. In addition there is a broad requirement in article 6(2)(i) that employers, as part of their normal activity in managing and directing the work, should provide appropriate instructions to their workers, while Article 6(3)(d) makes it clear that employers must take appropriate steps to ensure that "only workers who have received adequate instructions" are to have access to areas where there is "serious and specific danger".

5.   *Duties to Inform, Consult and Involve the Workforce.*

These duties are extremely widely cast by the Directive. The details of the information which is to be given by employers to workers and/or their representatives are set out in Article 10, while broad duties to provide for consultation and participation of workers are set out in Article 6(3)(c) and, most importantly, in Article 11.

---

21. For further detail concerning the structure and approach of the framework Directive, see *Alan C. Neal*, "The European Framework Directive on the Health and Safety of Workers: Challenges for the United Kingdom?", (1990) 6 *International Journal of Comparative Labour Law and Industrial Relations* 80, especially at pp.82-85.

6.   *Recording and Notification Duties.*

A variety of obligations on the part of employers and undertakings are set out in Article 9. These include duties to list certain accidents and occupational illnesses at the undertaking or establishment, and the preparation and provision of reports, for the benefit of relevant national authorities, on occupational accidents and occupational illnesses suffered by the workforce.

Small and Medium-sized Undertakings

In recent years, the Community authorities have begun to recognize the burdens which may be placed upon small and medium-sized undertakings (SMEs) in living up to ever-rising standards of employment protection — not least in relation to health and safety at work. As a direct response to concern that the employment creation benefits of SMEs within the European Community should not be stifled, the special status of SMEs is now safeguarded in the revised Treaty of Rome itself. Thus, Article 118a provides, in sub-paragraph (2), that the directives adopted with the objective of harmonizing conditions in the working environment as regards the health and safety of workers "shall avoid imposing administrative, financial and legal constraints in a way which would hold back the creation and development of small and medium-sized undertakings".

The 1989 framework Directive provides similar safeguards. It is stated that the size of the undertaking and/or establishment is a relevant matter in relation to determining the sufficiency of resources for dealing with the organization of protective and preventive measures, and in the opportunity for Member States to establish whether an employer is to be considered competent to discharge various Article 7 obligations himself. It is also a factor to be considered in relation to obligations concerning first aid, fire fighting, and evacuation of workers. Furthermore, Article 9 includes power for differential requirements to be imposed upon varying sizes of undertakings as regards documentation to be provided. Finally, in relation to the provision of information, Article 10(1) states that national measures" may take account, *inter alia*, of the size of the undertaking and/or establishment".

Information, Consultation and Participation

The development of rights to information, consultation and participation for workers and their representatives has been a long-standing ambition of the European commission. In recent years, the preference for formal institutionalized channels of information, consultation and participation has given way to a more flexible approach, with a much greater sensitivity being shown towards the many and varied experiences at Member State level in matters of collective bargaining, worker representation, and trade union structures.

In relation to health and safety at work, the recognition of how important a contribution can be made to promoting improvement in this area has been evident throughout the Community's periods of programmatization described earlier. Thus, for example, "the wishes expressed by management and labour" are specifically acknowledged in the Preamble to the Council's 1974 Resolution concerning a social action programme,[22] while the Advisory Committee on Safety, Hygiene and Health Protection

---

22. Council Resolution of 21st January 1974 concerning a social action programme of the European Communities (74/c 13/1).

at Work, established in 1974, is self-evidently built upon notions of tripartism, with "for each Member State two representatives of the Government, two representatives of trade unions, and two representatives of employers' organizations".[23] The Preamble to the 1978 Action Programme stressed that "it is essential to encourage the increasing participation of management and labour in the decisions and initiatives in the field of safety, hygiene and health protection at work at all levels, particularly at the level of the undertaking",[24] while the Preamble to the 1984 Action Programme reiterated that same need for encouragement of participation of management and labour.[25]

By the time of the passing of the Single European Act, the promotion of information, consultation and participation rights for workers and their representatives was being seen as a matter standing alongside the new impetus given by the "social dialogue" provision included in Article 118b of the Treaty of Rome. Thus, the Commission was undertaking, in its 1988 Action Programme,[26] to "ensure adequate involvement of the trade unions in European standardization work and related activities", while the "social dialogue" itself had become an accepted vehicle for the promotion of safety, hygiene and health at work. In the words of the Commission's Communication setting out the 1988 Action Programme,

> "In its joint opinion on information and consultation, the Val Duchesse Working Party stated 'When technological changes which imply major consequences for the workforce are introduced into the firm, workers and/or their representatives should be informed and consulted in accordance with the laws, agreements and practices in force in the Community countries'. The Commission considers this objective to be particularly important where such practices have a potential impact on health and safety".

Indeed, the 1988 Action Programme included a special section (Part "F") under the heading of "Social dialogue", in which the Advisory Committee on Safety, Hygiene and Health Protection at Work was identified as "a highly appropriate forum for consultation between the two sides of industry" in the context of the Commission's stated intention "to develop the dialogue between the two sides of industry in this field pursuant to Article 118b of the Single Act".

Consequently, the inclusion within the Community Social Charter of specific provisions relating to information, consultation and participation for workers and their representatives marked a continuation of a process which had already been evident for a decade and a half.[27]

The most evident expression of this promotion for information, consultation and participation in decisions and initiatives within the field of health and safety at work is to

---

23. Council Decision of 27th June 1974 on the setting up of an advisory committee on Safety Hygiene and Health Protection at Work (74/325/EEC).

24. Council Resolution of 29th June 1978 on an action programme of the European Communities on Safety and Health at Work (78/c 165/1).

25. Council Resolution of 27th February 1984 on a second programme of action of the European Communities on Safety and Health at Work (84/c 67/02).

26. Commission Communication on its programme concerning safety, hygiene and health at work (88/c 28/02).

27. Community Charter of Fundamental Social Rights (COM(89) 471 final). The proposed means for implementing the Social Charter's provisions are set out in the Communication from the Commission concerning its Action Programme relating to the implementation of the Community Charter of Basic Social Rights for Workers (COM(89) 568 final).

be found in the provisions of Articles 10 (concerning the provision of "information") and 11 (which deals with "consultation and participation") of the 1989 framework Directive. These radical provisions met with a positive reaction from all twelve Member States when the framework Directive was eventually adopted by the Council, and it is noteworthy that the full import of these measures has been incorporated into each of the so-called "daughter directives" introduced under powers set out in Article 16 of the 1989 Directive.

"Daughter Directives" Under the 1989 Framework

Under the umbrella of the 1989 framework Directive, a number of so-called "daughter directives" have also been adopted, all of which incorporate the general provisions of the framework, and all of which refer, specifically, to the dramatically extended "social dialogue" provisions. In particular, "daughter Directives" have been adopted on the minimum safety and health requirements for the workplace,[28] on the minimum safety and health requirements for the use of work equipment by workers at work,[29] on the minimum health and safety requirements for the use by workers of personal protective equipment at the workplace,[30] on the minimum health and safety requirements for the manual handling of loads where there is a risk particularly of back injury to workers,[31] and on the minimum safety and health requirements for work with visual screen equipment.[32]

Extending the Social Dimension *via* Article 118a

Alongside the developments under the umbrella of the 1989 framework Directive, however, there has also been a much more controversial approach discernible on the part of the Commission. This is evidenced by a variety of initiatives, all taking their inspiration from the Social Charter and from the Commission's Programme of Action of November 1989, and many of them involving what some would claim to be highly creative use of the Treaty base provided for in Article 118a.

In order to appreciate some of the arguments which have been sparked by these developments in the field of health and safety at work, it is necessary briefly to sketch the nature of the treaty powers in question.

The extent of powers exercised by the institutions of the European Community (in particular, the Commission, the Parliament, and the Council) is limited by what is provided for under the terms of the Treaty of Rome. Thus, a fairly strict adherence to the doctrine of "express grant of powers" is maintained, in line with continental legal theory. Consequently, it is considered necessary to find the relevant power to act in the Treaty, and, commonly, instruments deriving from the Community's organs make explicit

28. Council Directive of 30th November 1989 concerning the minimum safety and health requirements for the work place (89/654/EEC).

29. Council Directive of 30th November 1989 concerning the minimum safety and health requirements for the use of work equipment by workers at work (89/655/EEC).

30. Council Directive of 30th November 1989 on the minimum health and safety requirements for the use by workers of personal protective equipment at the workplace (89/656/EEC).

31. Council Directive of 29th May 1990 on the minimum health and safety requirements for the manual handling of loads where there is a risk particularly of back injury to workers (90/269/EEC).

32. Council Directive of 29th May 1990 on the minimum safety and health requirements for work with visual screen equipment (90/270/EEC).

reference to the Treaty basis upon which the power to act is claimed to rest.

At the initiation of the EEC, in the provisions set out in the original Treaty of Rome in 1957, the power granted to the Commission and the Council to act was clearly intended to be Article 100, which provided that "The Council shall, acting unanimously on a proposal from the Commission, issue directives for the approximation of such provisions laid down by law, regulation or administrative action in Member States as directly affect the establishment or functioning of the common market".

However, from a very early stage, what had commonly been considered as a minor "residual" source of power, was utilized successfully on an increasing scale; namely, the power given in Article 235 of the Treaty. This laid down that "If action by the Community should prove necessary to attain, in the course of the operation of the common market, one of the objectives of the Community, and this Treaty has not provided the necessary powers, the Council shall, acting unanimously on a proposal from the Commission and after consulting the European Parliament, take the appropriate measures". Whatever criticisms may have been levelled at the Community's organs for using Article 235 as a basis for acting were not, however, likely to provoke much more than a theoretical legal storm, since the political will to act was always required under both Article 100 and Article 235 by virtue of the need for the Council to act *unanimously* on any such matter.

This general picture changed with the passing of the Single European Act in 1986, since a new provision was inserted into the Treaty of Rome whereby the Council could, in certain circumstances, and after following a specially laid down "co-operation procedure", act by what was described as "qualified majority". the new treaty base permitting this use of "qualified majority" decision-making was that established by a new Article 100a. However, in the face of strong opposition from the then British Prime Minister, a degree of restriction was included in the Article 100a provision, by virtue of Article 100a (2), which expressly set out the restriction that the new powers "shall not apply to fiscal provisions, to those relating to the free movement of persons not to those relating to the rights and interests of employed persons".

At the same time, a number of the new Articles inserted into the Treaty of Rome by the Single European Act expressly provided for "qualified majority" decision-making in the legislative process, so that such procedures automatically constitute the appropriate method for legislation. Amongst the specific provisions which expressly give rise to such "qualified majority" decision-making procedures is that upon which the Commission has chosen to rest its post-1988 policies on health and safety at work; *ie.* Article 118a.

Consequently, so long as a health and safety matter really is a health and safety matter falling under Article 118a, and does not affect "the rights and interests of employed persons" as envisaged under Article 100a(2), the procedure for reaching agreement is by way of "qualified majority" voting in the Council. This means, therefore, that the adverse vote of one Member State (or, indeed, of just two Member States) cannot be enough to prevent such a measure gaining majority support and thus being enacted as Community legislation. Much of the controversy stirred by the recent use by the Commission of "qualified majority" procedure treaty bases for various initiatives in the field of health and safety at work has centred upon whether an initiative ostensibly falling under Article 118a in fact involves "the rights and interests of employed persons" as provided for in Article 100a(2) of the revised Treaty of Rome.

The arguments about this broadened approach of the Commission have arisen in particular in relation to a proposed Directive on health and safety for temporary workers,[33] another on the organization of working time,[34] and a further proposed Directive on health and safety for pregnant workers and women in the period following birth of a child.[35] Each of these proposals has taken as its treaty base of power to act the new Article 118a, with consequent "qualified majority" procedures in the same manner as provided for under Article 100a.

The first of these initiatives, on health and safety for temporary workers, formed one of a trio of measures designed to improve the employment conditions of "a-typical" workers in the European Community.[36] It is notable that, of the three proposals first put forward — each of which was founded upon a different Treaty base for its legitimacy — only that based on Article 118a has survived, and, indeed, looks set to receive *unanimous* support (notwithstanding the possibility for a "qualified majority") in its final stages before the Council.

The second and third initiatives, both of which were put forward by the Commission towards the end of the Summer 1990, have received opposition to the legal base adopted. Indeed, despite their both having been discussed at a specially convened "informal" Council meeting on 6th May 1991, neither looks set to make further progress in the foreseeable future.

There are evident dangers associated with the "creative" use by the Commission of Article 118a, and it is to be hoped that none of the adverse responses to this perceived "abuse" of the Treaty powers will rub off onto the broader range of measures currently working their way towards completion in the field of health and safety at work. These cover a significant range of topics, most of which were spelled out in the Commission's Action Programme of November 1989, and many of them reflect a response at European Community level to recognized areas of risk at work throughout the Communities and the European labour market.

Further Initiatives and Current Challenges

Of other measures introduced to date, one has already found final form; *i.e.* the Recommendation on a European Schedule of Industrial Diseases.[37] In addition, a proposal for a Directive to amend the 1983 Directive on Asbestos has now passed all of its legislative stages and been adopted.[38] A further proposal, updating earlier Directives

---

33. See now, Council Directive supplementing the measures to encourage improvements in the safety and health at work of workers with a fixed-duration employment relationship or a temporary employment relationship (91/383/EEC).

34. Proposed directive concerning certain aspects of the organization of working time (OJ/C 254/90, amended version OJ/C 124/91).

35. Proposed directive concerning measures to encourage improvements in the safety and health of pregnant workers, women workers who have recently given birth and women who are breastfeeding (OJ/C 281/90, amended version 91/C 25/04).

36. The other two are a proposed directive on the approximation of the laws of the Member States relating to certain employment relationships with regard to working conditions (OJ/C 224/90), and a proposed directive relating to certain employment contracts and employment relationships involving distortions of competition (OJ/C 224/90, amended version OJ/C 305/90).

37. Commission Recommendation of 22nd May 1990 to the Member States concerning the adoption of a European schedule of occupational diseases (90/326/EEC).

38. Council Directive amending Directive 83/447/EEC on the protection of workers from the risks related to exposure to asbestos at work (91/382/EEC).

on Safety and Health Signs at the workplace, is also likely to make swift progress.[39] Of other initiatives, covering medical assistance on board vessels,[40] health and safety requirements at temporary or mobile work sites,[41] and health and safety protection in the extractive industries (containing two separate items in the Commission's 1989 Action Programme, dealing with *(a)* drilling industries and *(b)* quarrying and open-cast mining),[42] it seems unlikely that objection in principle will be forthcoming, or that any challenge will be made to the clearly applicable Treaty base of Article 118a.

One additional initiative might be noted in this context, however. This is a proposal for a Directive introducing measures to promote improvements in the travel conditions of workers with motor disabilities — a measure which is based upon Article 118a, and was introduced in February 1991.[43]

It is clear, therefore, that, whatever political problems may lie in wait for them, the initiatives will continue to flow thick and fast in this vital area of social protection in which the European Community has committed itself to act. Areas such as agriculture, fishing and construction have been targeted as high risk sectors where further legislation is necessary. Aspects relating specifically to agriculture will be incorporated into the general directives, while other directives specific to agriculture are to be introduced — relating *inter alia* to plant protection products, animal husbandry, farm buildings, and other specific areas of activity. A further proposed directive will lay down minimum health and safety requirements on board new or old fishing vessels — a measure intended to enhance the preventive approach in training as regards accidents and ill-health onboard, as well as covering rescue and survival equipment, personal protective equipment, and the use of fishing gear.

With significant activity also to be witnessed in a number of fields closely related to that of occupational health and safety — notably, in the field of the environment and in relation to measures concerning public health — the period ushered in by 1992 looks set to witness ever greater concern at the level of the European Communities for the safeguarding of the health and safety of workers at work in the completed Single European Market.

Acknowledgement

We gratefully acknowledge permission from the Office for Official Publications of the European Communities to reproduce the texts of the legislative material in this book.

39. Proposed directive on the minimum requirements for safety and health signs at the workplace (OJ/C 53/91), intended to amend substantially the existing Directive 77/576/EEC (as already amended by Directive 79/640/EEC).

40. Proposed directive on the minimum health and safety requirements to encourage improved medical assistance on board vessels (OJ/C 183/90, amended version OJ/C 74/91).

41. Proposed directive on the minimum health and safety requirements for work at temporary or mobile work sites (OJ/C 213/90, amended version OJ/C 74/91).

42. Proposed directive concerning minimum requirements for improving the health and safety protection of workers in the extractive industries (OJ/C 32/91).

43. Proposed directive on the introduction of measures aimed at promoting an improvement in the travel conditions of workers with motor disabilities (OJ/C 68/91).

TREATY CREATING THE EUROPEAN ECONOMIC COMMUNITY

—————— EEC Treaty 1957 (as amended 1985) ——————

[Extracts]

Article 1

By this Treaty, the HIGH CONTRACTING PARTIES establish among themselves a EUROPEAN ECONOMIC COMMUNITY.

Article 2

The Community shall have as its task, by establishing a common market and progressively approximating the economic policies of Member States, to promote throughout the Community a harmonious development of economic activities, a continuous and balanced expansion, an increase in stability, an accelerated raising of the standard of living and closer relations between the States belonging to it.

Article 3

For the purposes set out in Article 2, the activities of the Community shall include, as provided in this Treaty and in accordance with the timetable set out therein

(a)   the elimination, as between Member States, of customs duties and of quantitative restrictions on the import and export of goods, and of all other measures having equivalent effect;

(b)   the establishment of a common customs tariff and of a common commercial policy towards third countries;

(c)   the abolition, as between Member States, of obstacles to freedom of movement for persons, services and capital;

(d)   the adoption of a common policy in the sphere of agriculture;

(e)   the adoption of a common policy in the sphere of transport;

(f)   the institution of a system ensuring that competition in the common market is not distorted:

(g)   the application of procedures by which the economic policies of Member States can be co-ordinated and disequilibria in their balances of payments remedied;

(h)   the approximation of the laws of Member States to the extent required for the proper functioning of the common market;

(i)   the creation of a European Social Fund in order to improve employment opportunities for workers and to contribute to the raising of their standard of living;

(j)   the establishment of a European Investment Bank to facilitate the economic expansion of the Community by opening up fresh resources;

(k)   the association of the overseas countries and territories in order to increase trade and to promote jointly economic and social development.

Article 100

The Council shall, acting unanimously on a proposal from the Commission, issue directives for the approximation of such provisions laid down by law, regulation or administrative action in Member States as directly affect the establishment or functioning of the common market.

The European Parliament and the Economic and Social Committee shall be consulted in the case of directives whose implementation would, in one or more Member States, involve the amendment of legislation.

Article 100a

1. By way of derogation from Article 100 and save where otherwise provided in this Treaty, the following provisions shall apply for the achievement of the objectives set out in Article 8a. The Council shall, acting by a qualifed majority on a proposal from the Commission in co-operation with the European Parliament and after consulting the Economic and Social Committee, adopt the measures for the approximation of the provisions laid down by law, regulation or administrative action in Member States which have as their object the establishment and functioning of the internal market.

2. Paragraph 1 shall not apply to fiscal provisions, to those relating to the free movement of persons nor to those relating to the rights and interests of employed persons.

3. The Commission, in its proposals envisaged in paragraph 1 concerning health, safety, environmental protection and consumer protection, will take as a base a high level of protection.

4. If, after the adoption of a harmonization measure by the Council acting by a qualified majority, a Member State deems it necessary to apply national provisions on grounds of major needs referred to in Article 36, or relating to protection of the environment or the working environment, it shall notify the Commission of these provisions.

    The Commission shall confirm the provisions involved after having verified that they are not a means of arbitrary discrimination or a disguised restriction on trade between Member States.

    By way of derogation from the procedure laid down in Articles 169 and 170, the Commission or any Member State may bring the matter directly before the Court of Justice if it considers that another Member State is making improper use of the powers provided for in this Article.

5. The harmonization measures referred to above shall, in appropriate cases, include a safeguard clause authorizing the Member States to take, for one or more of the non-economic reasons referred to in Article 36, provisional measures subject to a Community control procedure.

Article 100b

1. During 1992, the Commission shall, together with each Member State, draw up an inventory of national laws. regulations and administrative provisions which fall under Article 100a and which have not been harmonized pursuant to that Article.

   The Council. acting in accordance with the provisions of Article 100a, may decide that the provisions in force in a Member State must be recognized as being equivalent to those applied by another Member State.

2. The provisions of Article 100a (4) shall apply by analogy.

3. The Commission shall draw up the inventory referred to in the first sub-paragraph of paragraph 1 and shall submit appropriate proposals in good time to allow the Council to act before the end of 1992.

Article 117

Member States agree upon the need to promote improved working conditions and an improved standard of living for workers, so as to make possible their harmonization while the improvement is being maintained.

They believe that such a development will ensue not only from the functioning of the common market, which will favour the harmonization of social systems, but also from the procedures provided for in this Treaty and from the approximation of provisions laid down by law, regulation or administrative action.

Article 118

Without prejudice to the other provisions of this Treaty and in conformity with its general objectives, the Commission shall have the task of promoting close co-operation between Member States in the social field, particularly in matters relating to:

— employment;

— labour law and working conditions:

— basic and advanced vocational training:

— social security:

— prevention of occupational accidents and diseases:

— occupational hygiene:

— the right of association, and collective bargaining between employers and workers.

To this end, the Commission shall act in close contact with Member States by making studies, delivering opinions and arranging consultations both on problems arising at national level and on those of concern to international organizations.

Before delivering the opinions provided for in this Article, the Commission shall consult the Economic and Social Committee.

Article 118a

1. Member States shall pay particular attention to encouraging improvements, especially in the working environment, as regards the health and safety of workers, and shall set as their objective the harmonization of conditions in this area, while maintaining the improvements made.

2. In order to help achieve the objective laid down in the first paragraph, the Council, acting by a qualified majority on a proposal from the Commission, in co-operation with the European Parliament and after consulting the Economic and Social Committee, shall adopt, by means of directives, minimum requirements for gradual implementation having regard to the conditions and technical rules obtaining in each of the Member States.

    Such directives shall avoid imposing administrative, financial and legal constraints in a way which would hold back the creation and development of small and medium-sized undertakings.

3. The provisions adopted pursuant to this Article shall not prevent any Member State from maintaining or introducing more stringent measures for the protection of working conditions compatible with this Treaty.

Article 118b

The Commission shall endeavour to develop the dialogue between management and labour at European level which could, if the two sides consider it desirable, lead to relations based on agreement.

Article 235

If action by the Community should prove necessary to attain in the course of the operation of the common market, one of the objectives of the Community and this Treaty has not provided the necessary powers, the Council shall, acting unanimously on a proposal from the Commission and after consulting the European Parliament, take the appropriate measures.

CHAPTER III
HEALTH AND SAFETY

Article 30

Basic standards shall be laid down within the Community for the protection of the health of workers and the general public against the dangers arising from ionizing radiations.

The expression 'basic standards' means:

(a)   maximum permissible doses compatible with adequate safety;

(b)   maximum permissible levels of exposure and contamination;

(c)   the fundamental principles governing the health surveillance of workers.

Article 31

The basic standards shall be worked out by the Commission after it has obtained the opinion of a group of persons appointed by the Scientific and Technical Committee from among scientific experts, and in particular public health experts, in the Member States. The Commission shall obtain the opinion of the Economic and Social Committee on these basic standards.

After consulting the European Parliament the Council shall, on a proposal from the Commission, which shall forward to it the opinions obtained from these Committees, establish the basic standards; the Council shall act by a qualified majority.

Article 32

At the request of the Commission or of a Member State, the basic standards may be revised or supplemented in accordance with the procedure laid down in Article 31.

The Commission shall examine any request made by a Member State.

Article 33

Each Member State shall lay down the appropriate provisions, whether by legislation, regulation or administrative action, to ensure compliance with the basic standards which have been established and shall take the necessary measures with regard to teaching, education and vocational training.

The Commission shall make appropriate recommendations for harmonizing the provisions applicable in this field in the Member States.

To this end, the Member States shall communicate to the Commission the provisions applicable at the date of entry into force of this Treaty and any subsequent draft provisions of the same kind.

Article 34

Any Member State in whose territories particularly dangerous experiments are to take place shall take additional health and safety measures, on which it shall first obtain the opinion of the Commission.

The assent of the Commission shall be required where the effects of such experiments are liable to affect the territories of other Member States.

Article 35

Each Member State shall establish the facilities necessary to carry out continuous monitoring of the level of radio-activity in the air, water and soil and to ensure compliance with the basic standards.

The Commission shall have the right of access to such facilities; it may verify their operation and efficiency.

Article 36

The appropriate authorities shall periodically communicate information on the checks referred to in Article 35 to the Commission so that it is kept informed of the level of radio-activity to which the public is exposed.

Article 37

Each Member State shall provide the Commission with such general data relating to any plan for the disposal of radioactive waste in whatever form as will make it possible to determine whether the implementation of such plan is liable to result in the radioactive contamination of the water, soil or airspace of another Member State.

The Commission shall deliver its opinion within six months, after consulting the group of experts referred to in Article 31.

Article 38

The Commission shall make recommendations to the Member States with regard to the level of radioactivity in the air, water and soil.

In cases of urgency, the Commission shall issue a directive requiring the Member State concerned to take, within a period laid down by the Commission, all necessary measures to prevent infringement of the basic standards and to ensure compliance with regulations.

Should the State in question fail to comply with the Commission directive within the period laid down, the Commission or any Member State concerned may forthwith, by way of derogation from Articles 141 and 142, bring the matter before the Court of Justice.

Article 39

The Commission shall set up within the framework of the Joint Nuclear Research Centre, as soon as the latter has been established, a health and safety documentation and study section.

This section shall in particular have the task of collecting the documentation and information referred to in Articles 33, 36 and 37 and of assisting the Commission in carrying out the tasks assigned to it by this Chapter.

PART TWO

Action Programmes, Decisions, Resolutions, Recommendations, etc.

COUNCIL DECISION

of 9 July 1957

concerning the terms of reference and rules of procedure of the
Mines Safety Commission
[487/57]

Having noted the recommendations adopted by the Conference on Safety in Coal Mines
and the proposals submitted by the High Authority in the light of the final report of this
Conference, which constitute a useful basis for the improvement of safety in coal mines;

Having regard to their decisions taken at the 36th and 42nd Sessions of the Council on
6 September 1956 and 9 and 10 May 1957 to create the Mines Safety Commission;

THE REPRESENTATIVES OF THE GOVERNMENTS OF THE MEMBER STATES,
MEETING WITHIN THE SPECIAL COUNCIL OF MINISTERS,

— define the terms of reference of that Safety Commission as follows:

1. The Safety Commission shall follow developments in safety in coal mines, including safety regulations made by the public authorities, and shall assemble
the necessary information on progress made and practical results obtained,
particularly in the field of accident prevention.

   In order to obtain the necessary information, the Safety Commission shall
apply to the Governments concerned.

   The Safety Commission shall make use of the information in its possession,
and shall submit proposals to the Governments for the improvement of
safety in coal mines.

2. The Safety Commission shall assist the High Authority in seeking a method
of compiling comparable statistics on accidents.

3. The Safety Commission shall ensure that the appropriate information
assembled by it is swiftly communicated to the quarters concerned (in
particular the administrative bodies in mines, and employers' and workers'
organisations).

4. The Safety Commission shall, by regular contact with the Governments,
keep itself informed of steps taken to follow up proposals made by the
Conference on Safety in Coal Mines, and also those drawn up by itself.

5. The Safety Commission shall propose such studies and research as seem to
it to be most appropriate for improving safety, and shall specify how best they
may be put into effect.

6. The Safety Commission shall facilitate the exchange of information and of
experience between the persons responsible for safety, and shall propose
measures appropriate to such end (*eg.* study visits, setting up of documentation services).

7. The Safety Commission shall propose suitable measures for establishing the necessary liaison between the rescue services of the Community countries.

8. The Safety Commission shall submit an annual report to the Governments meeting within the Council, and to the High Authority, on its activities and on developments in the field of safety in the coal mines of the various Member States. At such time it shall in particular study the statistics compiled on the subject of accidents and incidents in coal mines:

—   lay down the rules of procedure for this Commission which are annexed to the present Decision,

—   request the High Authority to ensure that this Commission commence work with the least possible delay.

This Decision was adopted at the 44th Session of the Council, held on 9 July 1957.

ANNEX

RULES OF PROCEDURE
of the Mines Safety Commission

Chairmanship
Article 1

The chairmanship of the Mines Safety Commission shall be held by a member of the High Authority of the European Coal and Steel Community.

Article 2

The Chairman shall direct the work of the Safety Commission in conformity with the present rules of procedure.

Composition
Article 3

The Safety Commission shall have twenty-four members appointed by the Governments (*ie.* for each country four members consisting of two representatives of the national governments and one representative of employers and workers respectively).

Each Government shall communicate, in writing, to the Chairman, the list of names of the members whom it has appointed. Any alterations to that list shall be notified to the Chairman.

For every meeting of the Safety Commission each Government may appoint one or two advisers, whose names shall be communicated to the Chairman.

Participation of the International Labour Organisation
Article 4

Representatives of the International Labour Organisation shall be invited to participate in the work of the Safety Commission in an advisory capacity.

Participation of the United Kingdom
Article 5

Delegates appointed by the Government of the United Kingdom may participate in the work of the Safety Commission as observers.

Organisation

A. SELECT COMMITTEE

Article 6

A Select Committee is hereby set up, composed of the Government representatives on the Safety Commission.

Article 7

The Chairman of the Safety Commission shall be Chairman of the Select Committee.

Article 8

The task of the Select Committee shall be to ensure permanent liaison between the governments of Member States on the one hand, and between the latter and the Safety Commission on the other, particularly for the purpose of achieving a useful exchange of information. It shall supervise the preparation of the work of the Safety Commission.

Article 9

The Chairman shall convene the Select Committee.

The Select Committee shall be convened by the Chairman whenever at least three Government representatives request a meeting.

B. WORKING PARTIES

Article 10

In order to examine certain questions of a technical nature, the Safety Commission or the Select Committee may set up working parties composed of experts.

Article 11

The working parties shall determine their own working methods.

Article 12

The results of the work of the working parties, presented in the form of reports, shall be put before the Select Committee. It shall submit such reports to the Safety Commission together with the opinions of the Committee members.

In the event of differences of opinion within the working parties, note shall be taken of the opinions expressed together with the names of the experts expressing them.

Secretarial services

Article 13

The High Authority shall provide secretarial services for the Safety Commission, the Select Committee and the working parties.

The secretarial services shall be directed by an official of the High Authority, appointed as secretary.

All documents shall be drafted in the four official languages of the Community.

Operation

Article 14

The Chairman shall draw up the draft agenda, and the dates of meetings, after consultations with the members of the Select Committee.

Article 15

When requested, the Chairman shall grant leave to speak to the members of the Safety Commission, to representatives of the International Labour Organisation and to United Kingdom observers.

The Chairman may grant advisers leave to speak.

Article 16

The members of the High Authority shall be entitled to participate in, and to speak at meetings of, the Safety Commission and of the Select Committee.

The Chairman may be accompanied by advisers, to whom he may grant leave to speak.

Article 17

When the Safety Commission or the Select Committee shall consider it desirable to assemble information relating to the various aspects of safety in mines, it shall address requests to this effect to the Governments of the Member States.

Article 18

Proceedings shall only be valid if at least sixteen members are present. Resolutions shall be passed by a majority of the members present.

However, proposals from the Safety Commission made in conformity with paragraph 1(3) of its terms of reference shall be approved by two-thirds of the members present; such proposals shall carry at least thirteen votes.

Any dissenting opinions shall be brought to the notice of the Governments, should the members concerned so request.

RECOMMANDATION DE LA COMMISSION
—————— du 23 juillet 1962 ——————
aux États membres concernant l'adoption d'une
liste européenne des maladies professionnelles
[2188/62]

Exposé des motifs

1. Le traité instituant la Communauté economique européenne dans son article 117, exprime la volonté des États membres de « promouvoir l'amélioration des conditions de vie et de travail de la main-d'œuvre permettant leur égalisation dans le progrès » et, dans son article 118, déclare expressément que la Commission de la C.E.E. a pour mission de promouvoir une collaboration étroite entre les États membres dans le domaine social, notamment dans les matières relatives à la sécurité sociale et à la protection contre les maladies professionelles.

2. En matière de maladies professionnelles, la législation des six pays de la Communauté repose sur le système dit « de liste » qui consiste à énumérer limitativement les maladies reconnues comme ayant une origine professionnelle. Ce système est recommandé par l'Organisation internationale du travail dans ses conventions de 1925 (n° 18) et de 1934 (n° 42). Les listes étant différentes, une harmonisation est nécessaire pour atteindre un réel progrès social. En outre la réalisation progressive de la libre circulation des travailleurs à l'intérieur de la Communauté prévue par les dispositions du traité et entamée par le règlement n° 15, nécessite également l'établissement d'une législation harmonisée pour assurer une protection de même nature à tous les travailleurs dans chacun des pays de la Communauté où ils seront amenés à établir leur résidence et leur lieu de travail. Une telle harmonisation facilitera l'application des règlements n° 3 et n° 4 relatifs à la sécurité sociale des travailleurs migrants, dont certaines dispositions visant le cas de travailleurs ayant été exposés à un même risque dans deux ou plusieurs pays s'appliquent difficilement si les législations ne reconnaissent pas la même affection comme maladie professionnelle.

3. Les listes de maladies figurant dans les six législations diffèrent pour diverses raisons: divergences de nomenclatures, différences dans les conditions d'application de l'assurance, peu d'importance ou même inexistence, suivant les pays, de certaines catégories d'activités industrielles ou agricoles. Ces diversités peuvent entraîner des différences importantes dans les garanties accordées aux travailleurs tant en ce qui concerne la prévention que la réparation des maladies professionnelles. Elles sont, en outre, un obstacle à l'établissement de comparaisons valables, notamment d'ordre statistique, quant à l'application des législations des pays de la Communauté.

4. Il paraît ainsi souhaitable que les États membres adoptent une *liste européenne uniforme* des maladies ou agents pouvant les provoquer, afin de réaliser une première étape vers l'harmonisation des prescriptions légales et réglementaires en matière de protection contre les maladies professionelles et de réparation de leurs conséquences dommageables. Les étapes suivantes pourraient porter tant sur les conditions d'octroi que sur les niveaux des prestations.

5. L'analyse approfondie des listes nationales, tant générales que spéciales à l'agriculture dans certains États membres, montre qu'il est possible de réunir dans une liste unique, en les classant selon leur nature, les maladies ou agents figurant dans une ou plusieurs listes nationales actuelles, les États membres étant à même d'adopter cette liste selon la procédure en vigueur dans chaque pays. Certaines mesures récemment intervenues dans ce domaine semblent d'ailleurs avoir déjà tenu compte des travaux préparatoires de cette liste.

6. Le système des listes a été considéré pendant longtemps comme constituant une garantie pour les travailleurs, grâce à la notion de présomption d'origine qui s'y attache; cependant, lorsque la liste comporte des conditions limitatives trop restrictives (travaux, symptômes, délais), les avantages présentés par la présomption d'origine ne jouent pas pour les travailleurs qui ne remplissent pas strictement les conditions de la loi et qui cependant ont indéniablement contracté une maladie dans l'exercice de leur profession. Les travailleurs seraient garantis d'une manière plus complète si la législation ouvrait en outre un droit à réparation pour des maladies ne figurant pas dans la liste nationale mais dont l'origine professionnelle serait suffisamment établie.

7. En vue de faciliter les échanges d'information tendant à l'harmonisation des listes nationales sur la base de la liste européenne, il convient que chacun des États membres dont la législation connait des agents nocifs ou des maladies professionnelles non encore inscrits dans les listes d'autres États, établisse à l'usage de ces derniers et sur leur demande transmise par la Commission, des fiches documentaires comportant des informations, aussi précises et complètes que possible, de caractère technique, médical et statistique portant sur ces cas concrets. Le nombre et les caractéristiques des cas décrits devront être suffisants pour permettre leur exploitation par des enquêtes et études sur le plan national.

8. Les études ont permis en outre l'établissement d'une liste de maladies ou agents ne figurant encore dans aucune des listes nationales, mais qu'il serait souhaitable d'introduire dans une liste moderne tenant compte des plus récentes acquisitions de la médecine et de la technique. Cette liste, annexée à la liste européenne, devrait être retenue par les États membres comme liste des maladies soumises à déclaration; de caractère simplement indicatif, elle permettra de recueillir une documentation intéressante du point de vue médical, statistique et économique en vue d'une mise à jour périodique de la liste européenne; elle stimulera les recherches sur les maladies ou agents y figurant.

9. Le corollaire de la prévention des risques auxquels est exposée la santé des travailleurs n'est pas obligatoirement leur réparation dans le cadre de la législation sur les accidents de travail et sur les maladies professionnelles: ainsi l'hygiène du travail protège la santé des travailleurs sur les lieux de leur emploi sans que toutes les atteintes possibles à cette santé soient réparées au titre d'un risque professionnel.

Cependant la reconnaissance qu'une affection est liée à un risque professionnel fait porter une attention particulière sur ce risque et entraîne une amélioration de la prévention puisque le danger est mis en lumière, que des mesures préventives sont préconisées et que des contrôles plus efficaces peuvent s'exercer lorsque le risque existe.

L'effort doit donc porter, en premier lieu, sur la prévention.

Or, le rôle de la prévention dans le domaine des maladies professionnelles est d'autant plus important qu'il existe à cet égard *une différence capitale* entre accidents du travail et maladies professionnelles:

— pour les accidents, quelle que soit leur cause – manque d'organisation, défaillances matérielles, défaillances humaines toujours possibles – leur survenance est toujours fortuite et on ne peut, quel que soit le développement de la prévention, avoir une certitude absolue de le faire disparaître;

— pour les maladies professionnelles, au contraire, étant donné qu'il est possible, tout au moins dans un grand nombre de cas, de connaître les causes des maladies et de prévoir leur évolution, les remèdes préventifs peuvent, en principe, conduire à des résultats de loin supérieurs et se rapprocher graduellement d'une efficacité totale.

Bien qu'intéressant en premier lieu la réparation, la liste européenne stimulera le développement de la prévention pour chacun des agents nocifs et des maladies professionnelles déjà reconnus.

Au cours des étapes ultérieures, la Commission provoquera la collaboration des États membres en vue de favoriser l'application des meilleures méthodes de prévention.

10. Toute législation ou réglementation sur les maladies professionnelles devant avoir un caractère général, s'applique egalement aux personnes et entreprises relevant de la compétence de la Communauté européenne du charbon et de l'acier et de la Communauté européenne de l'énergie atomique.

La Commission de la C.E.E. à donc tenu à consulter la Haute Autorité de la C.E.C.A. et la Commission de l'Euratom qui, chacune pour sa competence respective, ont donné leur plein appui à la présente recommandation, sans préjudice des actions qu'elles peuvent mener dans le cadre de leurs traités respectifs.

Recommandation

Pour ces raisons la Commission de la Communauté économique européenne, au titre des dispositions du traité instituant cette Communauté, et notamment de l'article 155, recommande aux États membres:

a) d'introduire dans leurs dispositions législatives, réglementaires et administratives relatives aux maladies professionnelles la liste européenne ci-jointe au titre de liste des maladies professionnelles susceptibles de donner lieu à réparation sur la base de leur législation, en complétant à cet effet leur liste nationale ou leurs tableaux de maladies professionnelles indemnisables;

b) de coopérer à cette harmonisation en procédant par l'intermédiaire de la Commission à des échanges d'informations d'ordre médical, scientifique et technique relatifs aux cas de maladies professionnelles ayant effectivement donné lieu à réparation dans un ou plusieurs États, de fournir notamment à cet effet toutes informations utiles sur les maladies ou agents reconnus dans leur législation nationale, à la date de la présente recommandation, lorsque la demande en sera faite par un autre État membre par l'intermédiaire de la Commission, en établissant des fiches documentaires conformes au modèle ci-joint;

c) d'introduire en outre dans leurs dispositions législatives, réglementaires et administratives un droit à réparation au titre de la législation sur les maladies professionnelles, lorsque la preuve sera suffisamment établie par le travailleur intéressé qu'il a contracté en raison de son travail, une maladie qui ne figure pas dans la liste nationale;

d) d'informer la Commission des adjonctions à la liste nationale de maladies professionnelles ne figurant pas dans la liste européenne, afin de permettre une mise à jour périodique de la dite liste;

e) d'utiliser la liste européenne comme document de base concernant la prévention et la déclaration des accidents du travail et des maladies professionnelles;

f) de développer et d'améliorer les diverses mesures de prévention des maladies mentionnées dans la liste européenne, en recourant, le cas échéant, à la Commission pour avoir connaissance des expériences acquises par les États membres de la Communauté;

g) de rendre obligatoire la déclaration des cas de maladies inscrites sur la liste annexe, de faire procéder à une étude particulière de ces cas et d'en communiquer les résultats periodiquement à la Commission;

h) d'adapter leurs statistiques à la classification et à la nomenclature de la liste européenne et de la liste annexe et de les communiquer à la Commission.

Fait à Bruxelles, le 23 juillet 1962

ANNEXE I

Liste européenne des maladies professionnelles

A. *Maladies professionnelles provoquées par les agents chimiques suivants:*

1. Arsenic et ses composés
2. Beryllium (glucinium) et ses composés
3. Oxyde de carbone – oxychlorure de carbone – acide cyanhydrique, cyanures et composés du cyanogène
4. Cadmium et ses composés
5. Chrome et ses composés
6. Mercure et ses composés
7. Manganèse et ses composés
8. Acide nitrique – oxydes d'azote – ammoniaque
9. Nickel et ses composés
10. Phosphore et ses composés
11. Plomb et ses composés
12. Anhydride sulfureux, acide sulfurique, hydrogène sulfuré, sulfure de carbone
13. Thallium et ses composés
14. Vanadium et ses composés
15. Chlore, brome et iode et leurs composés inorganiques – fluor et ses composés
16. Hydrocarbures aliphatiques saturés ou non, cycliques ou non, constituants de l'éther de pétrole et de l'essence
17. Dérivés halogénés des hydrocarbures aliphatiques saturés ou non, cycliques ou non
18. Alcools, glycols, éthers, cétones, esters organiques et leurs dérivés halogénés
19. Acides organiques, aldéhydes
20. Nitrodérivés aliphatiques, esters de l'acide nitrique
21. Benzène, toluène, xylènes et autres homologues du benzène, naphtalènes et homologues (l'homologue d'un hydrocarbure aromatique est défini par la formule $C_n H_{2n-6}$ pour les homologues du benzène et par la formule $C_n H_{2n-12}$ pour les homologues du naphtalène)
22. Derivés halogénés des hydrocarbures aromatiques
23. Phénols et homologues, thiophénols et homologues, naphtols et homologues et leurs dérivés halogénés: dérivés halogénés des alkylaryloxydes et des alkylarysulfures, benzoquinone
24. Amines (primaires, secondaires, tertiaires, hétérocycliques) et hydrazines aromatiques et leurs dérivés halogénés, phénoliques, nitrosés, nitrés et sulfonés
25. Nitrodérivés des hydrocarbures aromatiques et des phénols

B. *Maladies professionnelles de la peau causées par des substances et agents non compris sous d'autres positions*

1. Cancers cutanés et affections cutanées précancéreuses dues à la suie, au goudron au bitume, au brai, à l'anthracène, aux huiles minérales, à la paraffine brute et aux composés, produits et résidus de ces substances

2. Affections cutanées provoquées dans le milieu professionnel par des substances non considérées sous d'autres positions

C. *Maladies professionnelles provoquées par l'inhalation de substances et agents non compris sous d'autres positions*

1. Pneumoconioses:
   a) – Silicose, associée ou non à la tuberculose pulmonaire,
   b) – Asbestose, associée ou non à la tuberculose pulmonaire ou à un cancer du poumon
   c) – Pneumoconioses dues aux poussières de silicates
2. Affections broncho-pulmonaires dues aux poussières ou fumées d'aluminium ou de ses composés
3. Affections broncho-pulmonaires dues aux poussières de métaux durs
4. Affections broncho-pulmonaires causées par les poussières de scories Thomas
5. Asthme provoqué dans le milieu professionnel, par des substances non incluses sous d'autres positions

D. *Maladies professionnelles infectieuses et parasitaires*

1. Helminthiases, ankylostome duodénal, anguillule de l'intestin
2. Maladies tropicales dont: paludisme, amibiase, trypanosomiase, dengue, fièvre à pappataci, fièvre de Malte, fièvre récurrente, fièvre jaune, peste, leischmaniose, pian, lèpre, typhus exanthématique et autres rickettsioses
3. Maladies infectieuses ou parasitaires transmises à l'homme par des animaux ou débris d'animaux
4 Maladies infectieuses du personnel s'occupant de prévention, soins, assistance à domicile et recherches

E *Maladies professionnelles par carence*

1. Scorbut

F. *Maladies professionnelles provoquées par des agents physiques*

1. Maladies provoquées par les radiations ionisantes
2. Cataracte provoquée par l'énergie radiante
3. Hypoacousie ou surdité provoquée par le bruit
4. Maladies provoquées par les travaux dans l'air comprimé
5. Maladies ostéo-articulaires ou angio-neurotiques provoquées par les vibrations mécaniques
6. a) – Maladies des bourses péri-articulaires dues à des pressions, cellulites sous-cutanées
   b) – Maladie par surmenage des gaines tendineuses, du tissu péri-tendineux, des insertions musculaires et tendineuses
   c) – Lésions du ménisque chez les mineurs
   d) – Arrachements par surmenage des apophyses épineuses
   e) – Paralysies des nerfs dues à la pression
7. Nystagmus des mineurs

ANNEX II

Liste annexe indicative de maladies à soumettre à déclaration en vue d'une inscription éventuelle dans la liste européenne

A   *Maladies provoquées par les agents chimiques suivants*

1. Ozone
2. Esters des acides du soufre
3. Mercaptanes et thyoethers
4. Oxyde de zinc
5. Boranes
6. Composés organiques du chlore, du brome et de l'iode
7. Hydrocarbures aliphatiques autres que ceux visés sous la rubrique A. 16 de la liste européenne
8. Amines aliphatiques et leurs dérivés halogénés
9. Nitriles et esters isocyaniques
10. Vinylbenzène et divinylbenzène, diphényle, décaline, tétraline
11. Acides aromatiques, anhydrides aromatiques et leurs dérivés halogénés
12. Oxyde de diphényle, dioxane, tétrahydrophurane
13. Thiophène
14. Furfurol

Aa.   *Maladies provoquées par des agents divers*

1. Maladies provoquées par l'inhalation de poussières de nacre
2. Maladies provoquées par des substances hormonales

B.   *Maladies provoquées par l'inhalation de substances non comprises sous d'autres positions*

1. Pneumoconioses provoquées par les poussières de charbon, de carbone, de graphite, de sulfate de baryum, d'oxydes d'étain
2. Fibroses pulmonaires dues aux métaux non désignés dans la liste européenne
3. Maladies pulmonaires provoquées par l'inhalation de poussières de coton, de lin, de chanvre, de jute, de sisal et de bagasse
4. Asthmes et bronchites asthmatiques provoqués par l'inhalation de poussières de poils d'animaux, de gomme arabique, d'antibiotiques, de bois exotiques et d'autres substances allergènes

C.   *Maladies provoquées par des causes physiques*

1. Crampes professionnelles

ANNEXE III

MODELES
de fiches documentaires relatives aux cas de maladies professionnelles ayant donné lieu à indemnisation

NOTICE
Chaque cas fera l'objet d'une fiche distincte

Les cas seront choisis – dans toute la mesure du possible – de telle sorte qu'ils présentent par leur nombre et leurs caractéristiques variées, un ensemble d'informations susceptibles d'être utilement exploitées dans les pays intéressés. Dans le cas où un même agent nocif est capable d'engendrer des maladies très diverses intéressant, par exemple, la peau, les muqueuses, les appareils respiratoire et digestif, le système nerveux, etc. il serait très souhaitable qu'un échantillonnage fût fourni par chaque catégorie de cas constatés.

Les fiches seront autant que possible groupées en un même envoi pour un agent nocif ou une maladie professionnelle déterminés et référence sera faite à la dénomination et au classement de la maladie correspondante dans la liste européenne ou la liste annexe.

Les renseignements d'ordre statistique ne figureront qu'une fois pour l'ensemble des cas faisant l'objet d'un même envoi.

**COMMUNAUTÉ ÉCONOMIQUE
EUROPÉENNE**

Commission
Direction générale
des affaires sociales

Classement C.E.E.

(^{1})

RECOMMANDATION
de la Commission aux États membres
concernant l'adoption d'une
LISTE EUROPÉENNE DES MALADIES PROFESSIONNELLES

FICHE DOCUMENTAIRE

pour l'analyse de cas de maladie professionnelle
ayant donné lieu à indemnisation en vertu de la législation de:

.... (^{1})

I

Dénomination de la maladie	Agent l'ayant provoquée
....	....
....	....

Identification du cas

Initiales et date de naissance de la victime ou n° d'ordre	M	F
....		(^{2})

II
Base juridique

Lois ou autres textes généraux concernant les maladies professionnelles (titres et dates)	Dénomination de la maladie ou de l'agent et n° dans la liste ou série de tableaux de la législation nationale
	Date d'inscription dans la liste ou série de tableaux:
	Classement dans la liste européenne ou dans la liste annexe } (^{2})

(^{1}) Nom de l'État membre.
(^{2}) Biffer la mention inutile.

III

RENSEIGNEMENTS TECHNIQUES

Agent nocif

....

....

....

....

....

.... (¹)

Travaux exécutés

Nature	Durée continue ou non	Évaluation quantitative du risque	Circonstances ayant pu accroître le risque d'une façon particulière

(¹) Description scientifique et technique de l'agent, de ses utilisations et de ses emplois.

IV

Renseignements médicaux

Diagnostic: ....

Anamnèse	Description de la maladie — manifestations cliniques, localisation, évolution, pronostic, etc.	Méthodes de laboratoire ou autres pour confirmer le diagnostic
	Conséquences temporaires et-ou permanentes sur la capacité de travail de la victime	
Durée de l'exposition au risque antérieure à l'apparition des troubles		

V

Difficultés rencontrées

Pour la constatation du risque professionnel	Pour la reconnaissance du caractère professionnel de la maladie
Expertises médicales éventuelles	Expertises médicales éventuelles
Expertises techniques éventuelles	Expertises techniques éventuelles
Mesures judiciaires éventuelles	Mesures judiciaires éventuelles

**COMMUNAUTÉ
ÉCONOMIQUE EUROPÉENNE**

Commission
Direction générale
des affaires sociales

Classement C.E.E.

(^{1})

RECOMMANDATION

de la Commission

aux États membres concernant l'adoption

d'une

LISTE EUROPÉENNE DES MALADIES PROFESSIONNELLES

DONNÉES STATISTIQUES

relatives aux cas de maladies professionnelles

ayant donné lieu à indemnisation en vertu de la

législation de

.... (^{1})

Maladies professionnelles causées par .... ($^2_3$)

Statistiques relatives aux années .... (^{1})

Affection (^{3}): ....

	19......	19......	19......	19......	19......
1. Nombre de personnes exposées au risque considéré (^{5})					
2. Nombre d'établissements où le risque existe					
3. Nombre de cas ayant donné lieu à indemnisation pour la première fois au cours de l'année considérée: Nombre total .... dont: Nombre de cas d'incapacité permanente (^{6})					
4. Nombre de décès au cours de l'année considérée (^{7})					
5. Nombre total de cas indemnisés au cours de l'année considérée (^{8})					

(^{1}) Nom de l'État membre.
(^{2}) Dénomination de l'agent, références dans la liste nationale et dans la liste européenne.
(^{3}) Si des affections distinctes sont engendrées par le même agent, fournir des renseignements séparés par affection, en donnant chaque fois les références indiquées ci-dessus.
(^{4}) Ne pas remonter au-delà de 1955.
(^{5}) A défaut de données exactes, fournir si possible une estimation.
(^{6}) Les cas étant suivis, si possible, dans leur évolution au-delà de l'année au cours de laquelle ils sont intervenus.
(^{7}) Distinguer si possible les décès afférents à des cas survenus au cours de l'année considérée de ceux afférents à des cas survenus au cours d'années antérieures
(^{8}) C'est-à-dire cas indemnisés pour la première fois au cours de l'année considérée plus cas en cours.

Observations

(Fournir toutes indications, remarques, observations susceptibles d'apporter aux pays intéressés le maximum de renseignements sur l'expérience acquise dans le pays qui a déjà reconnu la maladie professionnelle.)

RECOMMANDATION DE LA COMMISSION
du 20 juillet 1966
aux États membres relative aux conditions d'indemnisation des victimes
de maladies professionnelles
[66/462/CEE]

I

Exposé des motifs

1. La Commission de la Communauté économique européenne à adressé aux États membres, le 23 juillet 1962, une recommandation concernant l'adoption d'une liste européenne des maladies professionnelles; cette recommandation préconisait, en outre, l'introduction, dans les législations nationales sur les maladies professionnelles, de dispositions permettant l'indemnisation des travailleurs atteints de maladies qui ne sont pas inscrites sur les listes nationales mais dont l'origine professionnelle est prouvée, ainsi que l'établissement, entre les pays de la Communauté, d'un échange d'informations sur les agents nocifs et sur les maladies professionnelles donnant droit à réparation dans un pays, mais non reconnues dans un ou plusieurs autres.

2. Le paragraphe 4 de l'exposé des motifs de la recommandation du 23 juillet 1962 évoquait les problèmes que posent encore les divergences existant dans les dispositions législatives, règlementaires et administratives en la matière et indiquait qu'après l'harmonisation des listes de maladies professionnelles, « les étapes suivantes pourraient porter tant sur les conditions d'octroi que sur les niveaux des prestations ».

   En outre, la réalisation progressive de la libre circulation des travailleurs à l'intérieur de la Communauté prévue par le traité nécessite également l'harmonisation des législations en vue d'assurer à tous les travailleurs une protection égale dans chacun des pays de la Communauté où ils seront amenés à établir leur résidence et leur lieu de travail. Une telle harmonisation facilitera l'application des règlements relatifs à la sécurité sociale des travailleurs migrants, dont certaines dispositions visant le cas de travailleurs ayant été exposés à un même risque dans deux ou plusieurs pays, s'appliquent difficilement en raison des différences existant entre les législations.

   La recommandation ci-après vise exclusivement les conditions mises à l'octroi des prestations qui, en raison de leur nature, sont propres aux maladies professionnelles.

3. Dans la mesure où elle repose sur le système dit « de la liste » (ou sur le système dit « mixte » qui comporte également une liste) — comme c'est le cas pour les législations des six États membres — toute législation relative à la réparation des maladies professionnelles fait bénéficier le travailleur d'une présomption légale quant à l'origine professionnelle de la maladie dont il est atteint, dès lors que cette maladie figure à la liste et que son activité professionnelle le met en contact avec l'agent nocif, générateur d'une telle maladie.

4. Les listes nationales d'agents nocifs ou de maladies professionnelles contiennent souvent pour chaque agent nocif ou pour certains d'entre eux, des indications complémentaires de differente nature.

    Ces indications peuvent consister:

    a) En une symptomatologie ou en une description plus ou moins complète des manifestations cliniques que doit présenter l'affection pour pouvoir être considérée comme maladie professionnelle, ou en une indication relative à son degré de gravité eu égard à la cessation du travail qu'elle doit avoir entrainée;

    b) En une énumération des activités, travaux ou milieux professionnels de nature à exposer le travailleur au risque considéré;

    c) Dans la mention d'une durée minimum de l'exposition au risque pour que celui-ci puisse être considéré légalement comme cause de la maladie;

    d) Dans la mention d'un délai maximum dit « de prise en charge », qui court à partir de la cessation de l'exposition au risque, et avant l'expiration duquel la maladie doit être constatée pour être encore légalement imputée à ce risque.

5. Quant à leur effet juridique, ces mentions peuvent avoir un caractère simplement indicatif ou être, au contraire, impératives.

    Dans le premier cas, elles n'ont qu'une valeur de renseignement pour le médecin expert et l'organisme assureur et ne devraient normalement pas être reprises dans des dispositions de droit positif.

    Dans le second cas, elles constituent des conditions limitatives fixées pour l'attribution des prestations, conditions à défaut desquelles la maladie ne peut être considérée comme ayant une origine professionnelle ni, par conséquent, donner lieu à indemnisation à ce titre.

6. Le jeu de la présomption légale établie par l'existence de la liste des maladies professionnelles, et les conditions d'octroi de prestations dont sont assorties celles-ci, permettent une application quasi automatique des dispositions législatives créées d'ailleurs à défaut d'une définition générale de la maladie professionnelle. Mais, compte tenu de l'état actuel des connaissances dans le domaine de la médecine du travail ainsi que des moyens d'investigation toujours plus développés mis à la disposition des experts, il est devenu nécessaire d'éliminer la plupart des conditions limitant de manière impérative le droit à indemnisation.

    Les réalites médicales ne peuvent être inscrites dans un cadre de limites impératives, car les manifestations cliniques et l'évolution des maladies peuvent présenter des variations importantes suivant la constitution et la manière de réagir de chaque malade.

    En outre, l'évolution technique entraîne des modifications des conditions et, le cas échéant, des délais dans lesquels un travailleur peut subir les effets de certains agents nocifs générateurs de maladies professionnelles.

    Aussi les conditions restrictives actuelles sont-elles généralement arbitraires comme le prouve d'ailleurs le fait que, lorsque, pour une même maladie professionnelle, de telles conditions existent dans plusieurs législations nationales, elles n'y sont en aucune manière identiques. Par ailleurs, ces conditions, de limitatives qu'elles étaient à l'origine, sont devenues très souvent de simples énumérations n'ayant plus qu'une valeur indicative.

7.   Néanmoins, il en subsiste qui revêtent encore un caractère impératif et créent de ce fait une situation préjudiciable à l'égard des travailleurs: d'une part, en effet, si l'organisme assureur peut, même lorsque les conditions sont remplies, faire tomber la présomption légale en apportant la preuve qu'il n'y a pas de relation de cause à effet entre l'activité professionnelle et la maladie constatée, d'autre part, en revanche, le tratailleur n'est pas admis, lorsque tout ou partie des conditions ne sont pas remplies, à fournir la preuve de cette relation de cause à effet.

8.   Il existe cependant un petit nombre d'affections pour lesquelles certaines conditions doivent être remplies, mais il n'existe aucune raison d'ordre médical ou autre pour que la liste de ces affections et lesdites conditions ne soient pas les mêmes dans les différentes législations des États membres de la Communauté.

Cette « liste d'exceptions » qui figure en annexe, devra, comme la liste des maladies professionnelles, être révisée par décision de la Commission, au fur et à mesure de l'évolution des connaissances en la matière.

9.   La présente recommandation vise donc essentiellement à faire supprimer, dans la mesure du possible, le caractère limitatif des conditions mentionnées au paragraphe 4 ci-dessus, auxquelles peut être subordonné le jeu d'une présomption légale et à donner son plein effet à une appréciation par les médecins compétents en la matière de la relation de cause à effet sur laquelle est fondée l'attribution des prestations.

Cependant, les indications que contiennent ces conditions doivent être laissées à la disposition des experts, à titre d'information. A cet effet, une série de notices sur les travaux et les milieux de travail exposant au risque, sur les circonstances de la naissance des affections, sur les critères du diagnostic de celles-ci et, dans une certaine mesure, de leur pronostic, relativement aux agents nocifs et maladies professionnelles de la liste européenne seront publiées sous forme de compléments à la présente recommandation.

Ces notices résulteront de la confrontation scientifique sur le plan communautaire des expériences déjà réalisées dans les États membres; en favorisant une meilleure connaissance des risques, elles aideront indirectement, mais de façon non négligeable, la prévention des maladies professionnelles et faciliteront la tâche des médecins de travail.

10.   L'appréciation par un médecin compétent, visée à l'alinéa 1 du paragraphe précédent, doit, le cas échéant, s'appuyer sur une enquête faite sur le lieu du travail avec le concours notamment des représentants de la direction de l'entreprise, de représentants du personnel, du médecin d'usine ou du médecin du service de médecine du travail auquel l'entreprise est affiliée.

11.   Certains pays ont prévu, à côté d'une liste de maladies professionnelles valable pour l'ensemble des catégories professionnelles, une liste spéciale pour l'agriculture et, le cas échéant, pour l'horticulture. Or, la généralisation de l'usage d'engrais chimiques et de pesticides, la modernisation et la mécanisation des procédés de culture, rapprochent de plus en plus les conditions de travail de l'agriculture de celles de l'industrie en ce qui concerne le risque de maladie

professionnelle. Ces listes spéciales ont en réalité un effet équivalant à celui d'une condition limitative quant au secteur d'application. Pour rester dans la logique du système préconisé ci-dessus et ne pas défavoriser les travailleurs agricoles, il convient donc de supprimer ces listes spéciales et d'incorporer dans la liste générale les maladies professionnelles qui y étaient énumérées; il doit en être de même pour les listes spéciales concernant d'autres catégories.

12. Enfin, pour compléter l'ensemble des objectifs ainsi constitué par la recommandation et aboutir à ce qu'en aucun cas une personne, victime d'une maladie à laquelle son activité professionnelle l'a exposée à un degré plus élevé que l'ensemble de la population, ne puisse pas être indemnisée, il convient de rappeler et preciser le système dit « mixte » déjâ préconisé dans la première recommandation sur les maladies professionnelles, car le risque de nouvelles maladies professionnelles peut toujours se présenter et des cas peuvent surgir avant que la liste européenne et les listes nationales n'aient été révisées en vue de tenir compte des acquisitions scientifiques les plus récentes.

13. Toute législation ou réglementation sur les maladies professionnelles ayant un caractère général, elle s'applique également aux personnes et entreprises relevant de la compétence de la Communauté européenne du charbon et de l'acier et de la Communauté européenne de l'énergie atomique.

La Commission de la C.E.E. a donc tenu, ainsi qu'elle l'avait fait pour la recommandation concernant la liste européenne des maladies professionnelles de 1962, à consulter la Haute Autorité de la C.E.C.A. et la Commission de l'Euratom qui, chacune dans sa sphère de compétence, ont donné leur entier appui à la présente recommandation, sans préjudice des actions qui peuvent être menées en application de leurs traités respectifs.

Pour ces motifs, la Commission de la Communauté économique européenne, au titre des dispositions du traité instituant cette Communauté, et notamment des articles 118 et 155, et après avoir consulté le Parlement européen et le Comité economique et social, recommande aux États membres, sans préjudice des dispositions nationales plus favorables:

1. Sans porter atteinte à la présomption légale d'origine résultant de l'inscription d'une maladie sur la liste des maladies professionnelles, de supprimer dans leurs dispositions législatives, réglementaires ou administratives, relatives aux maladies professionnelles, les conditions limitatives mises à l'octroi des prestations, à l'exception des conditions qui sont indiquées pour certaines maladies professionnelles dont la liste figure en annexe à la présente recommandation sous le nom de « liste d'exceptions » ; devront être supprimées les conditions qui portent sur la description des manifestations cliniques des affections, les activités, les travaux ou les milieux professionnels, les délais d'exposition au risque et les délais concernant la constatation de la maladie après la cessation de l'exposition au risque. Si des doutes sérieux subsistent quant à la relation de cause à effet entre l'activité professionnelle et la maladie, la constatation concernant la relation de cause à effet doit se fonder essentiellement sur l'appréciation d'un médecin spécialisé, appuyée éventuellement par l'avis d'un technicien qualifié.

2. D'incorporer dans la liste générale des maladies professionnelles les listes spéciales quir pourraient exister, notamment pour l'agriculture;

3. Lorsqu'une maladie ne figurant pas encore dans la liste européenne est ajoutée dans une liste nationale, de ne prévoir de conditions limitatives, en ce qui la concerne, que s'il s'agit d'une maladie pouvant également être observée avec une certaine fréquence en dehors d'un milieu professionnel determiné, mais à laquelle certains travailleurs, de par leurs activités professionnelles, sont exposés à un degré plus élevé que l'ensemble de la population;

dans ce cas, les conditions doivent être limitées à celles qui sont réellement indispensables pour pallier la difficulté d'établir avec certitude dans chaque cas d'espèce l'origine professionnelle de la maladie et pour garantir l'intervention de solutions identiques pour des cas semblables;

ces conditions ne devront porter que sur:

— la cessation, entraînée par l'affection, de l'activité professionnelle exercée antérieurement;

— les activités, travaux ou milieux professionnels dans lesquels peut exister le risque de la maladie considérée;

— la durée minimum d'exposition au risque;

4. De faire publier les notices sur les maladies professionnelles de leur liste nationale sur la base des notices sur les maladies professionnelles de la liste européenne, notices qui seront établies ultérieurement par la Commission de la C.E.E., afin de fournir, à titre d'information, aux médecins et autres experts techniques des indications sur la symptomatologie de ces maladies, sur les activités, travaux et milieux qui y exposent, sur la durée moyenne d'exposition au risque, ainsi que sur les délais qui s'écoulent généralement entre la cessation de l'activité exposant au risque et la constatation de la maladie;

5. D'introduire dans leur législation une disposition permettant d'indemniser, au titre de la réparation des maladies professionnelles, les travailleurs atteints de maladies contractées du fait de leur travail mais ne pouvant bénéficier de la présomption légale d'origine de la maladie, soit parce que cette maladie n'est pas inscrite sur la liste nationale, soit parce que les conditions établies par la législation ne sont pas remplies ou ne sont remplies qu'en partie; il ne pourra s'agir que de maladies dont le risque est inhérent à l'activité professionnelle et auquel certains travailleurs sont exposés à un degré plus élevé que l'ensemble de la population.

Il y a lieu de prévoir que la preuve de l'origine professionnelle de la maladie est apportée dans chaque cas par l'intéressé, ou établie par son organisme assureur, qui doit, en tout état de cause, prendre d'office toutes initiatives nécessaires à la recherche de l'origine professionnelle de la maladie.

L'indemnisation, dans ces cas particuliers, n'impliquera pas la reconnaissance générale de la maladie comme maladie professionnelle, mais les États membres devront, dès qu'un certain nombre de cas d'une même maladie, dans la même profession, auront bénéficié de cette disposition, entamer la procédure nécessaire en vue de l'inscription de cette maladie sur la liste nationale et en informer la Commission de la C.E.E.

II

En conclusion, la Commission:

— recommande aux gouvernements des États membres d'adopter, dans les meilleurs délais, les mesures nécessaires en vue de réaliser les objectifs indiqués ci-dessus;

— suggère que les administrations nationales compétentes assurent une large diffusion de cette recommandation et des notices sur les maladies professionnelles tant à l'intérieur de leurs propres services qu'auprès des organismes spécialisés — quel que soit le caractère public, semi-public ou privé de ces derniers — ainsi qu'auprès des organisations professionnelles d'employeurs et de travailleurs, des chaires, instituts et des services et associations de médecine du travail;

— invite les gouvernements des États membres à l'informer tous les deux ans, et pour la première fois lors de la prochaine communication relative aux suites données à la recommandation du 23 juillet 1962 concernant la liste européenne des maladies professionnelles, des mesures adoptées en vue de l'application de la présente recommandation;

— rappelle la procedure d'echange d'informations instituée entre les États membres par la recommandation précitée du 23 juillet 1962.

Fait à Bruxelles, le 20 juillet 1966.

ANNEXE

LISTE D'EXCEPTIONS

énumérant les agents nocifs et maladies professionnelles pour lesquels les conditions limitatives indiquées peuvent être prévues

(Liste visée au paragraphe 1 alinéa 1 de la recommandation)

Numéro Correspondant de la liste européenne	Agent nocif ou maladie professionnelle	Conditions
B – 2	Affections cutanées provoquées dans le milieu professionnel, à l'exception de celles engendrées par des agents nocifs désignés ex- pressément dans la liste en vigueur	Affections graves ou à récidives répétées qui ont entraîné la cessation des activités pro- fessionnelles ou l'abandon de toute activité lucrative
C – 5	Troubles respiratoires de caractère asthmatiforme provoqués dans le milieu professionnel, à l'exception de l'asthme provoqué par des agents nocifs désignés expressément dans la liste en vigueur	L'affection doit avoir entraîné la cessation des activités professionnelles ou l'abandon de toute activité lucrative
D – 1	Ankylostomiose	Travaux souterrains, travaux dans des terrains marécageux ou argileux
D – 3	Tétanos	Travaux dans les égouts; travaux pouvant mettre en contact avec des animaux ou des débris d'animaux
D – 4	Maladies contagieuses	Personnes exerçant leurs activités dans les hôpitaux, dans des services de cure et de soins, dans les mater nités et dans d'autres services s'oc cupant de soigner des personnes; personnes exerçant leurs activités dans des services et institutions d'assistance sociale, publiques et privées, dans des services de santé, dans des laboratoires de diagnostic et de recherche médicaux

F – 6 – a	Maladies des bourses périarticulaires dues à des pressions, à l'exception des maladies provoquées par l'emploi des outils pneumatiques	Affections chroniques
F – 6 – b	Maladies par surmenage des gaines tendineuses du tissu péritendineux, des insertions musculaires et tendineuses	L'affection doit avoir entraîné la cessation des activités profession nelles ou l'abandon de toute activité lucrative
F – 6 – c	Lésions du ménisque	Travaux exécutés dans les mines, travaux souterrains pendant au moins trois ans
F – 7	Nystagmus	Travaux exécutés dans les mines

RECOMMANDATION DE LA COMMISSION

du 27 juillet 1966

adressée aux États membres et concernant le contrôle médical des travailleurs
exposés à des risques particuliers
[66/464/CEE]

Exposé des motifs

1. La Commission de la Communauté économique européenne a déjà approuvé et recommandé une liste européenne des maladies professionnelles et une liste annexe indicative de maladies à soumettre à déclaration en vue d'une inscription éventuelle dans la liste européenne, ainsi que la généralisation des services de médecine du travail dans les entreprises.

2. De nombreux travailleurs sont exposés au risque de maladies professionnelles et ce risque peut être considérablement réduit grâce au contrôle médical des travailleurs. Ce contrôle médical constitue depuis longtemps l'un des principes fondamentaux de la médecine du travail qui tend à prévenir ainsi les maladies des travailleurs en général et les maladies professionnelles en particulier.

3. Tous les États membres ont déjà inclus dans leur législation sur la protection du travail le principe du contrôle médical des travailleurs exposés à des risques particuliers, mais avec des modalités d'application différentes et en particulier avec des listes des risques qui varient parfois sensiblement d'un État à l'autre. Il est donc utile d'harmoniser ces dispositions et en particulier d'adopter pour tous les États une liste de base, aussi uniformisée que possible, des risques en question, pour assurer à tous les travailleurs exposés une protection médicale égale.

4. Il est opportun de prendre la liste européenne des maladies professionnelles comme base d'une liste européenne des risques spécifiques comportant l'obligation du contrôle médical périodique des travailleurs, et il est souhaitable que le contrôle médical soit également étendu aux risques possibles considérés dans la liste annexe en vue, notamment, de recueillir des informations utiles.

5. Il est opportun que le contrôle médical — à effectuer par des médecins spécialisés en médecine du travail — consiste en visites médicales d'embauche et en visites médicales périodiques et que ces visites périodiques soient effectuées à intervalles plus ou moins longs selon la nature du risque, la gravité de l'exposition et l'état physique du travailleur. Les autorités médicales de surveillance compétentes des États membres devront avoir la faculté d'étendre le contrôle médical à d'autres cas, de varier la fréquence des visites périodiques ou de les faire compléter par tout autre examen, et, même, sous certaines conditions, d'exempter l'employeur de l'obligation de ce contrôle.

6. Toute réglementation sur le contrôle médical des travailleurs devant avoir un caractère général, s'applique également aux personnes et entreprises relevant de la compétence de la Communauté européenne du charbon et de l'acier. Le domaine de la compétence de la Communauté européenne de l'energie atomique, qui est reglé par des normes de base de l'Euratom, n'est pas couvert par la présente recommandation. La Commission de la Communauté économique européenne a donc tenu à consulter la Haute Autorité de la C.E.C.A. qui a donné son plein appui à la présente recommandation sans préjudice des actions qu'elle peut mener dans le cadre de son traité.

Recommandation

Pour ces motifs, et en vertu des dispositions du traité instituant la Communauté économique européenne, et notamment des articles 118 et 155, la Commission, vu la recommandation pour l'adoption d'une liste européenne des maladies professionnelles, vu la recommandation relative à la médecine du travail dans l'entreprise, et en particulier le point 24 alinéa 5 qui recommande la mise en place immédiate des services de médecine du travail dans les entreprises relevant de branches d'activité dans lesquelles la fréquence des risques est en général très élevée ou celles où la santé des travailleurs est exposée à des risques particuliers, après consultation du Parlement européen et du Comité économique et social, recommande aux États membres de prendre, au plus tard dans un délai de deux ans, les dispositions législatives, réglementaires, administratives et toute autre initiative appropriées en vu d'assurer la réalisation des objectifs suivants:

a) Introduire l'obligation du contrôle médical des travailleurs salariés occupés à des travaux exposant à des risques particuliers;

b) Étendre progressivement cette obligation au moins à tous les salariés occupés aux travaux énumérés, à titre indicatif, au tableau annexé, dans la mesure où ces travaux les exposent effectivement aux risques visés dans la liste européenne des maladies professionnelles;

c) Faire consister ce contrôle en:

   1. Visite médicale d'embauche à répéter à l'occasion du changement de travail dès que le nouveau travail comporte l'exposition aux risques considérés au point b);

   2. visites médicales periodiques à effectuer aux intervalles indiqués au tableau;

d) Effectuer une visite médicale, dans le cas d'absences répétées et de courte durée pour maladie, à la reprise du travail après une absence prolongée à cause d'une maladie ou d'un accident ou à la reprise après maladie professionnelle, quelle qu'en ait été la durée; effectuer une visite médicale à la demande du travailleur quand celui-ci, sur la base de symptômes subjectifs, estime être atteint d'une maladie professionnelle;

e) Faire completer la visite médicale d'embauche par l'examen radiologique du thorax (radiophotographie ou radiographie), l'examen des urines et l'examen de l'acuité visuelle et de l'audition;

f) Faire compléter la visite d'embauche et les visites périodiques par des examens complémentaires: de spécialistes, radiologiques ou de laboratoire, jugés nécessaires pour le diagnostic d'une maladie professionnelle ou pour l'appréciation de la capacité de travail, le médecin qui a effectué la visite restant en outre libre d'en demander éventuellement d'autres qui ne sont pas normalement prévus, pourvu qu'ils soient indispensables;

g) Donner à l'autorité de surveillance compétente la faculté:

1. D'étendre l'obligation du contrôle médical à d'autres risques que ceux considérés dans la liste européenne des maladies professionnelles ou à d'autres travaux que ceux énumérés au tableau annexé;

2. de faire répéter les visites périodiques à des intervalles autres que ceux indiqués en annexe, compte tenu des conditions d'hygiène dans lesquelles se déroule le travail, des mesures techniques de prévention adoptées, et compte tenu des conditions psychiques et physiques du travailleur, selon l'appréciation du médecin responsable des visites;

3. d'étendre également l'obligation du contrôle médical à d'autres catégories de travailleurs qui, étant occupés dans le même local, sont exposés, bien que dans une moindre mesure, au même risque;

4. d'exempter l'employeur de l'obligation du contrôle médical périodique des travailleurs lorsque, par suite de la faible quantité des matières et des agents nocifs traités et de l'efficacité des mesures préventives adoptées ou du caractère occasionnel du travail insalubre, on peut raisonnablement considérer comme inexistant le risque couru par la santé des travailleurs;

5. de prescrire, pour compléter les examens complémentaires cités à l'alinéa e), d'autres examens de spécialistes, radiologiques ou de laboratoire, s'ils sont estimés indispensables au diagnostic à des fins préventives;

h) Confier l'exécution des visites médicales à des médecins spécialisés en médecine du travail et publier, par l'intermédiaire des autorités médicales compétentes en matière d'inspection du travaiol, des directives pour leur exécution;

i) Envoyer aux services de la C.E.E., en même temps que les informations annuelles pour l'établissement de l'exposé sur l'évolution de la situation sociale dans la Communauté, toute information utile pour la révision périodique biennale du tableau indicatif annexé.

La Commission signale enfin qu'il serait souhaitable que le contrôle médical soit également étendu aux travailleurs exposés aux risques indiqués dans l'annexe II de la recommandation européenne sur les maladies professionnelles, et plus précisément aux risques considérés dans la « Liste indicative des maladies à soumettre à déclaration en vue d'une inscription éventuelle dans la liste européenne », afin de recueillir les éléments relatifs à l'existence, à la fréquence et à la nature des maladies professionnelles qu'ils provoquent, en vue de prévenir ces maladies.

Fait à Bruxelles, le 27 juillet 1966.

ANNEXE

Tableau indicatif des risques – selon la classification établie dans la liste européenne des maladies professionnelles – et des catégories de travailleurs pour lesquels est recommandé le contrôle périodique de la santé et la périodicité des visites médicales

A. MALADIES PROFESSIONNELLES PROVOQUÉES PAR DES AGENTS CHIMIQUES

Risques	*Travailleurs employés aux travaux suivants dans la mesure où ils sont exposés à l'action nocive de la substance*	*Périodicité des visites*
1. Arsenic et ses composés	a) Production de l'arsenic b) grillage des pyrites arsenicales c) préparation d'alliages et de composés d) préparation de couleurs, d'émaux et d'autres produits contenant des composés de l'arsenic e) travaux de peinture, de vernissage et d'émaillage f) préparation de mélanges pour la production du verre g) teinture des fils et des tissus h) autres emplois de l'arsenic, d'alliages et de ses composés dans l'industrie chimique i) emploi dans l'agriculture de produits antiparasitaires contenant de l'arsenic, dans la mesure où cet emploi a un caractère professionnel	3–6–12 mois
2. Béryllium (glucinium) et ses composés	a) Production de béryllium b) préparation d'alliages et de composés c) fabrication de lampes, écrans et autres matériaux fluorescents d) fabrication de cristaux, de céramiques et de produits réfractaires qui expose à l'inhalation de poudres, fumées ou vapeurs contenant du béryllium ou ses composés	12–24 mois
3. Oxyde de carbone	a) Production, distribution et emploi industriel de l'oxyde de carbone et de mélanges gazeux contenant de l'oxyde de carbone b) réparation de conduites de gaz contenant CO c) conduites thermiques des fours de l'industrie métallurgique et de céramique d) second traitement de verre à la flamme e) travaux effectués dans des salles d'essais industriels de moteurs à combustion interne ou à explosion	6–12 mois
3. a)Oxychlorure de carbone	a) Production et utilisation de l'oxychlorure de carbone	mensuelle
3. b)Acide cyanhydrique, cyanures et composés du cyanogène	a) Production d'acide cyanhydrique de cyanures et d'autres composés de cyanogène b) dératisation et désinfestation avec des produits, dans la mesure où ces travaux ont un caractère professionnel c) destruction de parasites nocifs pour l'agriculture avec ces produits, dans la mesure où elle prend un caractère professionnel d) dépuration chimique du gaz d'éclairage avec ces produits e) opération de trempe et de cémentation avec des cyanures	3–6 mois
4. Cadmium et ses composés	a) Production du cadmium b) préparation d'alliages et de composés c) fabrication de colorants	6–12 mois

Risques	Travailleurs employés aux travaux suivants dans la mesure où ils sont exposés à l'action nocive de la substance	Périodicité des visites
	d) cadmiage	
	e) métallisation au pistolet	
	f) fabrication des accumulateurs au nickel-cadmium	
	g) soudure, taille, chauffage au rouge blanc, à l'arc électrique ou à la flamme oxyacétylénique ou oxhydrique d'objets recouverts de cadmium ou en contenant	
	h) production d'étoffes étanches avec des préparations de cadmium	
	i) coloration du verre	
	j) fabrication de feux d'artifice	
	k) fabrication de lampes à vapeurs de cadmium	
5. Chrome et ses composés	a) Production du chrome	semestrielle
	b) préparation d'alliages et de composés	
	c) régénération des chromates alcalins	
	d) fabrication de colorants	
	e) travaux de teinture avec des colorants de chrome dans l'industrie textile, des tapis, du verre et de la porcelaine	
	f) chromage	
	g) tannage des peaux	
	h) fabrication de feux d'artifice	
	i) production des allumettes	
	j) opérations comportant l'emploi de chromates alcalins dans l'industrie de la photographie et de la presse	
	k) imprégnation du bois avec des produits à base de chrome	
6. Mercure et ses composés	a) Production de mercure	4–6–12 mois
	b) préparation d'amalgames et de composés	
	c) fabrication réparation et entretien d'appareils et instruments à mercure	
	d) traitement du poil pour chapeaux (secrétage)	
	e) production et traitement en blanc du feutre obtenu par secrétage par des préparations mercurielles	
	f) travail en noir du feutre secrété	
	g) opérations d'électrolyse avec cathode au mercure	
	h) dorure et argenture au feu avec utilisation de mercure	
	i) fabrication de capsules	
	j) traitement des minerais aurifères et argentifères de récupération	
	k) emploi de pompes à mercure	
	l) emploi d'antiparasitaires contenant des composés organiques du mercure, dans la mesure où cet emploi a un caractère professionnel	
	m) préparation et emploi de vernis contenant du mercure et ses composés	
7. Manganèse et ses composés	a) Production du manganèse	6–12 mois
	b) travaux aux moulins de péroxyde de manganèse	
	c) préparation d'alliages et de composés	
	d) fabrication de piles à sec	
	e) préparation de mélanges du verre et des émaux	
	f) production des allumettes	
8. Acide nitrique, oxydes d'azote, ammoniaque	a) Production de l'acide nitrique	3–6 mois
	b) production de la nitrocellulose	

Risques	Travailleurs employés aux travaux suivants dans la mesure où ils sont exposés à l'action nocive de la substance	Périodicité des visites
	c) production des explosifs par processus de nitration	
	d) production de colorants azoïques	
	e) décapage et gravure des métaux	
9. Nickel et ses composés	a) Production et emploi du nickel carbonyle	annuelle
10. Phosphore et ses composés	a) Production du phosphore et de ses composés	4–6 mois
	b) emploi du phosphore blanc	
	c) travaux qui exposent à l'inhalation d'hydrogène phosphoré	
	d) travaux agricoles comportant l'emploi d'antiparasitaires contenant des composés organiques du phosphore, dans la mesure où ces travaux ont un caractère professionnel	
11. Plomb et ses composés	a) Production du plomb	3–6 mois
	b) préparation d'alliages et de composés	
	c) fabrication et préparation de couleurs, émaux, vernis et mastics contenant du plomb	
	d) fabrication de lames, tubes, projectiles et autres objets en plomb; triage et récupération de matériaux plombifères	
	e) opérations de peinture et de revêtement avec des mastics et des couleurs au plomb; enlèvement de vernis plombifères	
	f) composition typographique (à la main, à la linotype, à la monotype, à la stéréotypie)	
	g) chromolithographie effectuée avec des couleurs et des poudres plombifères	
	h) trempage au plomb	
	i) grillage des pyrites plombifères	
	j) travaux de soudure avec des alliages plombifères et taille à l'arc électrique ou au chalumeau de pièces de métal contenant du plomb ou vernies avec ces produits	
	k) fabrication et réparation d'accumulateurs	
	l) métallisation au plomb au pistolet	
	m) travaux de vernissage à chaud de pièces métalliques couvertes, même partiellement, de peintures à base de pigments plombifères	
	n) travaux de nettoyage à la limaille de plomb	
	o) taille de diamants en utilisant les «dops» plombifères	
11. a) Plomb tétraéthyle	a) Production de plomb tétraéthyle	hebdomadaire
	b) éthylation de l'essence	mensuelle
	c) nettoyage et réparation de réservoirs contenant du plomb tétraéthyle et de l'essence éthylée	trimestrielle
12. Anhydride sulfureux	a) Production de soufre	semestrielle
	b) production de l'anhydride sulfureux	
	c) blanchissage de la paille, du papier et des fibres textiles	
	d) soufrage des fruits et des substances alimentaires en général	
	e) dératisation et désinfestation, dans la mesure où ces travaux ont un caractère professionnel	
12. a) Acide sulfurique	a) Production de l'acide sulfurique	semestrielle
	b) carbonissage des laines	
	c) décapage des métaux	
	d) production du zinc électrolytique	
	e) purification et raffinage des graisses et des huiles	
	f) emploi dans les synthèses organiques	

Risques	Travailleurs employés aux travaux suivants dans la mesure où ils sont exposés à l'action nocive de la substance	Périodicité des visites
12. b) Hydrogène sulfuré	a) Raffinage des huiles minérales b) filage de la viscose	semestrielle
12. c) Sulfure de carbone	a) Production du sulfure de carbone b) emploi comme solvant c) traitement de l'alcalicellulose et opérations consécutives jusqu'au séchage du produit d) vulcanisation du caoutchouc e) dératisation et désinfestation, dans la mesure où ces travaux ont un caractère professionnel	trimestrielle
13. Thallium et ses composés	a) Production du thallium b) préparation d'alliages et de composés c) désinfestation au sulfate de thallium, dans la mesure où ce travail a un caractère professionnel	4 mois
14. Vanadium et ses composés	a) Production du vanadium b) préparation d'alliages et de composés c) nettoyage des installations de combustion de la naphte et récupération des cendres d) emploi comme matières premières dans les processus industriels e) production de mélanges pour la fabrication du verre	semestrielle
15. Chlore et ses composés inorganiques	a) Production du chlore et de l'acide chlorhydrique b) emploi comme matières premières dans les processus industriels c) décapage des métaux à l'acide chlorhydrique d) blanchiment des fibres textiles avec l'acide chlorhydrique	trimestrielle
15. a) Brome et ses composés inorganiques	a) Production du brome et de composés inorganiques b) emploi du brome comme matière dans les processus industriels	trimestrielle
15. b) Iode et ses composés inorganiques	a) Production de l'iode et de composés inorganiques	trimestrielle
15. c) Fluor et ses composés	a) Production du fluor et de l'acide fluorhydrique b) préparation de composés c) gravure du verre d) préparation de la cryolithe artificielle e) électrolyse de l'alumine avec l'emploi de cryolithe f) fabrication des superphosphates dans le groupe des fluorures de sodium . g) emploi des fluorures dans l'industrie des émaux h) imprégnation du bois avec des composés du flour	trimestrielle
16. Hydrocarbures aliphatiques saturés ou non, cycliques, ou non, constituants de l'éther de pétrole et de l'essence	a) Distillation du pétrole b) préparation industrielle de mélanges d'essence c) préparation et emploi industriel de solvants à base d'essence	semestrielle
17. Dérivés halogénés des hydrocarbures aliphatiques saturés ou non, cycliques ou non	a) Production b) emploi comme matières premières dans les processus industriels c) emploi de solvants contenant des dérivés halogénés des hydrocarbures aliphatiques	semestrielle

Risques	Travailleurs employés aux travaux suivants dans la mesure où ils sont exposés à l'action nocive de la substance	Périodicité des visites
18. Alcools, glycols, éthers cétones, esters organiques et leurs dérivés halogénés	a) Production b) emploi comme matières premières dans les processus industriels c) emploi comme solvants dans les processus industriels	semestrielle
19. Acides organiques, aldéhydes	a) Production	annuelle
20. Nitrodérivés aliphatiques, esters de l'acide nitrique	a) Production b) emploi comme matières premières dans les processus industriels	semestrielle
21. Benzène, toluène, xylène et autres homologues du benzène, naphtaline et homologues	a) Production des hydrocarbures benzéniques et homologues b) rectification du benzène et de ses homologues c) emploi du benzène et de ses homologues comme matières premières dans les processus industriels d) préparation et emploi de solvants contenant du benzène ou ses homologues e) rotocalcographie	trimestrielle
22. Dérivés halogénés des hydrocarbures aromatiques	a) Production b) emploi comme matières premières dans les processus industriels	trimestrielle
23. Phénols et homologues, thiophénols et homologues, naphtols et homologues et leurs dérivés halogénés: dérivés halogènes des alkylaryloxydes et des alkylarysulfures, benzoquinone	a) Production b) emploi comme matières premières dans les processus industriels	semestrielle
24. Amines (primaires, secondaires, tertiaires, hétérocycliques) et hydrazines aromatiques et leurs dérivés halogénés, phénoliques, nitrosés, nitrés et sulfonés	a) Production b) emploi dans les processus industriels	semestrielle
25. Nitrodérivés des hydrocarbures aromatiques et des phénols	a) Production b) emploi comme matières premières dans les processus industriels	trimestrielle

B. MALADIES PROFESSIONNELLES DE LA PEAU CAUSÉES PAR DES SUBSTANCES ET DES AGENTS NON COMPRIS SOUS D'AUTRES POSITIONS

Risques	Travailleurs employés aux travaux suivants dans la mesure où ils sont exposés à l'action nocive des substances indiquées	Périodicité des visites
1. Cancers cutanés et affections cutanées precancéreuses dues à la suie, au goudron, au bitume, au brai, à l'anthracène, aux huiles minérales, à la paraffine brute et aux composés, produits et résidus de ces substances.	a) Production du goudron, du bitume, des huiles minérales, de la paraffine brute, du brai b) production d'agglomérés d'anthracite c) asphaltage des routes d) emploi habituel du goudron pour le crépissage ou comme isolant	annuelle

Risques	Travailleurs employés aux travaux suivants dans la mesure où ils sont exposés à l'action nocive des substances indiquées	Périodicité des visites
2. Affections cutanées provoquées dans le milieu professionnel par des substances non considérées sous d'autres positions	a) Travaux de peinture, de vernissage, replâtrage, émaillage, comportant l'emploi fréquent ou prolongé d'huile de lin, de térébenthine et d'autres solvants ou diluants irritant la peau b) production de la laine de verre c) production du ciment et travaux au mortier d) production et emploi professionnel de détergents e) production et traitement des résines naturelles f) fabrication de conserves alimentaires et de confiserie g) travaux dans les salines et les mines de sel	annuelle

C. MALADIES PROFESSIONNELLES PROVOQUÉES PAR L'INHALATION DE SUBSTANCES ET AGENTS NON COMPRIS SOUS D'AUTRES POSITIONS

Risques	Travailleurs employés aux travaux suivants dans la mesure où ils sont exposés à l'action nocive des substances indiquées	Périodicité des visites
1. Pneumoconioses a) Silicose, associée ou non à la tuberculose pulmonaire	Travaux qui exposent à l'inhalation de poussières de silice	annuelle
b) Asbestose, associée ou non à la tuberculose pulmonaire ou à un cancer du poumon	a) Extraction et traitement de l'amiante b) fabrication d'objets en amiante	annuelle
c) Pneumonconioses dues aux poussières de silicates	Travaux qui exposent à l'inhalation de poussières de silicates	annuelle
2. Affections broncho-pulmonaires dues aux poussières ou fumées d'aluminium ou de ses composés	a) Production de poussières d'aluminium b) extraction de l'oxyde d'aluminium de la bauxite	2 ans
3. Affections broncho-pulmonaires dues aux poussières de métaux durs	Travaux qui exposent à l'inhalation de poussières de métaux durs carbure de wolfram, cobalt, titanium et tantalium	tous les deux ans
4. Affections broncho-pulmonaires causées par les poussières de scories Thomas	a) Fabrication, broyage, stockage et expédition de scories Thomas	annuelle
5. Asthme provoqué dans le milieu professionnel par des substances non incluses sous d'autres positions		quand le travailleur présente ou rapporte des symptômes de la maladie annuelle

D. MALADIES PROFESSIONNELLES INFECTIEUSES ET PARASITAIRES

Risques	*Travailleurs employés aux travaux suivants dans la mesure où ils sont exposés aux maladies professionnelles indiquées*	*Périodicité des visites*
1. Helminthiase, ankylostome duodénal, anguillule de l'intestin	a) Travaux dans les galeries b) travaux dans les fours à briques c) travaux dans les mines d) maraîchers e) travaux agricoles dans les zones où l'ankylostomiase est endémique	annuelle
2. Maladies tropicales dont: paludisme, amibiase, trypanosomiase, dengue, fièvre pappataci, fièvre de Malte, fièvre récurrente, fièvre jaune, peste, leischmaniose, pian lèpre, typhus exanthématique et autres rickettsioses		quand le travailleur rapporte ou présente des symptômes de la maladie ou après séjour dans des régions où ces maladies sont endémiques
3. Maladies infectieuses ou parasitaires transmises à l'homme par des animaux ou débris d'animaux	a) Travaux dans les infirmeries animales b) travaux dans les abattoirs c) travaux dans l'industrie et destruction d'abats d) tannage des peaux e) travail du crin f) ramassage et traitement des résidus animaux pour la fabrication d'engrais, de colle et d'autres produits industriels	semestrielle
4. Maladies infectieuses du personnel s'occupant de prévention, soins, assistance à domicile et recherches	a) Travaux de prophylaxie, de soins et d'assistance sanitaire b) recherches	annuelle

F. MALADIES PROFESSIONNELLES PROVOQUÉES PAR DES AGENTS PHYSIQUES

Risques	*Travailleurs employés aux travaux suivants dans la mesure où ils sont exposés à l'action nocive de l'agent indiqué*	*Périodicité des visites*
1. Maladies provoquées par les radiations ionisantes	Voir normes Euratom	
2. Cataracte provoquée par l'énergie radiante	a) Travaux aux fours de fusion des aciéries, des forges, des fonderies, et aux fours des verreries et du carbure de calcium	annuelle
3. Hypoacousie ou surdité provoquée par le bruit	a) Travaux de chaudronnerie b) rivetage des boulons c) corroyage et perçage des tôles avec des poinçons d) essai des moteurs à explosion et à réaction e) production de poudres métalliques avec des machines à pilon f) fabrication de clous g) travail sur des métiers mécaniques de tissage h) travaux avec des marteaux pneumatiques i) granitage, également avec de la grenaille métallique, et travaux de nettoyage industriel à l'air comprimé j) travaux au moulin à aube ou au tambour rotatif k) travaux dans les salles de turbines l) essais des armes à feu	annuelle

Risques	Travailleurs employés aux travaux suivants dans la mesure où ils sont exposés à l'action nocive de l'agent indiqué	Périodicité des visites
4. Maladies provoquées par les travaux dans l'air comprimé	a) Travaux dans les caissons b) travaux de scaphandrier	variables selon la pression
5. Maladies estéo-articulaires ou angio-neurotiques provoquées par les vibrations mécaniques	a) Utilisation de marteaux pneumatiques b) utilisation de meules flexibles	annuelle
6. a) Maladies des bourses péri-articulaires dues à des pressions cellulites sous-cutanées b) maladies par surmenage des gaines tendineuses du tissu péritendineux des insertions musculaires et tendineuses c) lésions du ménisque chez les mineurs d) arrachements par surmenage des apophyses épineuses e) paralysie des nerfs dues à la pression		quand le travailleur rapporte ou présente des symptômes de la maladie
7. Nystagmus des mineurs	Mineurs de fond	annuelle

COUNCIL RESOLUTION
—— of 21 January 1974 ——
concerning a social action programme
[74/c 13/1]

THE COUNCIL OF THE EUROPEAN COMMUNITIES,

Having regard to the Treaties establishing the European Communities;

Having regard to the draft from the Commission;

Having regard to the Opinion of the European Parliament;

Having regard to the Opinion of the Economic and Social Committee;

Whereas the Treaties establishing the European Communities assigned to them tasks with relevance to social objectives;

Whereas, pursuant to Article 2 of the Treaty establishing the European Economic Community, the European Economic Community shall have as a particular task to promote throughout the Community a harmonious development of economic activities, a continuous and balanced expansion, an increase in stability and an accelerated raising of the standard of living;

Whereas the Heads of State or of Government affirmed at their conference held in Paris in October 1972 that economic expansion is not an end in itself but should result in an improvement of the quality of life as well as of the standard of living;

Whereas the Heads of State or of Government emphasized as one of the conclusions adopted at the abovementioned conference that they attach as much importance to vigorous action in the social field as to the achievement of Economic and Monetary Union and invited the Community institutions to draw up a Social action programme providing for concrete measures and the corresponding resources particularly in the framework of the European Social Fund on the basis of suggestions put forward by the Heads of State or of Government and the Commission at the said Conference;

Whereas such a programme involves actions designed to achieve full and better employment, the improvement of living and working conditions and increased involvement of management and labour in the economic and social decisions of the Community, and of workers in the life of undertakings;

Whereas actions described in the above programme should be implemented in accordance with the provisions laid down in the Treaties, including those of Article 235 of the Treaty establishing the European Economic Community;

Having regard to the wishes expressed by management and labour;

Whereas, irrespective of serious threats to employment which may arise from the situation obtaining at the time of adoption of this Resolution, and without prejudice to the results of any future studies or measures, the Community should decide on the objectives and priorities to be given to its action in the social field over the coming years;

Takes note of the Social Action Programme from the Commission,

Considers that vigorous action must be undertaken in successive stages with a view to realising the social aims of European union, in order to attain the following broad objectives: full and better employment at Community, national and regional levels, which is an essential condition for an effective social policy; improvement of living and working conditions so as to make possible their harmonization while the improvement is being maintained; increased involvement of management and labour in the economic and social decisions of the Community, and of workers in the life of undertakings;

Considers that the Community social policy has an individual role to play and should make an essential contribution to achieving the aforementioned objectives by means of Community measures or the definition by the Community of objectives for national social policies, without however seeking a standard solution to all social problems or attempting to transfer to Community level any responsibilities which are assumed more effectively at other levels;

Considers that social objectives should be a constant concern of all Community policies;

Considers that it is essential to ensure the consistency of social and other Community policies so that measures taken will achieve the objectives of social and other policies simultaneously;

Considers that, to achieve the proposed actions successfully, and particularly in view of the structural changes and imbalances in the Community, the necessary resources should be provided, in particular by strengthening the role of the European Social Fund;

Expresses the political will to adopt the measures necessary to achieve the following objectives during a first stage covering the period from 1974 to 1976, in addition to measures adopted in the context of other Community policies:

Attainment of full and better employment in the Community

— to establish appropriate consultation between Member States on their employment policies guided by the need to achieve a policy of full and better employment in the Community as a whole and in the regions;

— to promote better co-operation by national employment services;

— to implement a common vocational training policy, with a view to attaining progressively the principal objectives thereof, especially approximation of training standards, in particular by setting up a European Vocational Training Centre;

— to undertake action for the purpose of achieving equality between men and women as regards access to employment and vocational training and advancement and as regards working conditions including pay, taking into account the important role of management and labour in this field;

— to ensure that the family responsibilities of all concerned may be reconciled with their job aspirations;

— to establish an action programme for migrant workers and members of their families which shall aim in particular:

— to improve the conditions of free movement within the Community of workers from Member States, including social security, and the social infra-structure of the

Member States, the latter being an indispensable condition for solving the specific problems of migrant workers and members of their families, especially problems of reception, housing, social services, training and education of children;

— to humanize the free movement of Community workers and members of their families by providing effective assistance during the various phases, it being understood that the prime objective is still to enable workers to find employment in their own regions;

— to achieve equality of treatment for Community and non-Community workers and members of their families in respect of living and working conditions, wages and economic rights, taking into account the Community provisions in force;

— to promote consultation on immigration policies vis-à-vis third countries;

— to initiate a programme for the vocational and social integration of handicapped persons, in particular making provisions for the promotion of pilot experiments for the purpose of rehabilitating them in vocational life, or where appropriate, of placing them in sheltered industries, and to undertake a comparative study of the legal provisions and the arrangements made for rehabilitation at national level;

— to seek solutions to the employment problems confronting certain more vulnerable categories of persons (the young and the aged);

— to protect workers hired through temporary employment agencies and to regulate the activities of such firms with a view to eliminating abuses therein;

— to continue the implementation of the Council's conclusions on employment policy in the Community and particularly those concerning the progressive integration of the labour markets including those relating to employment statistics and estimates;

Improvement of living and working conditions as to make possible their harmonization while the improvement is being maintained

— to establish appropriate consultations between Member States on their social protection policies with the particular aim of their approximation on the way of progress;

— to establish an action programme for workers aimed at the humanization of their living and working conditions, with particular reference to:

— improvement in safety and health conditions at work;

— the gradual elimination of physical and psychological stress which exists in the place of work and on the job, especially through improving the environment and seeking ways of increasing job satisfaction;

— a reform of the organization of work giving workers wider opportunities, especially those of having their own responsibilities and duties and of obtaining higher qualifications;

— to persevere with and expedite the implementation of the European Social Budget;

— gradually to extend social protection, particularly in the framework of social security schemes, to categories of persons not covered or inadequately provided for under existing schemes;

— to promote the co-ordination of social security schemes for self-employed workers with regard to freedom of establishment and freedom to provide services;

— to invite the Commission to submit a report on the problems arising in connection with co-ordination of supplementary schemes for employed persons moving within the Community;

— progressively to introduce machinery for adapting security benefits to increased prosperity in the various Member States;

— to protect workers' interests, in particular with to the retention of rights and advantages in the case of mergers, concentrations or rationalization operations;

— to implement, in co-operation with the Member States, specific measures to combat poverty by drawing up pilot schemes;

Increased involvement of management and labour in the economic and social decisions of the Community, and of workers in the life of undertakings

— to refer more extensively to the Standing Committee on Employment for the discussion of all questions with a fundamental influence on employment;

— to help trade union organizations taking part in Community work to establish training and information services for European affairs and to set up a European Trade Union Institute;

— progressively to involve workers or their representatives in the life of undertakings in the Community;

— to facilitate, depending on the situation in the different countries, the conclusion of collective agreements at European level in appropriate fields;

— to develop the involvement of management and labour in the economic and social decisions of the Community;

Lays down the following priorities among the actions referred to in this Resolution:

Attainment of full and better employment in the Community

1. The establishment of appropriate consultation between Member States on their employment policies and the promotion of better co-operation by national employment services.

2. The establishment of an action programme for migrant workers who are nationals of Member States or third countries.

3. The implementation of a common vocational training policy and the setting up of a European Vocational Training Centre.

4. The undertaking of action to achieve equality between men and women as regards access to employment and vocational training and advancement and as regards working conditions, including pay.

Improvement of living and working conditions so as to make possible their harmonization while the improvement is being maintained

5. The establishment of appropriate consultations between Member States on their social protection policies.

6. The establishment of an initial action programme, relating in particular to health and safety at work, the health of workers and improved organization of tasks, beginning in those economic sectors where working conditions appear to be the most difficult.

7. The implementation, in co-operation with the Member States, of specific measures to combat poverty by drawing up pilot schemes.

Increased involvement of management and labour in the economic and social decisions of the Community, and of workers in the life of undertakings

8. The progressive involvement of workers or their representatives in the life of undertakings in the Community.

9. The promotion of the involvement of management and labour in the economic and social decisions of the Community.

Takes note of the Commission's undertaking to submit to it, during 1974, the necessary proposals concerning the priorities laid down above;

Takes note of the Commission's undertaking to submit to it, before 1 April 1974, proposals relating to:

an initial action programme with regard to migrant workers;

the setting up of a European Vocational Training Centre;

a directive on the harmonization of laws with regard to the retention of rights and advantages in the event of changes in the ownership of undertakings, in particular in the event of mergers;

Notes that the Commission has already submitted to it proposals relating to:

—   assistance from the European Social Fund for migrant workers and for handicapped workers;

—   an action programme for handicapped workers in an open market economy;

—   the setting-up of a European General Industrial Safety Committee and the extension of the competence of the Mines Safety and Health Commission;

—   a Directive providing for the approximation of legislation of Member States concerning the application of the principle of equal pay for men and women;

—   the designation as an immediate objective of the overall application of the principle of the standard 40-hour working week by 1975, and the principle of four weeks annual paid holiday by 1976;

— the setting up of a European Foundation for the improvement of the environment and of living and working conditions;

— a Directive on the approximation of the Member States' legislation on collective dismissals.

Undertakes to act, at the latest five months after the Commission has informed the Council of the results of its deliberations arising from the opinions given by the European Parliament and the Economic and Social Committee, if such consultations have taken place, or, if such consultations have not taken place, at the latest nine months from the date of the transmission of the proposals to the Council by the Commission;

Takes note of the Commission's undertaking to submit to it before 31 December 1976 a series of measures to be taken during a further phase.

COUNCIL DECISION

of 27 June 1974

on the setting up of an Advisory Committee
on Safety, Hygiene and Health Protection at Work
[74/325/EEC]

THE COUNCIL OF THE EUROPEAN COMMUNITIES,

Having regard to the Treaty establishing the European Economic Community, and in particular Article 145 thereof;

Having regard to the draft of the Commission;

Having regard to the Opinion of the European Parliament;

Having regard to the Opinion of the Economic and Social Committee;

Whereas the profound transformation in production methods in all sectors of the economy and the spread of dangerous techniques and materials have created new problems for the safety, hygiene and health protection of workers at their place of work;

Whereas the prevention of occupational accidents and diseases, as well as occupational hygiene, are among the objectives of the Treaty establishing the European Economic Community;

Whereas the Council resolution of 21 January 1974 concerning a social action programme envisages an action programme for workers which aims *inter alia* at improvement in safety and health conditions at work;

Whereas a standing body should be envisaged to assist the Commission in the preparation and implementation of activities in the fields of safety hygiene and health protection at work and to facilitate co-operation between national administrations trades unions and employers' organizations;

Whereas this Decision does not conflict with Article 118 of the Treaty establishing the European Economic Community,

HAS DECIDED AS FOLLOWS:

Article 1

An Advisory Committee on Safety, Hygiene and Health Protection at Work (hereinafter called the 'Committee') is hereby established.

Article 2

1.   The Committee shall have the task of assisting the Commission in the Preparation and implementation of activities in the fields of safety, hygiene and health protection at work.

This task shall cover all sectors of the economy except the mineral extracting industries falling within the responsibility of the Mines Safety and Health Commission and except the protection of the health of workers against the dangers arising from ionizing radiations which is subject to special regulations pursuant to the Treaty establishing the European Atomic Energy Community.

2.   The Committee shall have the task in particular, of undertaking the following activities:

(a)   conducting, on the basis of information available to it, exchanges of views and experience regarding existing or planned regulations;

(b)   contributing towards the development of a common approach to problems existing in the fields of safety, hygiene and health protection at work and towards the choice of Community priorities as well as measures necessary for implementing them;

(c)   drawing the Commission's attention to areas in which there is an apparent need for the acquisition of new knowledge and for the implementation of appropriate educational and research projects;

(d)   defining, within the framework of Community action programmes, and in co-operation with the Mines Safety and Health Commission:

—   the criteria and aims of the campaign against the risk of accidents at work and health hazards within the undertaking;

—   methods enabling undertakings and their employees to evaluate and to improve the level of protection;

(e)   contributing towards keeping national administrations, trades unions and employers' organizations informed of Community measures in order to facilitate their co-operation and to encourage initiatives promoted by them aiming at exchanges of experience and at laying down codes of practice.

Article 3

1.   The Committee shall produce an annual report on its activities.

2.   The Commission shall forward that report to the European Parliament, the Council, the Economic and Social Committee and the Consultative Committee of the European Coal and Steel Community.

Article 4

1.   The Committee shall consist of 54 full members, there being for each Member State two representatives of the Government, two representatives of trade unions, and two representatives of employers' organizations.

2. An alternate member shall be appointed for each full member.

   Without prejudice to Article 6(3), the alternative member shall attend Committee meetings only when the member for whom he deputizes is unable to be present.

3. Full members and alternate members of the Committee shall be appointed by the Council which, in respect of representatives of trade unions and employers' associations, shall endeavour to achieve a fair balance in the composition of the Committee between the various economic sectors concerned.

4. The list of the members and the alternate members shall be published by the Council in the *Official Journal of the European Communities* for information purposes.

Article 5

1. The term of office of full members and alternate members shall be three years. Their appointments shall be renewable.

2. On expiry of their term of office, the full members and alternate members shall remain in office until they are replaced or their appointments are renewed.

3. A member's term of office shall end before the expiry of the three year period with his resignation or following a communication from the Member State concerned indicating that the term of office is terminated.

   For the remainder of the term of office, a member shall be replaced in accordance with the procedure laid down in Article 4.

Article 6

1. The Committee shall be chaired by a member of the Commission or, where such member is prevented from so doing and as an exception, by a Commission official to be nominated by him. The Chairman shall not vote.

2. The Committee shall meet when convened by the Chairman, either at the latter's initiative or at the request of a least one-third of its members.

3. The Chairman may, on his own initiative, invite up to two experts to participate in Committee meetings.

   Each Committee member may be accompanied by an expert, provided that he so informs the Chairman at least three days before the Committee meeting.

4. The Committee may establish working parties under the chairmanship of a Committee member.

   They shall submit the results of their proceedings in the form of a report at a meeting of the Committee.

5. Representatives of the Commission's department concerned shall participate in meetings of the Committee and of working parties.

   Secretarial services shall be provided for the Committee and for working parties by the Commission.

Article 7

1. An opinion delivered by the Committee shall not be valid unless two-thirds of its members are present.

2. Opinions of the Committee shall state the reasons on which they are based; they shall be delivered by an absolute majority of the votes validly cast. They shall be accompanied by a written statement of the views expressed by the minority, when the latter so requests.

Article 8

The Committee shall adopt its rules of procedure which shall enter into force after the Council, having received an opinion from the Commission, has given its approval.

Article 9

Without prejudice to Article 214 of the Treaty, Committee members shall be required not to disclose information to which they have gained access through Committee or working party proceedings, if the Commission informs them that the opinion requested or the question raised is of a confidential nature.

In such cases, only Committee members and representatives of the Commission's department shall attend the meetings concerned.

Article 10

This Decision shall enter into force on the fifth day following its publication in the Official Journal of the European Communities.

Done at Luxembourg, 27 June 1974

COUNCIL DECISION

of 27 June 1974

on the extension of the responsibilities of the Mines Safety and Health
Commission to all mineral-extracting industries
[74/326/EEC]

THE COUNCIL OF THE EUROPEAN COMMUNITIES,

Having regard to the Treaty establishing the European Economic Community and in particular Article 145 thereof;

Having regard to the draft of the Commission;

Having regard to the Opinion of the European Parliament;

Having regard to the Opinion of the Economic and Social Committee;

Whereas the representatives of the Governments of the Member States meeting within the special Council of Ministers, by Decision of 9 and 10 May 1957, set up a Mines Safety and Health Commission whose terms of reference as laid down by decision of 9 July 1957 of the representatives of the Governments of the Member States meeting within the Special Council of Ministers, amended by Decision of 11 March 1965 are to follow developments in safety and in the prevention of occupational risks to health in coal mines and to draw up proposals appropriate for the improvement of safety and health in coal mines;

Whereas this body has proved to be an effective and suitable instrument for safeguarding the health and safety of workers in coal mines;

Whereas problems of safety similar to those in coal mines also exist in other mineral-extracting industries;

Whereas the prevention of occupational accidents and diseases, as well as occupational hygiene, are among the objectives of the Treaty establishing the European Economic Community;

Whereas the Council resolution of 21 January 1974 concerning a social action programme envisages an action programme for workers which aims inter alia at improvement in safety and health conditions at work;

Whereas the Safety and Health Commission should be assigned the task of extending to all mineral extracting industries the preventive action which has hitherto been confined to coal mines;

Whereas the representatives of the Governments of the Member States meeting within the Council agreed to assign this task to the Safety and Health Commission,

HAS DECIDED AS FOLLOWS:

Article 1

1. Preventive action against risks of accident and occupational risks to the safety and health of workers in all mineral-extracting industries except simple excavation, excluding the protection of the health of workers against the dangers arising from ionizing radiations which is subject to special regulations pursuant to the Treaty establishing the European Atomic Energy Community shall be the responsibility of the Mines Safety and Health Commission within the terms of reference laid down by Decision of 11 March 1965 of the representatives of the Governments of the Member States meeting within the special Council of Ministers.

2. Mineral extracting industries shall be taken to mean the activities of prospecting and of extraction in the strict sense of the word as well as of preparation of extracted materials for sale (crushing, screening, washing), but not the processing of such extracted materials.

3. Simple excavation shall be taken to mean work whose purpose is not the extraction of materials for use.

Article 2

1. This Decision shall enter into force on the fifth day following its publication in the Official Journal of the European Communities.

2. It shall apply:

   — to the underground activities of the mineral-extracting industries: as from the day laid down in paragraph 1

   — to the other activities of the mineral-extracting industries: as from 1 January 1976.

Done at Luxembourg, 27 June 1974.

COUNCIL RESOLUTION

of 29 June 1978

on an action programme of the European Communities on safety and health
at work
[78/c 165/1]

THE COUNCIL OF THE EUROPEAN COMMUNITIES,

Having regard to the Treaties establishing the European Communities,

Having regard to the draft resolution submitted by the Commission,

Having regard to the opinion of the European Parliament,

Having regard to the opinion of the Economic and Social Committee,

Whereas the Council resolution of 21 January 1974 concerning a social action
programme provides for the establishment of an action programme on safety and
health at work;

Whereas, under Article 2 of the Treaty establishing the European Economic
Community, the Community shall have among its tasks, by establishing a common
market and progressively approximating the economic policies of Member States,
that of promoting throughout the Community a harmonious development of
economic activities, a continuous and balanced expansion and an accelerated
raising of the standard of living;

Whereas at the Conference held in Paris in October 1972 the Heads of State or of
Government affirmed that the first aim of economic expansion, which is not an
end in itself, should be to enable disparities in living conditions to be reduced and
that it should result in an improvement in the quality of life as well as in standards
of living;

Whereas moreover, in Article 117 of the said Treaty, the Member States agree upon
the need to promote improved working conditions and an improved standard of
living for workers, so as to make possible their harmonization while the improvement is being maintained;

Whereas prevention of occupational accidents and diseases and also occupational
hygiene fall within the fields and objectives referred to in Article 118 of the said
Treaty; whereas in this context collaboration should be strengthened between the
Member States and the Commission and between the Member States themselves;

Whereas suitable health protection for the public and effective prevention of
accidents at work and occupational diseases would meet these general objectives;

Whereas in spite of sustained efforts the continuing high level of accidents at work
and of occupational diseases remains a serious problem;

Whereas efforts made in the field of accident prevention and health protection at
the work place have beneficial effects which are reflected in the economic sphere
and in industrial relations;

Whereas a considerable effort is needed at Community level to search for and implement suitable means for maintaining or creating a working environment tailored to the needs of man and his legitimate aspirations;

Whereas both the effectiveness of the measures and their cost should be taken into account in the choice of action at Community level to be undertaken and of the measures to be taken to implement it;

Whereas the improvement of working conditions and the working environment must be envisaged in overall terms and must concern all sectors of the economy;

Whereas the actions should be implemented in accordance with the provisions of the Treaties, including those of Article 235 of the Treaty establishing the European Economic Community;

Whereas it is essential also to encourage the increasing participation of management and labour in the decisions and initiatives in the field of safety, hygiene and health protection at work at all levels, particularly at the level of the undertaking;

Whereas the Advisory Committee on Safety, Hygiene and Health Protection at Work, set up by Council Decision 74/325/EEC of 27 June 1974, must be closely associated with this work;

Whereas the European Foundation for the Improvement of Living and Working Conditions and the European Centre for the Development of Vocational Training may have a role to play in the implementation of certain aspects of the programme;

Whereas, in implementing the actions, account must be taken of work undertaken in other fields, notably in the context of the Council resolution of 17 December 1973 on industrial policy and of the Declaration of the Council of the European Communities and of the representatives of the Governments of the Member States meeting in the Council of 22 November 1973 on the programme of action of the European Communities on the environment, in order to ensure the closest possible co-ordination of actions and proposals;

Whereas, in order to carry out the actions, it is important to ensure that concepts, terminology and also methods of identification, measurement and assessment relating to safety and health risks are harmonized; whereas such a task is of major importance in the context of these actions;

Notes   the action programme from the Commission annexed hereto and approves its general objective which is to increase protection of workers again occupational risks of all kinds by improving the means and conditions of work, knowledge and human attitudes;

Expresses   the political will to take, in keeping with the urgency of the matter and bearing in mind what is feasible at national and Community level, the measures required so that between now and the end of 1982 the following actions in particular can be undertaken:

Accident and disease aetiology connected with work – Research

1. Establish, in collaboration with the Statistical Office of the European Communities, a common statistical methodology in order to assess with sufficient accuracy the frequency, gravity and causes of accidents at work, and also the mortality, sickness and absenteeism rates in the case of diseases connected with work.

2. Promote the exchange of knowledge, establish the conditions for close co-operation between research institutes and identify the subjects for research to be worked on jointly.

Protection against dangerous substances

3. Standardize the terminology and concepts relating to exposure limits for toxic substances.
   Harmonize the exposure limits for a certain number of substances, taking into account the exposure limits already in existence.

4. Develop a preventive and protective action for substances recognized as being carcinogenic, by fixing exposure limits, sampling requirements and measuring methods, and satisfactory conditions of hygiene at the work place, and by specifying prohibitions where necessary.

5. Establish, for certain specific toxic substances such as asbestos, arsenic, cadmium, lead and chlorinated solvents, exposure limits, limit values for human biological indicators, sampling requirements and measuring methods, and satisfactory conditions of hygiene at the workplace.

6. Establish a common methodology for the assessment of the health risks connected with the physical, chemical and biological agents present at the work place, in particular by research into criteria of harmfulness and by determining the reference values from which to obtain exposure limits.

7. Establish information notices on the risks relating to and handbooks on the handling of a certain number of dangerous substances such as pesticides, herbicides, carcinogenic substances, asbestos, arsenic, lead, mercury, cadmium and chlorinated solvents.

Prevention of the dangers and harmful effects of machines

8. Establish the limit levels for noise and vibrations at the work place and determine practical ways and means of protecting workers and reducing sound levels at places of work.

   Establish the permissible sound levels of building-site equipment and other machines.

9. Undertake a joint study of the application of the principles of accident prevention and of ergonomics in the design, construction and utilization of the plant and machinery, and promote this application in certain pilot sectors, including agriculture.

10. Analyse the provisions and measures governing the monitoring of the effectiveness of safety and protection arrangements and organize an exchange of experience in this field.

Monitoring and inspection — improvement of human attitudes

11. Develop a common methodology for monitoring both pollutant concentrations and the measurement of environmental conditions at places of work; carry out inter-comparison programmes and establish reference methods for the determination of the most important pollutants.

   Promote new monitoring and measuring methods for the assessment of individual exposure, in particular through the application of sensitive biological indicators. Special attention will be given to the monitoring of exposure in the case of women, especially of expectant mothers, and adolescents.

   Undertake a joint study of the principles and methods of application of industrial medicine with a view to promoting better protection of workers' health.

12. Establish the principles and criteria applicable to the special monitoring relating to assistance or rescue teams in the event of accident or disaster, maintenance and repair teams and the isolated worker.

13. Exchange experience concerning the principles and methods of organization of inspection by public authorities in the fields of safety, hygiene at work and occupational medicine.

14. Draw up outline schemes at a Community level for introducing and providing information on safety and hygiene matters at the work place to particular categories of workers such as migrant workers, newly recruited workers and workers who have changed jobs.

Notes   that the Commission will take the necessary initiatives for the implementation of this resolution;

Invites   the Commission to submit an annual report to it on the progress made in implementing this resolution.

ANNEX

ACTION PROGRAMME OF THE EUROPEAN COMMUNITIES ON HEALTH AND SAFETY AT WORK

Introduction

A high percentage of the population of the nine Member States is exposed to varying degrees of many and widely divergent occupational risks which could threaten their health and personal safety. Occupational pathology is habitually concerned with accidents and diseases resulting from work, the prevention or diagnosis of which have been the subject of action within the Community for several years, and the harmful effects of which are partly or totally compensated through various schemes.

Despite the efforts made in the Member States of the Community, the number of accidents and diseases resulting from work remains high. Quite apart from their financial importance, the human and social consequences of occupational accidents and diseases are incalculable, since it is not easy to assess the psychological damage done or to take into account the long-term factors connected with accidents and disease. Thus there is good reason to believe that the total social and financial cost of occupational accidents and diseases is far greater than the quantitative estimates at our disposal suggest.

Modern technology uses increasingly advanced processes which present new dangers. They produce or use chemical substances which are inadequately tested for their harmful effects on man. All chemical, physical, mechanical and biological agents and the psychosocial factors connected with work must be readily recognizable and brought under control or eliminated by suitable means so as to avoid any damage to health or a significant reduction in safety.

The prevention, limitation and, where possible, elimination of occupational risks constitute major elements of a policy to protect the health and safety of the workers.

Of course, the Member States have a long tradition in the organization of industrial safety and health but they must also agree to shoulder a joint programme of positive and effective actions to improve the conditions under which man performs his job and do everything possible to ensure his well-being and guarantee the quality of his working environment. In order to implement such a programme, it is necessary not only to harmonize ideas and basic principles, but also to plan and guide technical progress and the organization of work in such a way as to take account of the requirement of health and safety.

In view of the persisting gravity of the problem, the Commission must initiate, promote and develop a common preventive policy with regard to all occupational risks, especially by obtaining fresh knowledge, by encouraging co-operation and co-ordination and by developing appropriate actions at different levels of responsibility or competence. In addition to promoting exchanges and the improvement of reciprocal information, such a programme should aim to persuade responsible authorities in the Member States and the social partners to join forces against risks of all kinds which the working environment brings to bear on the health and safety of workers and on society at large.

The present programme takes account of the guidelines proposed by the Commission and of several studies made and consultations held over the past two years. It also takes into consideration the experience gained by the Commission in the coal and steel industries and in the nuclear field where, under the terms of the ECSC and Euratom Treaties, research programmes and work on harmonization and standardization in accident and disease prevention and protection with regard to specific risks in these three sectors have been carried out for many years.

This programme does not effect other programmes such as those for the elimination of technical barriers in trade and for the protection of the environment. By proposing specific actions within the framework of the programme, the work undertaken by other research programmes, notably in the environmental field will be taken into account, so that maximum co-ordination is ensured.

Some action could be taken in collaboration or conjunction with other organizations, such as the European Foundation for the Improvement of Living and Working Conditions and the European Centre for the Development of Vocational Training.

I. General objectives of the action programme on safety and health

The main aim of the programme is to increase the level of protection against occupational risks of all types by increasing the efficiency of measures for preventing, monitoring and controlling these risks.

One of the primary conditions for the implementation of such a programme is the full participation of both sides of industry in preventive and protective measures.

Each of the actions proposed in the programme must be seen as an element contributing to the better organization of preventive and protective measures for workers and to closer collaboration between the social partners towards that end. Furthermore, in order to take account of the experience obtained by international organizations and to avoid duplication of effort in the surveys or actions undertaken, liaison between Member States must be improved with a view to organizing joint action in international agencies responsible for occupational health and safety.

Such a programme should make it possible to achieve the following general objectives:

(a) Improvement of the working situation with a view to increased safety and with due regard to health requirements in the organization of the work. Such an improvement should cover not only the existing situation but also new technical developments. Technical progress which contributes to the creation of a new working situation or to the improvement of an existing situation is not always conceived and directed in line with the dictates of safety and health; where machinery, premises and plant are concerned, safety aspects should be considered at the design stage and integrated into the subsequent stages of their production and commissioning. Due attention must also be paid to health considerations at every stage in the production and use of chemical substances.

There is a close link between occupational accident and disease prevention on the one hand, and the organization of work and safety and health training and information at the place of work on the other. There is an urgent need to review and redefine a more effective accident and disease prevention strategy in order to up-date traditional methods.

Where it is not possible to eliminate it, exposure to occupational risks must be kept to permissible levels applicable to all workers within the Community and based on common concepts and references.

So as to monitor more effectively the application of preventive measures, surveillance of health and working conditions must be intensified, notably in line with the exigencies of occupational medicine, hygiene and safety appropriate to present-day conditions.

(b)   Improvement of knowledge in order to identify and assess risks and perfect prevention and control methods.

In view of the complexity and diversity of the factors it embraces, aetiology is a priority subject for research and analysis. Valid and comparable statistics must be prepared and existing research co-ordinated. The promotion of new research is an essential corollary to any Community action in occupational medicine, hygiene and safety.

(c)   Improvement of human attitudes in order to promote and develop safety and health consciousness.

Alongside the technical aspects of accident prevention and health protection, a real system of safety instruction and health education must be created. This has yet to be introduced and will be taught in different ways at the various educational levels and at the various levels of responsibility and action within undertakings.

II. Description of the initiatives to be taken at Community level

Attainment of the general objectives requires many initiatives involving various scientific disciplines. Such initiatives pre-suppose the effective participation of individuals in managing their own health and safety and should encourage the social partners and the various professional associations and bodies to take a more active part in the formulation and implementation of a policy for the prevention of dangers at the workplace.

The following six concrete initiatives are planned within various time limits for the attainments of these general objectives:

1.   incorporation of safety aspects into the various stages of design, production and operation;

2.   determination of exposure limits for workers with regard to pollution and harmful substances present or likely to be present at the workplace;

3.   more extensive monitoring of workers safety and health;

4.   accident and disease aetiology and assessment of the risks connected with work;

5.   co-ordination and promotion of research on occupational safety and health;

6.   development of safety and health consciousness by education and training.

INITIATIVE 1

Incorporation of safety aspects into the various stages
of design, production and operation

Aim

In order to promote this incorporation the Commission will consider actions aimed essentially at harmonizing, from the safety point of view, the principles and designs of workplaces, machinery, equipment and plant and at the formulation or co-ordination of rules for their use and guidance on the use of dangerous substances.

The principle of integrated safety is today generally regarded as essential for all preventive measures and it is receiving increasing attention at national and international level. In all decisions with regard to undertakings (planning and construction of the undertaking, purchase and operation of plant, organization of production, working methods, *etc.*) more attention must be paid to safety. Similarly operational safety should be studied in advance for the design and manufacture of machinery and tools so as to guarantee protection of the worker's health as far as possible. As concerns the production and distribution of dangerous substances, the same principles have to be taken into account.

The principles of ergonomics are not yet sufficiently well applied in the search for better safety. In particular design ergonomics which is already widespread in the Community has not been sufficiently adopted, as compared with the work carried out in the Scandinavian countries and in the United States.

The results of research carried out over several years in the coal and steel industries indicate the measures which should be planned at Community level in others sectors of industry.

In this field the Commission is planning to propose a certain number of measures which will encourage the application of the principles and which could progressively form a basis of legal, regulatory and administrative provisions or of up-to-date technical guides drawn up at Community level in order to improve the current situation in many industrial or agricultural spheres. These measures concern in particular:

(a)   *Setting up of undertakings and planning of layout and equipment*

The Commission has selected the following points from amongst the numerous factors which must be taken into consideration: ventilation and lighting, temperature, protection against falling from heights and against falling heavy objects, protection against fire, noise and vibrations, gases, vapours and dusts, design of general and emergency thoroughfares and location of doors and windows.

(b)   *Organization of work within undertakings or between several undertakings*

The following points are to receive special attention: equipment and layout of workplaces, outdoor workplaces, warning signs, dangerous jobs, no-access and limited access areas, transport within the undertaking, inspections, maintenance work, plant testing, co-ordination of work within the undertaking, co-ordination of the work of various departments belonging to the same undertaking or to different undertakings, *etc.*

(c) *Manufacture and use of machinery, equipment and tools*

This is the chief area for the application or technical accident and disease prevention which is of paramount social and economic importance. In this sector harmonization measures require lengthy preparation. With regard to the manufacture of machinery and equipment the concept of their safety was already considered in the general programmce of 28 May 1969 on the elimination of technical barriers to trade. However, there exist inherent dangers in the use of machinery and equipment and a procedure should be introduced for the exchange of experience and information so that such dangers are recognized and identified. Furthermore, since 1969 the Council had already pointed out that it would be possible, if necessary, to lay down rules on use supplementing Community Directives on harmonization with regard to the manufacture of machinery and equipment. Guidelines and rules must be drawn up with a view to determining appropriate legislation at Community level.

(d) *Handling of dangerous substances and preparations*

In this field Community harmonization action must be taken with regard to the handling of dangerous substances and preparations, with a view to improving the practical organization of safety, that is, handling at the workplace, storage, marking of containers and pipes. Technical and health protection measures, working restrictions and prohibitions, the number of hours worked and medical protection measures should also be harmonized at Community level. The distribution of dangerous substances (classification, identification and packaging) is taken into account in the programmes for the 'elimination of the technical barriers to trade' and 'environment'.

Contents

Some of the objectives set out above can be achieved only in the medium and long-term. The problems will be selected for study on the basis of the wishes expressed or guidance given by relevant bodies who should above all bear in mind practical considerations and on the basis of urgent needs which may arise from unforeseen dangerous situations such as accidents or disasters, or which may be recognized as a result of the acquisition of fresh knowledge on the effects of chemical substances and the need to control their use with a view to protecting health.

The Commission plans to begin work in this field by studying the following matters:

(a) *Setting up of undertakings and planning of layout and equipment*

1. Organization and layout of agricultural holdings. There is reason to consider that modern agricultural holdings should meet requirements similar to those imposed upon industrial enterprises. So far these requirements have generally not been taken into account in national regulations and it would be appropriate to take the necessary steps at Community level.

2. Noise and vibration control. This requires special medium and long-term attention. The main task consists in setting an optimum machine-noise level on the basis of health data and an assessment of results obtained to date by research and the examination of practical experiments (for example the use of machinery with a low-noise level, which has already been perfected). Noise emission levels, designed to take account especially of the practical problems involved in occupational

protection, will be established after national experts have been consulted and will be published in the form of Directives.

(b) *Organization of work within undertakings*

1. Transport within undertakings. Internal transport, particularly the safe organization of general thoroughfares, needs to be examined and suitable practical instructions should be drawn up. This sector has a particularly high accident rate.

2. Safety signs at workplaces. Council Directive 77/576/EEC of 25 July 1977 on safety signs at workplaces provides that these signs must be able to keep up with technical progress and meet recommendations for harmonization at international level.

   In this connection provision is made for a committee to meet at regular intervals. This action was initiated in 1977 and will be continued in 1978 and 1979 by means of proposals for Directives.

3. Co-ordination of the work of principal and secondary undertakings. The internal and external collaboration of principal and secondary undertakings (sub-contracts) requires special technical examination from the point of view of safety. In practice — especially for the co-ordination of collaboration between several independent undertakings — there are many problems still to be solved. A Community examination of these questions leading to such co-ordination by means of suitable legal instruments is required.

(c) *Manufacture and use of machinery, equipment and tools*

In addition to the work completed within the context of the elimination of technical barriers to trade which is concerned with the design and manufacture of machines, equipment and tools it seems essential to examine in the short and medium-term the need for joint rules on the use of the following: agricultural machinery, lifting gear, machinery used in construction, metal scaffolding and woodworking machines. Depending on the circumstances and on the results of the collaboration to be organized such rules would take the form of guidelines or Directives

(d) *Handling of dangerous substances and preparations*

An urgent study must be made of the handling of dangerous or toxic substances and agreement reached on common standards which will then be proposed to the Member States. An essentially practical approach is required and attention will initially be directed towards the problems of health protection connected with the use of pesticides and herbicides in agriculture. Similar problems arise with other products, *eg.* arsenic, lead, mercury, cadmium, chrome, nickel, vegetable dusts, biological pollution, *etc.*

As information is obtained on the toxicological effect of these substances, as outlined in paragraph 5 of Initiative 2, practical guidelines will be drawn up for all products which involve handling problems or health risks.

INITIATIVE 2

Determination of exposure limits for workers with regard to pollutants and harmful substances present or likely to be present at the workplace

Aim

With a view to the organization of disease prevention and to the monitoring of many occupational risks it is essential to have data on exposure limits for workers with regard to pollutants and harmful substances. It is therefore important for the Commission to achieve, at Community level, harmonization of the concepts, methodologies and references on the basis of which the Member States determine their permissible exposure limits.

There are already standards for protection against radiation at Community level which have been in force since 1959 (Directive) and which were recently revised by a Directive issued in June 1976. They are an example of a joint health policy concerned with an industrial risk facing workers and the general public and based on uniform standards for the whole Community. This example should be extended to other pollutants present at the workplace.

Moreover the studies carried out by the Commission over the past four years in particular in relation to the environment programme and the experience acquired with regard to certain environmental pollutants now make it possible to present concrete proposals for action with regard to certain specific pollutants affecting workplaces in particular.

In addition to these short-term actions, however, the Commission plans to make an objective analysis at Community level of the harmful or undesirable effects of exposure to pollutants in given circumstances — taking account of the results already obtained at international level, in particular by the WHO and the ILO. From this analysis it is proposed to deduce criteria of noxiousness on which to base acceptable exposure limits for workers. Such a project would cover a large number of substances and would be extended as industrial toxicity studies currently in progress are completed.

The protection of human health against chemical substances requires a complex toxicological evaluation which at present is incomplete. The Commission must take priority action with regard to carcinogens, since it is generally accepted that a high proportion of human cancer is caused by external factors including chemicals present at the workplace.

Contents

The Commission is planning the following initiatives:

1. *Non-ionizing radiation and other physical agents*

With regard to non-ionizing radiation, proposals for Directives will be submitted to the Council on microwaves, laser radiation, ultra-violet radiation and ultrasound, on the basis of the procedure followed for standards in protection against radiation.

2. *Harmonization of exposure limits*

The Commission plans, at the earliest opportunity, to make a comparative study of existing regulations and recommendations in Member States with regard to permissible exposure levels of workers to toxic substances or physically harmful substances.

The values adopted in different countries vary, the terminology used is not the same and the concepts used to determine the limits are not based on the same principles. Harmonization is therefore essential and a general Directive co-ordinating and harmonizing exposure levels, possibly updated later on in accordance with the latest scientific data and international information available to the Commission, could be prepared between now and 1979.

This short-term initiative would have the advantage of achieving harmonization at Community level and avoiding the delay of waiting for the completion of on-going research projects in the field of occupational toxicology, whether within the Commission or in the Member States.

3. *Directives on specific pollutants*

The general harmonization discussed in paragraph 2 must be supplemented by the preparation of specific Directives such as those proposed by the Commission for vinyl chloride monomer and those shortly to be put forward on asbestos, lead, mercury, solvents, carbon monoxide, noise and vibrations. The studies in progress within the Commission and the state of knowledge have now reached the stage where they can be used to determine the permissible exposure levels for the above-mentioned pollutants from the point of view of health protection.

4. *Carcinogens*

Specific Commission action with regard to carcinogens present at workplaces will consist in:

— collecting data on the distribution of carcinogens and their concentration at the workplace,

— collecting and analysing medical data,

— perfecting readily applicable detection,

— fixing the lowest possible levels or, if necessary, prohibit a certain number of carcinogens present at the workplace.

5. *Toxicological evaluation*

Toxicological evaluation is central to take assessment of the health risks due to the presence of many chemical and biological agents in the working environment and can be carried out only if sufficient knowledge is available on the effects of the agents under consideration on man. The methodology adopted by the Commission for assessing the dangers from environmental pollutants in general is based on research into criteria for noxiousness from which permissible human exposure levels may be deduced. The data already collected by the Commission on the effects on health of urban atmospheric pollutants and certain water pollutants provide a basis for the action planned in industry, but it needs to be considerably extended and developed. Priority will be given to the following substances: arsenic, cadmium, chromium, iron oxides, nickel, vegetable dusts, ozone, nitrogen oxides and biological pollutants.

The Commission, while taking account of studies already carried out and projects being planned at international level, is to give priority to the extension and development of information relating to the objective evaluation of risks associated with toxic substances present at the workplace. This action will lead to Directives on exposure levels for workers and also to the compilation of handbooks on the safe handling of such substances at the workplace. The Commission intends to carry out this action by means of a series of studies and scientific and technical consultations. It will be assisted in this action by a Scientific Committee on Toxicology planned for the end of 1977.

INITIATIVE 3

More extensive monitoring of workers' safety and health

Whereas exposure limits for workers and safety and health protection measures are essential factors in the organization of accident and disease prevention, various permanent and well-adapted methods are also required with which to monitor the measures adopted and the exposure levels prescribed for the workplace.

These monitoring methods must be harmonized and co-ordinated at Community level.

The monitoring of workers' health and safety depends upon several types of monitoring which complement each other:

(a) monitoring of the effectiveness of individual or group safety and protection measures with regard to machinery, equipment and plant;

(b) monitoring of hygiene and working conditions from which the types of exposure to different physical, chemical and biological agents present in the working environment are derived;

(c) monitoring of the state of health and behaviour of the worker as part of occupational medicine;

(d) special monitoring as a result of work entailing special risks;

(e) industrial toxico-vigilance;

(f) inspections.

The Commission feels that it is essential to harmonize at Community level principles and methods applicable to monitoring. Moreover, efforts should be made to interest workers in monitoring within the undertaking, either by direct means or by means of existing bodies or institutions.

Any proposed solutions must allow workers' and employers' representatives to play a fuller part in the practical organization of such monitoring at various levels of action and responsibility.

Contents

1. *Monitoring of the effectiveness of safety and protection measures*

Planning and execution of this form of monitoring varies at present from country to country and according to the regulations and activities concerned. Once the provisions currently governing such monitoring have been analysed, suitable proposals will be submitted to the Council for adoption in order to harmonize and strengthen the organization of this type of monitoring in which the workers' and employers representatives should play a greater role.

2. *Monitoring of hygiene and working conditions*

Monitoring of pollutant concentrations at workplaces and the intensity of environmental factors is essential for the organization of disease prevention and monitoring.

Measuring programmes do exist in Member States but they are based on different methods and sometimes different principles. These measures must be harmonized at Community level with regard to sampling, techniques and measuring intervals.

When the Commission has analysed these different methods, it will draw up intercomparison programmes and prepare reference methods for the determination of the major pollutants present at workplaces.

Special attention will be paid to promoting the development of new monitoring and measuring methods for individual exposure.

The Commission will make a similar effort to apply the human biological indicators already in existence and will carry out research for new indicators which will make it possible to detect any changes in the state of health at an early stage. The European list of occupational diseases will be used as a reference document for drawing up the priorities for this action scheduled to take place as from 1978. Account will have to be taken not only of individual sensitivity, which may be very high for some pollutants, and of workplaces so that groups with a high occupational exposure risk may be identified, but also of some special groups of workers such as adolescents and women.

3. *Monitoring of workers' health*

In accordance with the terms of Article 118 of the Treaty establishing the EEC, occupational medicine must be considered as an area in which the Commission has the task of promoting close co-operation between Member States in the social field, particularly in matters relating to working conditions and to the prevention of occupational accidents and diseases. The term 'occupational medicine', as stated in the 1962 recommendation on occupational medicine in the undertaking, refers to a service established in or near a place of employment for the purposes of:

(a) protecting the workers against any health hazard which may arise out of their work or conditions in which it is carried on;

(b) contributing towards the workers' physical and mental adjustment, in particular by the adaptation of the work to the workers and their assignment to jobs for which they are suited; and

(c) contributing to the establishment and maintenance of the highest possible degree of physical and mental well-being of the workers.

In addition, Recommendation 112 of the ILO stated that the role of occupational health services should be essentially preventive and defined their functions so as to include the prevention of accidents and occupational diseases, the rehabilitation of workers, job analysis in the light of physiological and psychological considerations, surveillance of hygiene, advice on the placement of workers, medical supervision, emergency treatment and research in occupational health.

Consideration must be given to closer harmonization of the methods used by occupational health services in undertakings in order that the work of the industrial medical officer may be more fully integrated into the system for monitoring workers' safety and health, as recommended in this programme.

This revision will be carried out with effect from 1978 by consultation with the relevant bodies and should culminate in a directive on the organization of occupational medicine in the Member States of the Community, to be proposed in 1979.

4.   *Special monitoring*

In many undertakings there are some jobs which present higher than average risks; certain types of casual work may also involve exposure to risk which is higher than that present in normal working conditions or than the exposure levels laid down. Such jobs are done, for example, by members of rescue teams or of maintenance and repair teams and by workers in virology laboratories and in institutes producing sera or viruses, etc.

Exchanges of information and experience for cases involving these aspects should be organized at Community level and should lead to a definition of the principles and criteria for this particular type of monitoring.

5.   *Industrial toxico-vigilance*

The Commission plans to set up an industrial toxico-vigilance system along the lines proposed by the ILO and which is aimed at establishing a central information system for all observations made in industrial activity concerning the harmful effects of toxic substances. This system should be based on a network of highly specialized centres which could analyse information received from occupational health services and transmit it when required to interested persons or institutions.

The Commission will make an appropriate proposal to the Council, after holding the necessary consultations.

6.   *Inspections*

Inspections carried out for the purposes of occupational safety, medicine and hygiene should be organized so that they assume full responsibility and control by placing the emphasis on preventive measures. With this end in view the necessary provisions must be made in close collaboration with the competent authorities in Member States for the strengthening and development of the work of inspection at national level. The Commission intends to review the role of the inspectorate responsible for implementing in each Member State the regulations of occupational health, hygiene and safety. This review will cover diplomas, certificates and other qualifications, and the powers and scope of their responsibilities in this field.

INITIATIVE 4

Accident and disease aetiology and assessment of risks connected with work

Aim

The risk of accident or disease may be estimated objectively only if reliable methods are available which make it possible to determine the scope, seriousness and development in time and, in a general way, to acquire greater knowledge of the various factors involved in the cause of accidents at work and of diseases due to work.

Statistics are essential tools for the analysis and interpretation of facts and for assessment of the results obtained from an accident and disease prevention policy.

The improvement of statistics and their comparability, the harmonization of methodologies and the more precise interpretation of the data they provide are important steps in the development of an improved organization of work with regard to accident and disease prevention. Since so many different approaches are used a distinction must be drawn between action in respect of accidents at work and action in respect of disease due to work.

Such actions must provide a clearer picture of the different causative factors of accidents at work and of diseases due to work and must use them as a basis for practical preventive and protective measures against hazards connected with work. It will then be possible to provide preventive-type protection for men at work, on an objective and realistic basis.

In addition, special attention will be paid to calculating the economic and social cost of accidents at work and diseases due to work so as to establish the order of priority for preventive measures.

Account will be taken of the harmonization work already carried out by other international organizations and of work completed or in progress, particularly by the ILO.

Contents

These initiatives deal separately with accidents at work and diseases resulting from work.

As regards accidents at work the two sectors for which community statistics are already available are the iron and steel industry and mining. Drawing on the experience gained in the sectors the Commission plans to draw up Community statistics concerning other sectors, to launch sectoral in-depth studies and to harmonize accident definitions and methods of reporting accidents in order to establish more precisely the aetiology of accidents. Preparatory surveys are in progress and the first results will be available in 1979.

With regard to diseases due to work, statistics collected at national level usually concern only occupational diseases and are drawn up on different bases so that it is not possible to compare them. There are no Community statistics in this field and it would be appropriate to devise a joint methodology as soon as possible so that existing national statistics may be processed. The Commission therefore plans to gather and analyse national statistical information and to draw up proposals for methodologies with a view to a common approach, so that calculations may be made of mortality, sickness and absenteeism rates and their evolution over a period of time.

Close collaboration must be instituted with the national statistical offices and the national social security offices with regard to these new problems.

This is a medium-term initiative and the first results will become available only after two or three years.

INITIATIVE 5

Co-ordination and promotion of research on occupational safety and health

Aim

The action planned in the programme must find its scientific support in a research programme which is co-ordinated and/or carried out jointly and which deals on the one hand with the measurement and effects on health of pollutants and harmful substances and, on the other hand, with the development within undertakings of safer, 'cleaner' technologies which do not threaten the general environment.

Collaboration must be organized and strengthened between the institutes and laboratories of Member States in order to avoid duplication of work, to derive greater benefit from the financial resources available and where necessary to bring together highly specialized laboratories to work on problems which cannot be solved in a single Member State.

Moreover, research must be carried out in fields where little or no work has been done, such as agriculture and the tertiary industries.

Contents

Two permanent inventories of research in progress or planned (occupational safety and medicine) at national level are already being prepared at Community level. From 1978 the inventories will make it possible to set up a reciprocal information system on responsible bodies in order to promote the exchange of knowledge and create conditions for close collaboration between research institutes. These permanent inventories will also mention fields in which there are gaps. Three pilot studies are in progress on inflammable substances, occupational risks in the building industry and certain carcinogens. During 1978 these studies will also indicate which subjects should be covered by joint research.

On the basis of these inventories the data bank being compiled within the Commission should be progressively supplemented and should include details of new research; account is taken of the fact that this data bank will subsequently be linked to the information system on medical research which is being set up at Commission level.

Research work aimed at closing the gaps in knowledge on toxic agents and their effects on health or at improving methods for measuring these agents is of major importance for the success of several parts of the programme — in particular the section on the determination of criteria for harmfulness. It will also help to determine as accurately as possible the potential and actual effect on health of pollutants and nuisances present or likely to be present at the workplace.

The results of the implementation of the various initiatives making up the programme will be analysed by the Commission with effect from 1979 and could form a basis for the preparation of a detailed and precise Community research and development programme which could be the subject of a future Commission proposal for adoption by the Council.

INITIATIVE 6

Development of safety and health consciousness by means of education and training

Aim

This initiative is aimed at developing safety and health consciousness by means of education and training. It is of paramount importance for the success of the promotion of safety and hygiene at workplaces. It is based on instruction and training and involves various levels of education and the undertaking itself. It also concerns in a general way occupational and social sectors involved in problems of accident prevention and health protection at work.

This is a medium and long-term initiative in view of the different sectors involved and of the absence to date of any real methodology and common principles. Various studies and consultations will be required before results and concrete proposals are obtained at Community level.

This action concerns educational bodies, undertakings and society in general.

As for education the basic principles of safety and of health education must be taught in schools. Knowledge of and the correct attitudes towards occupational safety and hygiene must be taught at various levels of education as an integral part of the curriculum and at the same time attention must be paid to the requirements of prevention in relation to real life situations. The question is one of establishing at Community level a safety training scheme which takes account of the differences between national characteristics and traditions but which is based on common principles and a common approach.

Within undertakings steps for the elimination of risks must be systematically organized and co-ordinated at all levels of responsibility and management. Principles of safety must be consolidated, developed and made public. Action designed to sharpen the awareness of industrialists and heads of undertakings must be taken together with the campaign aimed at workers.

For the training of society in general the action taken in education must be supplemented by action aimed at certain population groups. The use of audio-visual aids is one of the most modern and most effective means of informing the public of the importance and significance of accident and disease prevention.

Contents

1. *Education*

The Commission plans to carry out, together with the bodies responsible for national education, preparatory studies for the purpose of defining harmonized planning at Community level.

In general education — starting at the earliest age and continuing throughout school life — instruction must be on two levels:

— theoretical and practical instruction to give children and young people an awareness of the risk of accidents,

— instruction to develop a sense of moral and public responsibility with regard to safety and health protection.

In technical education relevant training in safety and health protection should accompany all levels of technical instruction and vocational training. Special attention should be paid to the training of persons particularly concerned with safety and health protection who have a specific task or responsibility in this field.

The Commission plans to propose Community training models for persons in certain occupations and concerned with specific tasks, such as industrial medical officer, occupational safety officer, engineer, architect, member of a company safety committee or union official.

2. *Undertaking*

Within an undertaking training in safety must be under the control of the undertaking itself since general and technical training cannot take the place of appropriate action at the workplace. This type of training must supplement the instruction received in schools and it must also be given to those who have not previously received any such instruction.

Such training, to be carried out within industry, will be more specialized and more detailed. In many cases it will be organized by specialist bodies whose work must be co-ordinated at Community level. It should be remembered that education covers a broad span of learning situations — for example instruction given by experienced workers and learning on-the-job.

Beginning in 1978, the Commission intends:

— to draw up Community models for safety training and refresher courses for certain categories of staff: administrative grades, executive grades, instructors for courses on safety and health education and safety delegates,

— to draw up Community models for presenting various aspects of safety to newly recruited workers, migrant workers and workers who have changed jobs,

— to draw up manuals and codes of practice with regard to sectoral activities or dangerous jobs,

— to organize safety campaigns of limited duration with a specific aim, in which workers will feel fully and actively involved,

— to extend the group training courses already in existence to other groups of persons concerned with accident prevention and safety measures.

This action will be furthered by making available to both management and labour knowledge or concepts acquired either by exchange of experience within specialist groups in the relevant sectors or by research projects jointly agreed and financed. Such knowledge could be included in instructions, regulations or codes of practice, to be distributed with commentaries in the appropriate quarters and to be kept constantly up to date.

The Commission will support this type of co-operation and promotion of safety by providing information gathered from specific aspects of the action programme, such as information on accidents and on technical progress in the design, manufacture and use of machinery and plant, and by making available the industrial toxico-vigilance results.

3. *Population groups*

In addition to the action taken in education, general information for certain population groups (such as parents' associations, professional bodies, women's associations) must be organized with regard to the importance of accident and disease prevention. Some steps have already been taken in this field at national level. Audio-visual aids are already used to provide this information. The Commission plans to co-ordinate these initiatives and develop them jointly, to produce films and set up a permanent file on audio-visual aids available on an exchange basis.

COUNCIL RESOLUTION

of 27 February 1984

on a second programme of action of the European Communities
on safety and health at work
[84/c 67/02]

THE COUNCIL OF THE EUROPEAN COMMUNITIES,

Having regard to the Treaties establishing the European Communities,

Having regard to the draft resolution submitted by the Commission,

Having regard to the opinion of the European Parliament,

Having regard to the opinion of the Economic and Social Committee,

Whereas the Council resolution of 21 January 1974 concerning a social action programme provides for the establishment of an action programme on safety and health at work;

Whereas, under Article 2 of the Treaty establishing the European Economic Community, the Community shall have in particular as its task, by establishing a common market and progressively approximating the economic policies of Member States, that of promoting throughout the Community a harmonious developing of economic activities, a continuous and balanced expansion and an accelerated raising of the standard of living;

Whereas, at the Conference held in Paris in October 1972, the Heads of State or of Government affirmed that the first aim of economic expansion, which is not an end in itself, should be to enable disparities in living conditions to be reduced and that it should result in an improvement in the quality of life as well as in standards of living;

Whereas, moreover, under Article 117 of the said Treaty the Member States agreed upon the need to promote improved working conditions and an improved standard of living for workers, so as to make possible their harmonization while the improvement was being maintained;

Whereas prevention of accidents at work and occupational diseases and also occupational hygiene fall within the fields and objectives referred to in Article 118 of the said Treaty; whereas in this context collaboration should be strengthened between the Member States and the Commission and between the Member States themselves;

Whereas suitable health protection for the public and effective prevention of accidents at work and occupational diseases would be in conformity with these general objectives;

Whereas, in spite of sustained efforts, the continuing high level of accidents at work and of occupational diseases remains a serious problem;

Whereas efforts made in the field of accident prevention and health protection at the work place have beneficial effects which are reflected in the economic sphere and in industrial relations;

Whereas a considerable effort is needed at Community level to search for and implement suitable means for maintaining or creating a working environment tailored to the needs of man and his legitimate aspirations;

Whereas both the effectiveness of the measures and the cost of implementing them should be taken into account in the choice of action to be undertaken at Community level and of the measures to be taken to implement it;

Whereas the improvement of working conditions and the working environment must be envisaged in overall terms and must concern all sectors of the economy;

Whereas it is also essential to encourage the increasing participation of management and labour, at all levels and particularly at the level of the undertaking, in decisions and initiatives in the field of safety, hygiene and health protection at work;

Whereas the Advisory Committee on Safety, Hygiene and Health Protection at Work, set up by Decision 74/325/EEC, must be closely associated with this work;

Whereas the European Foundation for the Improvement of Living and Working Conditions and the European Centre for the Development of Vocational Training may have a role to play in the implementation of certain aspects of the programme;

Whereas, in implementing the actions, account must be taken of work undertaken in other fields, especially in the context of the Council resolution of 17 December 1973 on industrial policy and of the Declaration of 22 November 1973 and resolutions of 17 May 1977 and of 7 February 1983 of the Council of the European Communities and of the representatives of the Governments of the Member States, meeting within the Council, concerning a European Community action programme on the environment, in order to ensure the closest possible co-ordination of actions and proposals;

Whereas, in order to carry out the actions successfully, it is important to ensure that concepts, terminology and methods of identification, measurement and assessment relating to safety and health risks are harmonized; whereas such a task is of major importance in the context of these actions;

Notes	that this second action programme takes into account the first action programme annexed to the Council resolution of 29 June 1978 on an action programme of the European Communities on safety and health at work;
Expresses	the political will to take, in keeping with the urgency of the matter and bearing in mind what is feasible at national and Community level, the measures required so that between now and the end of 1988 the following priority actions in particular can be undertaken:

I. PROTECTION AGAINST DANGEROUS SUBSTANCES

1. Continue with the establishment of Community provisions based on Council Directive 80/1107/EEC of 27 November 1980 on the protection of workers from the risks related to exposure to chemical, physical and biological agents at work[6],

2. Establish common methodologies for the assessment of the health risks of physical, chemical and biological agents present at the work place.

3. Develop a standard approach for the establishment of exposure limits for toxic substances by applying the methodologies referred to in point 2. Make recommendations for the harmonization of exposure limits for a certain number of substances, taking into account existing exposure limits.

4. Establish for toxic substances standard methods for measuring and evaluating work place air concentrations and biological indicators of the workers involved, together with quality control programmes for their use.

5. Develop preventive and protective measures for substances recognized as being carcinogenic and other dangerous substances and processes which may have serious harmful effects on health.

6. Establish Community rules for limiting exposure to noise and continue work on developing a basis for Community measures on vibrations and non-ionizing radiation.

II. ERGONOMIC MEASURES, PROTECTION AGAINST ACCIDENTS AND DANGEROUS SITUATIONS

7. Work out safety proposals, particularly for certain high-risk activities, including proposals on specific measures for the prevention of accidents involving falls, manual lifting, handling and dangerous machinery.

8. Examine the major accident hazards of certain industrial activities covered by Directive 82/501/EEC.

9. Work out ergonomic measures and accident prevention principles with the aim of determining the limits of the constraints imposed on various groups of the working population by the design of equipment, the tasks required and the working environment so that they are not prejudicial to their health and safety.

10. Work out proposals on lighting at the workplace.

11. Organize exchanges of experience with a view to establishing more clearly the principles and methods of organization and training of the departments responsible for inspection in the fields of safety, health and hygiene at work.

III. ORGANIZATION

12. Make recommendations on the organizational and advisory role of the departments responsible for dealing with health and safety problems in small and medium-sized undertakings, by defining, in particular, the role of specialists in occupational medicine, hygiene and safety.

13. Draw up the principles and criteria for monitoring workers whose health and safety are likely to be seriously at risk, such as certain maintenance and repair workers, certain migrant workers and certain workers in sub-contracting undertakings.

14. Draw up the principles for participation by workers and their representatives in the improvement of health and safety measures at the work place.

IV. TRAINING AND INFORMATION

15. Ensure that employers and workers who are liable to be exposed to chemicals and other substances at their work place have adequate information on those substances.

Prepare information notices and manuals on the handling of certain dangerous substances, particularly those covered by Community Directives. If need be, and taking account of existing Community regulations on the matter, draw up proposals on the establishment of systems and codes for the identification of dangerous substances at the work place.

16. (a) Draw up programmes aimed at better training as regards occupational hazards and measures for the safety of those at work (safety training) and

    (b) Training schemes intended for specific groups:

        — young workers,

        — groups who have special need of up-to-date information, *ie.* workers doing a job to which they are not accustomed or who have difficulty in acquiring information through the usual channels, or to whom it is difficult to communicate information.

        — workers in key positions, *ie.* people who lay down working conditions, communicate information, *etc.*

V. STATISTICS

17. Establish comparable data on mortality occupational diseases and collect data from existing sources on the frequency, gravity and causes of accidents at work and occupational diseases, including, as far as possible, data on vulnerable groups of workers and absenteeism due to illness.

18. Compile an inventory of existing cancer registers at local, regional and national level in order to assess the comparability of the data contained in them and to ensure better co-ordination at Community level.

VI. RESEARCH

19. Identify and co-ordinate topics for appropriate research in the field of safety and health at work which can be the subject of future Community action.

VII. CO-OPERATION

20. Within the framework of existing procedures continue to co-operate with international organizations such as the World Health Organization and the International Labour Office and national organizations and institutes outside the Community.

21. Continue to co-operate on other action by the Community and by Member States, where it proves worthwhile;

Requests the Commission to prepare annually, after consulting the Member States, a forward outline of the work it intends to carry out on the implementation of this resolution.

RESOLUTION OF THE COUNCIL AND THE REPRESENTATIVES OF THE GOVERNMENTS OF THE MEMBER STATES, MEETING WITHIN THE COUNCIL

of 7 July 1986

on a programme of action of the European Communities against cancer
[86/c 184/05]

THE COUNCIL OF THE EUROPEAN COMMUNITIES AND THE REPRESENTATIVES OF THE GOVERNMENTS OF THE MEMBER STATES, MEETING WITHIN THE COUNCIL,

Having regard to the Treaties establishing the European Communities,

Having regard to the draft resolution submitted by the Commission,

Having regard to the opinion of the European Parliament,

Having regard to the opinion of the Economic and Social Committee,

Whereas, pursuant to the Treaty establishing the European Economic Community, the Community has *inter alia* as its task, by establishing a common market and progressively approximating the economic policies of Member States, to promote throughout the Community a harmonious development of economic activities, a continuous and balanced expansion and an accelerated raising of the standard of living;

Whereas the European Council in Milan on 28 and 29 June 1985 emphasized the importance of launching a European programme of action against cancer;

Whereas that European Council also approved the proposals of the *ad hoc* Committee on a People's Europe calling for appropriate follow-up to the Commission Communication on co-operation at Community level on health problems;

Whereas various Community actions to prevent cancers arising from exposure to ionizing radiation or exposure to chemical carcinogens are already being carried out under the Treaties establishing the European Economic Community and the European Atomic Energy Community;

Whereas actions to reduce the risk of cancer from exposure to carcinogenic substances are included in a number of existing Community programmes on the environment, worker protection, consumer protection, nutrition, agriculture and the internal market;

Whereas the present programme would increase knowledge about the causes of the cancer and the possible means of preventing and treating it;

Whereas by ensuring a wider dissemination of knowledge of the causes, prevention and treatment of cancer, and an improvement in the comparability of information about those matters, in particular concerning the nature and degree of risk of cancer arising from exposure to given substances or processes, the programme will contribute to the achievement of Community objectives, in particular the removal of non-tariff barriers to trade, while contributing to the overall reduction of risks of cancer;

Whereas research into cancer and carcinogenic effects of physical and chemical agents is being undertaken in a number of Community research programmes;

Whereas the co-ordination of national research activities relating to the early detection and treatment of cancer is not the specific purpose of this resolution, whereas research should take place in the context of the promotion of research organized by the Commission and of the relevant medical and public health research programmes; whereas it is desirable to provide the necessary co-ordination and liaison between those activities and the activities undertaken pursuant to this resolution;

Whereas co-operation with international and national organizations carrying out work in this field will ensure a wider dissemination of knowledge of cancer and help to avoid duplication of effort,

1. *take note of* the conclusions of the *ad hoc* Committee of cancer experts (which met in Paris on 19 and 20 February 1986), forwarded to the Council by the Commission in its report of 10 March 1986;

2. *express* the political will to implement a five-year action programme of the European Communities against cancer;

3. *set* for this programme the objective of contributing to an improvement of the health and quality of life of citizens within the Community by:

   — reducing the number of illnesses due to cancer and the related mortality and

   — decreasing the potential years of life lost because of cancer;

4. *call on* the Commission to examine systematically whether and to what extent, Community legislation, measures and other activities are likely to constitute a hindrance to cancer prevention;

5. *take note of* the action programme proposed by the Commission and consider the following priority actions to be necessary:

(a) Tobacco

As a first priority, development of measures to limit and reduce the use of tobacco.

These measures should be based on the practical experience gained in the various Member States and should contribute to increasing the effectiveness of national programmes and actions.

Systematic examination of the various ways and means of limiting and reducing the use of tobacco, such as rules on advertising, rules on labelling, tax legislation, sponsorship, enforcement of no-smoking rules, extension of no-smoking areas and, if appropriate, the drafting of proposals for actions at Community level.

(b) Chemical substances

Development of harmonized criteria and procedures for evaluating the carcinogenic nature of chemical substances.

Identification and quantification of the carcinogens to which workers and/or the population are exposed and, if appropriate, drafting of new legislation and amendment of existing legislation in order to reduce exposure of workers and/or the population to carcinogens.

(c) **Nutrition and alcohol**

Assessment of the results of research and, where appropriate, preparation of measures, taking account of differing circumstances and habits in the Member States.

Attention should also be paid to abuses in the consumption of alcohol.

(d) **Prevention/early diagnosis**

Exchange of information and experience, particularly on Member States' preventive and early diagnosis programmes (type of examination, numbers participating, effectiveness, cost/benefit analysis excluding resources directly allocated to research) and, on this basis, the preparation of any appropriate measures.

(e) **Epidemiological data**

Exchange of information on Member States' structures and procedures for cancer epidemiology (*inter alia* through exchanges of experts) and evaluation of those structures and procedures with a view to improving them.

Improvement in the collection, availability and comparability of epidemiological data on cancer and factors causing cancer and, if appropriate, preparation of measures in this field taking particular account of the problems of data protection to assist in and help increase the effectiveness of epidemiological research.

(f) **Health education**

Exchange of information and experience, particularly on Member States' health education and information programmes and, on this basis, the preparation of any appropriate measures.

(g) **International collaboration**

Collaboration with international and national organizations active in the field covered by these actions to achieve maximum possible effectiveness;

consider   that a high degree of cohesion between current and future actions at national and Community level would help achieve the objective referred to in point 3 and therefore advocate the strengthening of cancer prevention measures already initiated at Community level and the co-ordination of such actions with those referred to in point 5;

call on   the Commission to submit within 12 months of the adoption of this resolution, and thereafter annually, in close collaboration with the Member States, a work programme detailing the work it intends to carry out in order to implement this resolution and, if necessary, to submit proposals to the Council to this effect.

The annual work programmes, together with a report on results achieved and activities carried out, will be forwarded to the European Parliament, the Council, and the Economic and Social Committee;

take note of the Commission's intention to:

— submit proposals on research and on a European Cancer Information Campaign and

— intensify work relating to cancer in all other Community programmes with a view to submitting appropriate proposals.

COUNCIL RESOLUTION

—— of 21 December 1987 ——

on safety, hygiene and health at work
[88/c 28/01]

THE COUNCIL OF THE EUROPEAN COMMUNITIES,

Whereas Article 118a of the Treaty establishing the European Economic Community sets the objective of harmonizing conditions, especially in the working environment, for protecting the health and safety of workers, while maintaining the improvements made;

Whereas, to this end, Article 118a stipulates that the Council, acting by a qualified majority on a proposal from the Commission, in co-operation with the European Parliament and after consulting the Economic and Social Committee, shall adopt, by means of directives, minimum requirements for gradual implementation, having regard to the conditions and technical rules obtaining in each of the Member States; whereas such directives shall avoid imposing administrative, financial and legal constraints in a way which would hold back the creation and development of small and medium-sized undertakings;

Whereas Article 118a will enable action at Community level as regards the improvement of the working environment to be intensified and expanded in order to protect the safety and health of workers;

Whereas the development of growth and the improvement of productivity at the level of both undertakings and the Community's economy depend *inter alia* on the quality of the working environment, the possibilities for workers to have an influence on the working environment in order to protect their safety and health and the motivation of workers;

Whereas Article 118a provides in particular for the improvement of safety and health conditions at work, which constitutes an essential feature of the social dimension of the internal market,

HAS ADOPTED THIS RESOLUTION:

I

The Council:

1. welcomes the Commission communication on its programme concerning safety, hygiene and health at work;

2. considers that this communication constitutes a useful framework for commencing implementation at Community level of Article 118a;

3. shares the Commission's opinion that the protection of the safety and health of workers must also include measures concerning ergonomics in connection with safety and health at work;

4.   stresses the need:

—   to place equal emphasis on achieving the economic and social objectives of the completion of the internal market,

—   to co-ordinate Community and national measures concerning the achievement of these two objectives,

notes in this context that the Commission has stated that it will take into account the social aspects of the proposals which it will submit with a view to the completion of the internal market.

II

1.   The Council takes note of the measures contemplated by the Commission in its communication on safety, hygiene and health at work and to this end:

(a)   suggests that the Commission draw up practical plans of work, preferably on an annual basis, in close co-operation with the Member States and after consulting the Advisory Committee on Safety, Hygiene and Health Protection at work;

(b)   takes note of the Commission's intention of submitting to it in the near future:

(i)   minimum requirements at Community level concerning:

—   the organization of the safety and health of workers at work including protection against risks resulting from the carrying of heavy loads by hand,

—   protection against risks resulting from dangerous substances, including carcinogenic substances; in this connection, the principle of substitution using a recognized non-dangerous or less dangerous substance should be taken as a basis,

—   the arrangement of the place of work;

(ii)   other activities:

—   harmonization of statistics on accidents at work and occupational diseases,

—   a study on the organization by the Member States of means of control and of sanctions;

(c)   also notes that the Commission intends to submit before 1992 a series of measures included in its programme on safety, hygiene and health at work.

2.   The Council considers that when drawing up the plans of work, account should be taken of the following criteria in particular:

—   the seriousness of the risks of accidents at work and/or occupational diseases,

—   the number of workers exposed to risks,

—   the possibilities for prevention.

3. The Council directives should lay stress on the enactment of the main provisions for the elimination of risks for workers at the place of work.

4. The Member States undertake to make available to the Commission the knowledge and experience available in the Member States *inter alia* to enable the Commission to improve the statements as regards the impact of its proposals on small and medium-sized undertakings.

   To this end, the Commission is asked to maintain close contacts with national experts.

5. The social partners will be involved in the preparation of the directives, particularly the Advisory Committee on Safety, Hygiene and Health Protection at Work.

   The social partners will also be involved, in accordance with national laws and practices, in:

   — the implementation of the Council directives at national level,

   — conceiving and implementing Member States' policies concerning the field covered by Article 118a,

   — the organization of the working environment in undertakings to protect the safety and health of workers and the implementation of the corresponding arrangements for the protection of workers.

6. The Council stresses that the information, increased awareness and, if necessary, the training of employers and workers will play a fundamental role in the success of the measures recommended in the Commission's communication on its programme on safety, hygiene and health at work.

7. In order to assist and expedite the implementation of the safety and health measures, the Member Sates will examine the possibility of measures in favour of undertakings in order to prompt the latter to implement preventive measures.

8. The Commission is requested to examine how the exchange of information and experience in the field covered by this resolution can be improved, particularly as regards the gathering and dissemination of data.

   At the same time, the Commission is invited to examine the advisability of setting up Community machinery to study the repercussions at national level of Community measures in the field of health and safety at work.

   Co-operation with and between bodies with tasks in the field covered by this resolution should be intensified.

9. The Council:

   — acknowledges the predominant role of the heightening of public awareness for the success of the measures recommended in the Commission's communication on its programme on safety, hygiene and health at work,

   — agrees to suggest that European year in this field be organized in 1992.

COMMISSION COMMUNICATION

of 21 December 1987

on its programme concerning safety, hygiene and health at work

[88/c 28/02]

SUMMARY

In order to remain coherent and achieve its full impact, the creation of the internal market, 'the heart of the strategy to relaunch the construction of Europe', must incorporate a significant element of social policy, within which the physical and mental protection of workers stands high on the list of priorities. Taking full advantage of the opportunities afforded by the provisions of the Article 118a of the Single Act concerning the improvement of health and safety at work, and confirming its commitment to making full and rapid use of all the resources put at its disposal by this legal provision, the Commission has adopted the following action programme.

1. INTRODUCTION

A. The situation in the Member States

1. Despite the absence of sufficiently reliable statistics for Europe as a whole, the data available at national level are adequate demonstration of the high cost in human and social terms of industrial accidents. The estimated level of compensation paid out in 1984 for occupational accidents and diseases was around 16,000 million ECU in the EEC as a whole, amounting to 7% of total sickness insurance payments.

2. An analysis of efforts made within the Member States to reduce occupational accidents and diseases shows that many means of increasing awareness of health and safety have been employed, not only at manager and worker level within firms, but also among the public at large. Co-operation between management, health and safety services and workers and their representatives within firms has been constantly improved and more efficiently organized. The inclusion of safety considerations, right from the planning stage, is recognized as a necessity.

3. National legislation is increasingly reflecting the work carried out at Community level; at the same time, there is a growing tendency to lift restrictions intended to protect women at work, in the interests of equal employment opportunities for men and women.

4. General measures can be divided into those which, while not necessarily legislative, are aimed at improving installations in respect of work safety, removing the hazards presented by the use of tools and machinery, and protecting workers engaged in particularly dangerous tasks; secondly, protective measures for handling dangerous substances and finally, those enabling the appropriate advisory committees to ensure proper implementation of the legal and administrative provisions.

B. Community action

Under the EEC Treaty, the Commission has implemented two action programmes on safety and health at work since 1978.

These programmes were the subject of two Council resolutions:

The first of these, of 29 June 1978, expressed the political will to enable a series of actions to be taken up to 1982 focussing on the aetiology substances, prevention of the dangers and harmful effects of machines, monitoring and inspection, and the improvement of human attitudes.

The second resolution, adopted on 27 February 1984, was a continuation of the first action programme.

In this context, the Commission drafted 10 directives — seven of which have been adopted by the Council — on the protection of workers exposed to physical and chemical agents at work and the prevention of major accident hazards related to chemicals.

C. Legal bases and content of the new work programme

In order to confirm its will to reinforce the social dimension of the completion of the internal market, the Commission intends to develop its initiatives in the field of safety, hygiene and health at work, on the basis of Articles 117 and 118 of the EEC Treaty concerning social policy, and the specific provisions of the Single Act given in Article 118a (1) on the harmonization of improvements in the conditions of protection of the health and safety of workers, and Article 118b which stresses the need to promote the dialogue between the two sides of industry.

The Commission has therefore decided, without awaiting the expiry of the second action programme, to draw up a new work programme concentrating chiefly on the following five subjects:

— safety and ergonomics,
— health and hygiene,
— information and training,
— initiatives specifically directed at small and medium-sized enterprises,
— social dialogue.

II. THE PROGRAMME

A. Safety and ergonomics at work

1. *Completion of the internal market – removal of technical barriers*

In the White Paper on completing the internal market, the Commission took into account the 'underlying reasons for the existence of barriers to trade' and recognized in particular 'the overall equivalence of Member States legislative objectives in the protection of health and safety'.

The Commission has established and will continue close co-operation in defining essential safety requirements at the design and construction stages of new equipment.

Legislative harmonization will enable principal safety requirements to be established progressively, the necessary technical specifications being entrusted to organizations competent to deal with standardization. In view of the importance of the technical specifications for meeting the essential requirements for products used at work and the need to fully guarantee the dialogue between the two sides of industry on this question, the Commission will ensure adequate involvement of the trade unions in European standardization work and related activities.

2. *Promotion of safety at work and application of ergonomic principles*

(a)   The Commission will prepare directives covering the organization of safety at work as well as the selection and use of appropriate plant, equipment, machinery and substances. In addition, the Commission will prepare a proposal for a Council decision regarding a system for the rapid exchange of information on specific safety hazards at work and the resulting restrictions placed on the use of dangerous substances, tools, equipment, *etc.*

(b)   The Commission will prepare directives on personal protective equipment provided at the work place by the employer, with particular regard to appropriate use, user acceptability, availability, maintenance and testing.

(c)   The Commission will revise the 1977 Directive on safety signs at work to bring it up to date and extend its scope.

(d)   The Commission will put forward recommendations on the selection and use of equipment resulting from the development of new technologies and process control systems, with particular regard to the intrinsic safety of the equipment and ergonomic factors in its use.

(e)   The Commission will prepare recommendations on good working practices aimed at avoiding back pain and back injury caused by bad work place design resulting in physical strain, faulty handling of materials, incorrect lifting and falls.

3. *Safety in high-risk sectors*

From the high-risk sectors the Commission has focussed its attention on the three with the highest accident rate and highest level of serious injury.

(a)   Work at sea:

Working and living conditions on board are particulary difficult: movement of the work area, limited space, long duration and high intensity of work, noise multiplicity of individual workers' tasks, geographic or meteorological isolation of the vessel, limiting the possibilities of assistance and thus exacerbating the consequences of accidents, all contribute to a higher fatal accident rate in the seafaring occupations than in other 'high-risk' jobs.

In view of this situation, which affects around 500,000 workers, urgent measures are envisaged in order to make safety a more integral part of the design of vessels and the definition of tasks and to ensure the availability of adequate medical assistance and emergency services at sea.

(b) Agriculture:

Agriculture employs around 10 million people in the Community. More than half of all work accidents occur in farmyards and farm buildings, in particular during the handling of animals, horizontal or vertical movements and the handling of tools, loads and pesticides. However, owing to their self-employed status, farmers are not covered or concerned by the regulations governing health and safety at work, even when such regulations apply to agriculture.

The Commission is drawing up a directive on plant-protection products and recommendations concerning the design of farm buildings and electrical installations in agriculture.

(c) The construction industry:

The construction industry (building and civil engineering) is an essential element of economic activity in the European Community and employs almost 10 million workers. The building sector is characterized by a high proportion of small firms, attracted by the low level of capital outlay needed. It is also an activity with a higher-than-average risk of accidents and occupational diseases. In addition, as the current system of bidding for contracts gives no specific indication of safety and health costs, it may encourage tenderers to propose working methods which are apparently cheaper but less safe, or to adopt such methods once the contract has been won. For these reasons, the traditional tripartite approach must be broadened to include designers and clients.

The Commission will prepare a directive on safety in the construction industry, which will stress the need to incorporate safety requirements right from the initial design stage, to make health and safety aspects clearer in the tenders, to closely define responsibility. On construction sites and to establish safety-related qualification requirements for certain tasks.

B. Occupational health and hygiene

1. In order to guarantee that exposure of workers to physical factors, biological organisms and chemical substances is as low as reasonably achieveable, and to enable the level of exposure to be monitored and measured, the Commission has forwarded to the Council a proposal for a directive establishing the basis for a Community list of exposure limit values for 100 agents.

The lists already drawn up by the Member States contain over 1,000 substances and the European Inventory of Existing Chemical Substances (EINECS) contains 100,000 entries.

The Commission intends to extend this list accordingly, and will carry out studies to collect and evaluate toxicological and health data for individual agents and their absorption pathways. The Commission will also examine ways and means of improving the collection of such data. In the case of special protective measures which may be required for those chemical agents which can be absorbed through the skin, the Commission will propose modifications to the existing directives.

2. In the case of agents likely to cause cancer, the Commission intends to submit to the Council a directive laying down general and specific measures relating to occupational carcinogens. Subsequent directives will be proposed for the other carcinogenic agents in line with ongoing work on the classification and labelling of chemical substances. The Commission will also submit proposals for directives on certain groups of compounds such as pesticides. A proposal for a directive will also have to be made on biological agents which cause ill health, such as pathogenic micro-organisms, and genetic engineering techniques which may present a risk to health.

3. Once the proposal for a Council directive on exposure limit values for 100 agents (see B (2)) has been adopted by the Council, detailed examination of the measures required — for example, technical analyses — must be carried out to ensure accurate determination of exposure levels. To this end, the Commission will request technical assistance from competent organizations such as CEN. Account will also be taken of the current work of the International Organizations Standards in this area. The Commission will also study ways of improving the measurement methods available.

4. For very dangerous agents or work activities, the Commission has already submitted a proposal for a directive to the Council, in which the conditions to be applied for the proscription of specific agents are set out. Studies will be carried out to determine the other agents and/or processes to be added to this directive.

5. The Commission is working on the technical aspects of the directive on noise, which will be implemented from 1990. A proposal will be submitted extending its field of application by including workers not currently covered and by re-evaluating the threshold values.

6. The proposed directive on the harmonizations of classification and labelling of dangerous preparations emphasizes the need for information on the composition of such preparations and the hazards they present. The Commission will investigate what supplementary measures are required for the health protection of workers under Article 118a.

7. In 1962 and 1966 the Commission made recommendations to the Member States concerning a European Schedule of Industrial Diseases. This list must be revised to take account of subsequent improvements in the diagnosis of occupational diseases. The competent advisory committee is considering what improvements should be made to the Schedule and the Commission will make recommendations to the Member States on the basis of its findings.

8. Legal provisions relating to occupational health services and their role in the protection of workers' health vary considerably between Member States. The appropriate advisory committee is currently preparing an opinion on the organization of these services and the respective roles of the various health and safety specialists, taking into account the previous work of the Economic and Social Committee. On the basis of this, the Commission intends to draft a recommendation on the subject.

C. Information

1. In its joint opinion on information and consultation, the Val Duchesse Working Party stated 'When technological changes which imply major consequences for the work-force are introduced into the firm, workers and/or their representatives should be informed and consulted in accordance with the laws, agreements and practices in force in the Community countries'. The Commission considers this objective to be particularly important where such practices have a potential impact on health and safety.

2. In order to overcome the disparity of available information on chemical substances, the Commission intends to provide information on all the substances for which directives are proposed in the field of health and safety. This information, together with that provided by the labelling system for dangerous substances and preparations, will be examined in order to determine its best use.

3. The protection of workers requires that research results and technical innovations aimed at improving working conditions are applied with the co-operation of all parties involved.

    To this end, the Commission will step up its work in the following fields:

    —   the evaluation of recent research, to select that most promising for application in pilot projects,

    —   the establishment of evaluation programmes with the co-operation in each case of two or more Member States,

    —   the development of methods of disseminating the results, particularly for high-risk activities such as deep-sea diving or off-shore exploration.

4. Finally, the Commission intends to increase information, training and exchange of experience between senior labour inspectors responsible for national implementation of regulations derived from Community directives.

    To this end, the regular meetings of the labour inspectors currently taking place at Community level will be formalized, seminars will be organized on specific topics, and the programme of exchange of inspectors between Member States will be expanded.

D. Training

1. The Commission recently submitted to the Council two communications on adult training in firms and vocational training for women. On the basis of the conclusions adopted by the Council, new action programmes will be drawn up in these areas, in which health and safety training at the workplace could be included.

    In addition, with the assistance of the European Centre for the Development of Vocational Training (CEDEFOP), the Commission will give special priority to the development of courses for the training of safety instructors.

2. Considerable differences exist between the Member States in the safety training and official recognition of those responsible for safety and health protection (company managers, safety officers, ergonomics and health specialists, first aiders, workers representatives, *etc.*). The Commission proposes to continue to encourage training initiatives for these various groups based upon generally accepted principles and practice.

3. When developing and during the course of special youth training schemes aimed particularly at the unemployed the Commission will study the provisions necessary to ensure the safety of participants, including those combined work/training schemes.

4. At university, or in higher level technical education, the Commission will investigate ways of providing a full course of training in the appropriate safety precautions required for the future specialization of those who will be responsible for the safety of others, *eg.* engineers, industrial chemists, and physicists.

5. In the high-risk sectors, the Commission has already developed a series of training modules for certain dangerous agricultural activities, and these have been tested in pilot projects. For sea fishing, financial and technical assistance has been provided for the development of the 'Medical Advice Centres Network' (Macnet), to extend the availability of medical assistance. The Commission intends to further develop these activities, which have a direct impact on these high-risk sectors.

6. To develop the training resources necessary to meet these various needs, the Commission intends to establish a network of collaboration centres involved in teaching the various disciplines and training workers and their representatives.

E. Small and medium-sized enterprises

1. The Community is devoting special attention to small and medium-sized enterprises, which are considered an essential element in economic recovery and job creation. The 'Action Programme for SMEs' stresses the need to keep regulations down to a necessary minimum. For its part, Article 118a of the Single European Act recognizes the special needs of SMEs in respect of safety and health problems.

   In order to fulfil both these essential requirements, and in order to keep the directives from imposing administrative, financial and legal constraints which may hold back the creation and development of SMEs, the Commission intends:

   — to undertake a study of how existing regulations on health and safety are interpreted and applied in a sample of SMEs,

   — to undertake a review of the special rules and exceptions which exist in national legislation regarding health, hygiene and safety at work, and to assess the need for harmonization of legislation in this field in accordance with Article 118a of the Single European Act.

2. When faced with activities which have a high health and safety risk, SMEs do not always possess the technical know-how in accident prevention, training and monitoring are difficult to carry out and accidents can have serious economic consequences.

Any impetus towards new patterns of working can pose additional problems for such enterprises. Moreover, longer working hours in SMEs may lead to increased fatigue and a slackening of vigilance, increasing the risk of accidents. In addition, the measurement of exposure limits to dangerous agents, normally calculated on an eight-hour working day, may have to be adjusted.

The Commission therefore intends to study the effect of new patterns of working on safety, hygiene and health in SMEs.

3.   The Commission is aware of the limited impact of information campaigns on the special rules and exceptions in health and safety legislation in SMEs. Furthermore, it would appear that efforts made within the Member States to provide advice and training on safety are not having the expected results.

To counter this, the Commission intends:

—   to consider how health and safety regulations can be made clearer for proprietors of SMEs,

—   to include advice on safety, hygiene and health at work in information manuals to be prepared for creators of SMEs,

—   to prepare training modules on safety specifically for creators of SMEs, develop pilot projects integrating these modules into general training and provide for specific safety counselling,

—   to develop a system for providing readily accessible information to SMEs on safety equipment and personal protective equipment.

F.   Social dialogue

Development of Community action on health and safety and the balance which must be achieved between economic and social policy, as the large internal market is developed, both necessitate close collaboration between employer and worker representatives during the stages leading up to Commission decisions.

The Commission therefore intends to develop the dialogue between the two sides of industry in this field pursuant to Article 118b of the Single Act.

The Advisory Committee on Safety, Hygiene and Health Protection at Work, which has existed since 1974, provides a highly appropriate forum for consultation between the two sides of industry. This Committee must play fully its part in assisting the Commission in defining the action it will take in this field. As in the past, the Commission will continue to consult the Committee on the proposals which it intends to present to the Council.

COMMISSION DECISION

——————————— of 24 February 1988 ———————————
providing for the improvement of information on safety, hygiene and
health at work
[88/383/EEC]

THE COMMISSION OF THE EUROPEAN COMMUNITIES,

Having regard to the Treaty establishing the European Economic Community, and in particular Article 118 thereof,

Whereas it is essential for the Commission to have the necessary information at its disposal before the adoption by the Member States of laws, regulations and administrative provisions on safety, hygiene and health at work; whereas, in certain cases, all the Member States must also be informed of the laws, regulations and administrative provisions contemplated by any one Member State;

Whereas Council Directive 83/189/EEC provided for a system of information in the field of technical standards and regulations, including those relating to safety, hygiene and health at work;

Whereas it would appear necessary to supplement this procedure by the improvement of information on other laws, regulations and administrative provisions provided for under this Decision;

Whereas it is necessary to set up a group of experts, the members of which will be nominated by the Member States, with the task of helping the Commission to examine draft national laws, regulations and administrative provisions,

HAS ADOPTED THIS DECISION:

Article 1

The Member States shall inform the Commission without delay on any laws, regulations and administrative provisions on safety, hygiene and health at work and of any draft laws, regulations and administrative provisions in this field, with the exception of draft technical regulations as defined in Article 1(6) of Directive 83/189/EEC. The Member States shall also submit any other provisions referred to in this field.

Article 2

1.   The Commission shall forward to the Member States any draft laws, regulations and administrative provisions which it receives under Article 1 and which it considers relevant for the purpose of the present Decision.

2.   The Commission and the other Member States may submit observations to the Member State whose draft provisions they have received under paragraph 1; the Member State concerned shall take these observations into consideration as far as possible. Member States shall forward their observations through the Commission.

Article 3

The Commission shall be assisted by a group of experts chaired by a representative of the Commission. The group shall be composed of 24 members, each Member State nominating two experts for appointment by the Commission. For each member an alternate member shall be appointed as described above.

Article 4

The Commission shall periodically inform the Advisory Committee on Safety, Hygiene and Health Protection at work of any activities arising from the implementation of this Decision, with the exception of any aspects deemed confidential by the Member States, and, where appropriate, the Standing Committee set up pursuant to Article 5 of Directive 83/189/EEC.

Article 5

This Decision is addressed to the Member States.

Done at Brussels, 24 February 1988.

DECISION OF THE COUNCIL AND OF REPRESENTATIVES OF THE GOVERNMENTS OF THE MEMBER STATES, MEETING WITHIN THE COUNCIL

——————— of 21 June 1988 ———————

adopting a 1988 to 1989 plan of action for an information and public awareness campaign in the context of the 'Europe against Cancer' programme

[88/351/EEC]

THE COUNCIL OF THE EUROPEAN COMMUNITIES AND THE REPRESENTATIVES OF THE GOVERNMENTS OF THE MEMBER STATES, MEETING WITHIN THE COUNCIL

Having regard to the Treaty establishing the European Economic Community,

Having regard to the proposal from the Commission,

Having regard to the opinion of the European Parliament,

Having regard to the opinion of the Economic and Social Committee,

Whereas in June 1985 in Milan and in December 1985 in Luxembourg the European Council underlined the advantages of launching a European programme against cancer;

Whereas in December 1986 in London the European Council decided that 1989 should be designated 'European Cancer Information Year' and specified that the aim would be to develop a sustained and concerted information campaign in all the Member States on the prevention, early screening and treatment of cancer;

Whereas the prevention and research measures already undertaken in the context of the 'Europe against cancer' programme should be supplemented by measures to increase awareness among members of the health professions and by an information campaign;

Whereas this campaign should be especially designed to increase the awareness of the public, teachers and members of the health professions regarding the fight against cancer, which includes in particular the fight against tobacco, protection against carcinogenic agents and, for the purposes of health promotion in general, improvement in nutrition;

Whereas information campaigns against cancer are a matter for the competent private organizations and public authorities of the Member States, but Community action may give a major boost towards success in the fight against cancer;

Whereas duplication of effort should be avoided by developing common basic material for the information of the public and for the training of members of the health professions, as well as by promoting exchanges of experience,

HAVE DECIDED AS FOLLOWS:

Article 1

1. The Commission and the Member States shall implement, in the period 1988 to 1989, the measures set out in the Annex.

2. The estimated amount of the Community distribution necessary for these measures is 10 million ECU.

Article 2

1. At intervals, and at least once a year, the Commission shall consult the competent authorities of the Member States during the various stages of implementation of the measures provided for in this Decision.

2. The Commission shall keep the European Parliament and the Council informed of progress.

Done at Luxembourg, 21 June 1988.

ANNEX

MEASURES TO BE IMPLEMENTED IN 1988 AND 1989

In 1988

A 1: Co-ordination, in 1988, of a 'European week' of information on the 'Europe against cancer' programme, with particular emphasis on the following aspects: 'Fight against tobacco', 'Improvement in nutrition' and 'Fight against carcinogenic agents', in preparation for the European Cancer Information Year.

In this context, organization of Community activities, encouragement of, and contribution to, the production of television information and health education programmes for the general public containing a European dimension in 1988 and in 1989 for the European Cancer Information Year.

A 2: Exchange of experience with teaching materials for health education on cancer for young people and on cancer awareness for the health professions and contributions to the dissemination and, if need be, adaptation of such materials in 1988 and 1989 for the European Cancer Information Year.

A 3: Campaign to interest teachers and the health professions in the 'Europe against cancer' programme.

In 1989

A 4: Organization and co-ordination at Community level in 1989 of an information campaign on the 'Europe against cancer' programme, designed to develop the campaign launched in 1988 and to fit into the framework of the European Cancer Information Year with regard to prevention, screening and treatment aspects.

COMMUNITY CHARTER

———— of 9 December 1989 ————

of the fundamental social rights of workers

[Social Charter 1989]

THE HEADS OF STATE AND GOVERNMENT OF THE MEMBER STATES OF THE EUROPEAN COMMUNITY MEETING AT STRASBOURG ON 9 DECEMBER 1989

Whereas, under the terms of Article 117 of the EEC Treaty, the Member States have agreed on the need to promote improved living and working conditions for workers so as to make possible their harmonization while the improvement is being maintained;

Whereas following on from the conclusions of the European Councils of Hanover and Rhodes the European Council of Madrid considered that, in the context of the establishment of the single European market, the same importance must be attached to the social aspects as to the economic aspects and whereas, therefore, they must be developed in a balanced manner;

Having regard to the Resolutions of the European Parliament of 15 March 1989 and 14 September 1989 and to the opinion of the Economic and Social Committee of 22 February 1989;

Whereas the completion of the internal market is the most effective means of creating employment and ensuring maximum well-being in the Community; whereas employment development and creation must be given first priority in the completion of the internal market; whereas it is for the Community to take up the challenges of the future with regard to economic competitiveness, taking into account, in particular, regional imbalances;

Whereas the social consensus contributes to the strengthening of the competitiveness of undertakings, of the economy as a whole and to the creation of employment; whereas in this respect it is an essential condition for ensuring sustained economic development;

Whereas the completion of the internal market must favour the approximation of improvements in living and working conditions, as well as to economic and social cohesion within the European Community while avoiding distortions of competition;

Whereas the completion of the internal market must offer improvements in the social field for workers of the European Community, especially in terms of freedom of movement, living and working conditions, health and safety at work, social protection, education and training;

Whereas, in order to ensure equal treatment, it is important to combat every form of discrimination, including discrimination on grounds of sex, colour, race, opinions and beliefs, and whereas, in a spirit of solidarity, it is important to combat social exclusion;

Whereas it is for Member States to guarantee that workers from non-Member countries and members of their families who are legally resident in a Member State of the European Community are able to enjoy, as regards their living and working conditions, treatment comparable to that enjoyed by workers who are nationals of the Member State concerned;

Whereas inspiration should be drawn from the Conventions of the International Labour Organization and from the European Social Charter of the Council of Europe;

Whereas the Treaty, as amended by the Single European Act, contains provisions laying down the powers of the Community, relating, *inter alia*, to the freedom of movement of workers (Articles 7, 48-51), to the right of establishment (Articles 52-58), to the social field under the conditions laid down in Articles 117-122 — in particular as regards the improvement of health and safety in the working environment (Article 118a), the development of the dialogue between management and labour at European level (Article 118b), equal pay for men and women for equal work (Article 119) — to the general principles for implementing a common vocational training policy (Article 128), to economic and social cohesion (Article 130a to 130e) and, more generally, to the approximation of legislation (Articles 100, 100a and 235); whereas the implementation of the Charter must not entail an extension of the Community's powers as defined by the Treaties;

Whereas the aim of the present Charter is on the one hand to consolidate the progress made in the social field, through action by the Member States, the two sides of industry and the Community;

Whereas its aim is on the other hand to declare solemnly that the implementation of the Single European Act must take full account of the social dimension of the Community and that it is necessary in this context to ensure at appropriate levels the development of the social rights of workers of the European Community, especially employed workers and self employed persons;

Whereas, in accordance with the conclusions of the Madrid European Council, the respective roles of Community rules, national legislation and collective agreements must be clearly established;

Whereas, by virtue of the principle of subsidiarity, responsibility for the initiatives to be taken with regard to the implementation of these social rights lies with the Member States or their constituent parts and, within the limits of its powers, with the European Community; whereas such implementation may take the form of laws, collective agreements or existing practices at the various appropriate levels and whereas it requires in many spheres the active involvement of the two sides of industry;

Whereas the solemn proclamation of fundamental social rights at European Community level may not, when implemented, provide grounds for any retrogression compared with the situation currently existing in each Member State,

HAVE ADOPTED THE FOLLOWING DECLARATION CONSTITUTING THE 'COMMUNITY CHARTER OF THE FUNDAMENTAL SOCIAL RIGHTS OF WORKERS':

TITLE I — FUNDAMENTAL SOCIAL RIGHTS OF WORKERS

FREEDOM OF MOVEMENT

1. Every citizen of the European Community shall have the right to freedom of movement throughout the territory of the Community subject to restrictions justified on grounds of public order, public safety or public health.

2. The right to freedom of movement shall enable any citizen to engage in any occupation or profession in the Community in accordance with the principles of equal treatment as regards access to employment, working conditions and social protection in the host country.

3. The right of freedom of movement shall also imply:

    — harmonization of conditions of residence in all Member States, particularly those concerning family reunification;

    — elimination of obstacles arising from the non-recognition of diplomas or equivalent occupational qualifications;

    — improvement of the living and working conditions of frontier workers.

EMPLOYMENT AND REMUNERATION

4. Every individual shall be free to choose and engage in an occupation according to the regulations governing each occupation.

5. All employment shall be fairly remunerated.

    To this effect, in accordance with arrangements applying in each country:

    — workers shall be assured of an equitable wage, *i.e.* a wage sufficient to enable them to have a decent standard of living;

    — workers subject to terms of employment other than an open-ended full time contract shall receive an equitable reference wage;

    — wages may be withheld, seized or transferred only in accordance with the provisions of national law; such provisions should entail measures enabling the worker concerned to continue to enjoy the necessary means of subsistence for himself and his family.

6. Every individual must be able to have access to public placement services free of charge.

IMPROVEMENT OF LIVING AND WORKING CONDITIONS

7. The completion of the internal market must lead to an improvement in the living and working conditions of workers in the European Community. This process must result from an approximation of these conditions while the improvement is being maintained, as regards in particular the duration and organization of working time and forms of employment other than open-ended contracts, such as fixed-term contracts, part-time working, temporary work and seasonal work.

The improvement must cover, where necessary, the development of certain aspects of employment regulations such as procedures for collective redundancies and those regarding bankruptcies.

8.   Every worker of the European Community shall have a right to a weekly rest period and to annual paid leave, the duration of which must be harmonised in accordance with national practices while the improvement is being maintained.

9.   The conditions of employment of every worker of the European Community shall be stipulated in laws, in a collective agreement or in a contract of employment, according to arrangements applying in each country.

SOCIAL PROTECTION

According to the arrangements applying in each country:

10.   Every worker of the European Community shall have a right to adequate social protection, and shall, whatever his status and whatever the size of the undertaking in which he is employed, enjoy an adequate level of social security benefits.

Persons who have been unable either to enter or re-enter the labour market and have no means of subsistence must be able to receive sufficient resources and social assistance in keeping with their particular situation.

FREEDOM OF ASSOCIATION AND COLLECTIVE BARGAINING

11.   Employers and workers of the European Community shall have the right of association in order to constitute professional organizations or trade unions of their choice for the defence of their economic and social interests.

Every employer and every worker shall have the freedom to join or not to join such organisations without any personal or occupational damage being thereby suffered by him.

12.   Employers or employers' organizations, on the one hand, and workers' organizations, on the other, shall have the right to negotiate and conclude collective agreements under the conditions laid down by national legislation and practice.

The dialogue between the two sides of industry at European level which must be developed, may, if the parties deem it desirable, result in contractual relations, in particular at inter-occupational and sectoral level.

13.   The right to resort to collective action in the event of a conflict of interests shall include the right to strike, subject to the obligations arising under national regulations and collective agreements.

In order to facilitate the settlement of industrial disputes the establishment and utilization at the appropriate levels of conciliation, mediation and arbitration procedures should be encouraged in accordance with national practice.

14.   The internal legal order of the Member States shall determine under which conditions and to what extent the rights provided for in Articles 11 to 13 apply to the armed forces, the police and the civil service.

VOCATIONAL TRAINING

15. Every worker of the European Community must be able to have access to vocational training and to receive such training throughout his working life. In the conditions governing access to such training there may be no discrimination on grounds of nationality.

    The competent public authorities, undertakings or the two sides of industry, each within their own sphere of competence, should set up continuing and permanent training systems enabling every person to undergo retraining, more especially through leave for training purposes, to improve his skills or to acquire new skills, particularly in the light of technical developments.

EQUAL TREATMENT FOR MEN AND WOMEN

16. Equal treatment for men and women must be assured. Equal opportunities for men and women must be developed.

    To this end, action should be intensified wherever necessary to ensure the implementation of the principle of equality between men and women as regards in particular access to employment, remuneration, working conditions, social protection, education, vocational training and career development.

    Measures should also be developed enabling men and women to reconcile their occupational and family obligations.

INFORMATION, CONSULTATION AND PARTICIPATION FOR WORKERS

17. Information, consultation and participation for workers must be developed along appropriate lines, taking account of the practices in force in the various Member States.

    This shall apply especially in companies or groups of companies having establishments or companies in several Member States of the European Community.

18. Such information, consultation and participation must be implemented in due time, particularly in the following cases:

    — when technological changes which, from the point of view of working conditions and work organization, have major implications for the workforce are introduced into undertakings;

    — in connection with restructuring operations in undertakings or in cases of mergers having an impact on the employment of workers;

    — in case of collective redundancy procedures;

    — when trans-frontier workers in particular are affected by employment policies pursued by the undertaking where they are employed.

HEALTH PROTECTION AND SAFETY AT THE WORKPLACE

19.   Every worker must enjoy satisfactory health and safety conditions in his working environment. Appropriate measures must be taken in order to achieve further harmonization of conditions in this area while maintaining the improvements made.

    The measures shall take account, in particular, of the need for the training, information, consultation and balanced participation of workers as regards the risks incurred and the steps taken to eliminate or reduce them.

    The provisions regarding implementation of the internal market shall help to ensure such protection.

PROTECTION OF CHILDREN AND ADOLESCENTS

20.   Without prejudice to such rules as may be more favourable to young people, in particular those ensuring their preparation for work through vocational training, and subject to derogations limited to certain light work, the minimum employment age must not be lower than the minimum school-leaving age and, in any case, not lower than 15 years.

21.   Young people who are in gainful employment must receive equitable remuneration in accordance with national practice.

22.   Appropriate measures must be taken to adjust labour regulations applicable to young workers so that their specific needs regarding development, vocational training and access to employment are met.

    The duration of work must, in particular, be limited — without it being possible to circumvent this limitation through recourse to overtime — and night work prohibited in the case of workers of under 18 years of age save in the case of certain jobs laid down in national legislation or regulations.

23.   Following the end of compulsory education, young people must be entitled to receive initial vocational training of a sufficient duration to enable them to adapt to the requirements of their future working life; for young workers, such training should take place during working hours.

ELDERLY PERSONS

According to the arrangements applying in each country:

24.   Every worker of the European Community must, at the time of retirement, be able to enjoy resources affording him or her a decent standard of living.

25.   Every person who has reached retirement age but who is not entitled to a pension or who does not have other means of subsistence, must be entitled to sufficient resources and to medical and social assistance specifically suited to his needs.

DISABLED PERSONS

26.   All disabled persons, whatever the origin and nature of their disablement, must be entitled to additional concrete measures aimed at improving their social and professional integration.

       These measures must concern, in particular, according to the capacities of the beneficiaries, vocational training, ergonomics, accessibility, mobility, means of transport and housing.

TITLE II – IMPLEMENTATION OF THE CHARTER

27.   It is more particularly the responsibility of the Member States, in accordance with the national practices, notably through legislative measures or collective agreements, to guarantee the fundamental social rights in this Charter and to implement the social measures indispensable to the smooth operation of the internal market as part of a strategy of economic and social cohesion.

28.   The European Council invites the Commission to submit as soon as possible initiatives which fall within its powers, as provided for in the Treaties, with a view to the adoption of legal instruments for the effective implementation, as and when the internal market is completed, of those rights which come within the Community's area of competence.

29.   The Commission shall establish each year, during the last three months, a report on the application of the Charter by the Member States and by the European Community.

30.   The report of the Commission shall be forwarded to the European Council, the European Parliament and the Economic and Social Committee.

COMMUNICATION FROM THE COMMISSION

of 29 November 1989

concerning its action programme relating to the implementation of the community
charter of basic social rights for workers

[COM (89) 568 final]

PART I

GENERAL INTRODUCTION

1.   On 27 September 1989 the Commission presented a draft Community Charter of fundamental social rights in which reference is made to an action programme. In its report to the European Council of 8 and 9 December 1989, the Presidency, at the end of the Social Affairs Council of 30 October 1989, has taken note of the Commission's intention to present an action programme relating to the concrete implementation of the rights defined in the Charter. Further, the Presidency "has invited the Commission to take into account the demands expressed by several delegations regarding, in particular, the determination of paid leave, the continued payment of remuneration during holidays and during sickness, the protection of children and adolescents, the situation of pregnant women and of mothers with children of an early age, the professional insertion of the disabled, protection of health and safety at the workplace, vocational guidance, mutual recognition of qualifications and temporary work".

This is the subject of this document which the Commission has prepared under its sole responsibility, pursuant to its right of initiative, with regard to proposals for Community instruments to be presented to the Council and recommendations under Article 155 of the EEC Treaty. The Commission will, however, present this document to the European Parliament, the Economic and Social Committee and the two sides of industry.

2.   The attached action programme contains a number of new measures which the Commission sees a need to develop in order to implement the most urgent aspects of the principles of the draft Charter.

These measures are grouped under thirteen short chapters, each covering an area relating to the development of the social dimension of the internal market and which, apart from the chapter on employment and the labour market, correspond to the various sections of the Charter in the context of completing the internal market and more generally, implementing the Treaty as amended by the Single European Act.

That being said, the social dimension has already become a fact. The reform of the structural funds, the improvement of the working environment in order to protect the health and safety of workers, occupational equality between men and women, the various exchange programmes, *etc.* are but a few examples of the important fields in which substantial advances have been made.

Each of the thirteen chapters also reviews measures already adopted by the Community with regard to the area concerned. Reference is also made to work that will be continued in each of these areas to adapt existing instruments to social

136

change or change in the Community (for example, adaptation of certain regulations concerning freedom of movement and social security for migrant workers) or technical change (for example, adaptation of certain directives concerning the safety and health of workers).

Each of the new measures is included in a Presentation in which the Commission emphasizes the reasons why it considers that action is needed at Community level and the essential components of the proposal it plans to draw up.

3. In accordance with the principle of subsidiarity whereby the Community acts when the set objectives can be reached more effectively at its level than at that of the Member States, the Commission's proposals relate to only part of the issues raised in certain articles of the draft Charter. The Commission takes the view that responsibility for the initiatives to be taken as regards the implementation of social rights lies with the Member States, their constituent parts or the two sides of industry as well as, within the limits of its powers, with the European Community.

4. Furthermore, in choosing the legal instruments it will propose, the Commission will take account of the fact that its proposals should be implemented in the form of laws or collective agreements, this making it possible to adapt to particular situations and enabling the two sides of industry to be actively involved.

5. The Commission has therefore limited its proposals for directives or regulations to those areas where Community legislation seems necessary to achieve the social dimension of the internal market and more generally, to contribute to the economic and social cohesion of the Community. It mainly concerns proposals relating to social security for migrant workers, freedom of movement, working conditions, vocational training, and improvements, particularly of the working environment, to protect the safety and health of workers. While the Commission is not making a proposal in respect of discrimination on the grounds of race, colour or religion, it nonetheless stresses the need for such practices to be eradicated, particularly in the workplace and in access to employment, through appropriate action by the Member States and by the two sides of industry.

6. In some cases, the Commission is not proposing any initiative. This applies in the case of that section of the draft Charter which is devoted to the right to freedom of association and collective bargaining.

The affirmation of these principles is vital in the field of industrial relations which largely control relations between the two sides of industry in firms, and more widely, on the labour market. Clearly, the problems deriving from the application of these principles must be settled directly by the two sides of industry, or where appropriate, by the Member States.

This section also refers to the social dialogue; in the spirit of Article 118b of the Treaty, the Commission will actively seek to develop it at Community, sectoral or inter-occupational level, but also at national and regional levels, it being understood that the dialogue could lead eventually to relations based on agreement at European level.

7. By seeking to make a distinction between the measures to be taken by the Community and those to be taken by the Member States or the two sides of industry, the Commission believes it is acting fully in consonance with the request made by the Heads of State and of Government at the European Council in Madrid

which emphasized that "the role to be played by Community standards, national legislation and contractual relations must be clearly established".

8. Although the draft Charter refers to employment policy and the necessary fight against unemployment only in the recitals, the Commission presents in this Action Programme some measures it plans to take to contribute on the one hand to improving knowledge about the labour market and measures to combat unemployment — thus responding to a request by the two sides of industry in the context of the social dialogue — and on the other hand to solve the problem of unemployment, particularly long-term unemployment, at Community level.

In addition, reference is made, albeit succinctly, in this Action Programme to European Social Fund operations, the main component of Community action in the field of vocational training for young people and for long-term unemployed workers, and thus an essential factor in the campaign against unemployment: the activities of the European Social Fund are henceforth a part of established Community practice.

9. The Commission believes that an Action Programme should include components concerning employment, training and workers' living and working conditions. Thus, by implementing a set of factors contributing to the development of the social dimension of the internal market a contribution will be made to the economic and social cohesion of the Community. It should be stressed that while priority must be given to job creation in the context of reinforcing firms' competitiveness, at the same time it is important to implement an overall policy aimed initially at workers' interests by affirming that the economic, industrial and social aspects form a whole. In this context, the Commission wants to stress the importance it attaches to the monitoring and assessment of the intervention of the structural funds as primary instruments for contributing to employment development and job creation, particularly from the point of view of regional imbalances.

10. In most cases, the Commission has indicated the nature of the proposals to be presented: proposals for directive, regulation, decision, recommendation or communications, or again opinions within the meaning of Article 118 of the Treaty.

However, it has not indicated the legal bases on which proposals will be based. The legal bases to which the Commission could refer are set out in one of the recitals in the Charter. It would be premature, at this stage, to make a statement with respect to the legal bases for proposals to be made in the course of the next three years.

11. With respect to the implementation of this Action Programme, the Commission will present all the proposals set out in the second part of this Action Programme. The first set of proposals, representing the most urgent priorities, will be put forward in the Commission's 1990 work programme. A second set will be included in the 1991 work programme. Any further proposals will be presented in 1992.

12. The Commission recalls that, in the context of the implementation of the Charter, it should also be instructed to present a regular report on the application of the Charter by the Member States and by the European Community.

The Commission therefore expects the Governments of the Member States to transmit an initial report by the end of 1990 stating how they have applied the principles of the Charter.

13. More generally, the Commission stresses that the Council should reach decisions without delay on the proposals it plans to present.

The Commission therefore asks the Council, as it did in 1974 at the time of the adoption of the social action programme, to undertake to adopt a decision concerning the Commission proposals within a period of 18 months, but in any case within two years at the outside, after transmission of the Commission proposals to the Council, the European Parliament, the Economic and Social Committee and the two sides of industry.

PART 10

HEALTH PROTECTION AND SAFETY AT THE WORKPLACE

A. INTRODUCTION

Protection of health and of safety in the working environment is ensured by means of technical regulations regarding products and equipments used by workers and by provisions regarding worker protection in the working environment.

Before the Single Act came into force there were already a number of directives applicable in the field of health and safety at work (notably protection against risks from asbestos, noise and lead).

Since October 1987, when the Commission adopted its programme concerning safety, hygiene and health at work which the Council welcomed in its Resolution of 21 December 1987, ten proposals for directives have been presented to the Council. Three of them have already been adopted, including Directive 89/391/EEC on improvements in the safety and health of workers at work, which is of particular importance.

Other proposals should be adopted by the end of the year or during the first half of 1990.

In parallel, the Community has developed the implementation of a new approach regarding technical regulation which entails, for example for industrial machines or for individual protective clothing, compulsory safety requirements for the protection of workers. The implementation of these requirements goes through European standards, to the definition of whose representatives of workers are from now on associated.

The Community already has, therefore, a series of binding provisions which ensure fairly broad protection for workers' health and safety at the workplace. It must be pointed out, moreover, that the Commission will propose, whenever necessary, amendments to the directives adopted to take account of developments occurring after their adoption (new substances, technical progress). Several such proposals will be presented in the next few years.

The Commission considers that priority should be given to new initiatives in areas where safety causes significant problems, notably, the building industry, fisheries, drilling rigs and open-cast mines, the improvement of medical assistance on board vessels and also workplaces excluded from the specific workplace directive.

In addition, when freedom of movement develops further and the labour market takes on a European dimension, the Commission believes that the Member States should endeavour to approximate their ideas concerning the schedule of industrial diseases. This would doubtless not be a question of introducing laws in an area closely connected with national social security systems. The Commission will accordingly put forward a recommendation emphasizing the importance of adopting a schedule of European industrial diseases.

B.   NEW INITIATIVES

— *Proposal for a Council Directive on the minimum health and safety requirements to encourage improved medical assistance on board vessels*

— *Proposal for a Council Directive on the minimum health and safety requirements for work at temporary or mobile work sites*

— *Proposal for a Council Directive on the minimum requirements to be applied in improving the safety and health of workers in the drilling industries*

— *Proposal for a Council Directive on the minimum requirements to be applied in improving the safety and health of workers in the quarrying and open-cast mining industries*

— *Proposal for a Council Directive on the minimum safety and health requirements for fishing vessels*

— *Recommendation to the Member States on the adoption of a European schedule of industrial diseases*

— *Proposal for a Council directive on the minimum requirements for safety and health signs at the workplace*

— *Proposal for a Council directive defining a system of specific information for workers exposed to certain dangerous industrial agents*

— *Proposal for a Council directive on the minimum safety and health requirements regarding the exposure of workers to the risks caused by physical agents*

— *Proposal for a Council directive amending Directive 83/447/EEC on the protection of workers from the risks related to exposure to asbestos at work*

— *Proposal for a Council directive on the minimum safety and health requirements for activities in the transport sector*

— *Proposal for the establishment of a safety, hygiene and health agency*

Proposal for a Council Directive on the minimum health and safety requirements to encourage improved medical assistance on board vessels

Work on board vessels involves specific risks. The consequences of accidents are heightened given that medical equipment on board is often inadequate and much time is required for help and intervention from elsewhere.

The proposed Directive aims to promote better worker safety and health on board vessels by improving medical assistance on board.

Proposal for a Council Directive on the minimum health and safety requirements for work at temporary or mobile work sites

Major risks are involved in work on temporary and mobile sites.

The Directive aims to incorporate health requirements from the initial stages of site design; it defines responsibilities as regards the safety and health of all persons operating on temporary and mobile work sites and lays down safety requirements for certain tasks.

Proposal for a Council Directive on the minimum requirements to be applied in improving the safety and health of workers in the drilling industries

No steps have so far been taken at Community level to promote improvement in the safety and health of workers in the drilling industries.

Following the disaster in the North Sea on the Piper Alpha oil and natural gas drilling rig, in which the explosions and fire caused the death of 167 persons on 6 July 1988, Parliament requested the Commission to take suitable measures as soon as possible.

Proposal for a Council Directive on the minimum requirements to be applied in improving the safety and health of workers in the quarrying and open-cast mining industries

There are no special Community measures covering the quarrying and open-cast mining industries.

The risks and accident rates are higher in these industries than in others and they are not covered by the first individual Directive on the workplace pursuant to Article 16(1) of Directive 89/391/EEC.

On this account steps should be taken at Community level to improve the safety and health protection of workers in these industries.

Proposal for a Council Directive on the minimum safety and health requirements for fishing vessels

The risks connected with work on board fishing vessels are greater than those in other "high-risk" occupations. The purpose of the proposed directive is to lay down minimum safety and health requirements in relation, in particular, to working procedures on board such vessels.

Recommendation to the Member States on the adoption of a European schedule of industrial diseases

The Commission recommendations of 23 July 1962 and 20 July 1966 established a European schedule of industrial diseases.

Since then, within each Member State the schedule of the various industrial diseases which can give right to compensation has gradually developed on account of many factors, such as changing techniques, the use of new substances, different activities and varying constraints at the workplace.

The number of diseases known as "industrial diseases" (that is where there is good reason to believe that they are closely connected with certain activities, but which the Member States have not yet recognized as giving any right to compensation) has constantly changed.

The Commission takes the view that in such a complex field it must, as in the past, make use of a recommendation to encourage the Member States to bring about the greatest possible convergence among themselves.

The recommendation would therefore consist in an updating of the European schedule of industrial diseases with the aim of harmonizing at Community level requirements in this area.

Proposal for a Council directive on the minimum requirements for safety and health signs at the workplace

The individual directive on workplaces establishes the minimal requirements for workplaces, but does not specifically cover the posting of signs. Some provisions on this subject already appear in the Council Directive 77/575/EEC and Commission Directive 79/640/EEC. This proposal for a directive aims to revise and extend the above-mentioned directives, updating the previous texts and adding a number of measures which are the result of technical progress.

Proposal for a Council directive defining a system of specific information for workers exposed to certain dangerous industrial agents

The proposal concerns the preparation of information sheets on dangerous agents. These sheets should be available whenever new substances are introduced.

This proposal defines the minimum requirements for the protection of workers and takes account of the work carried out by the ILO on chemical substances.

Information sheets on chemical substances are also required by the Council directives on the placing of chemical substances on the market, and these sheets are taken into consideration in this proposal for a directive.

Proposal for a Council directive on the minimum safety and health requirements regarding the exposure of workers to the risks caused by physical agents

Physical agents, such as vibration and electro-magnetic radiation, give rise to risks which are often considered to be unacceptable. It often takes some time before effects which are damaging to health become apparent. A proposal will be made to introduce the preventive and corrective measures necessary to reduce the possibility of over-exposure, accident and illness.

Proposal for a Council directive amending Directive 83/447/EEC on the protection of workers from the risks related to exposure to asbestos at work

Certain provisions are laid down by Directive 83/447/EEC to the effect that the Council, acting on a proposal from the Commission, must review this directive before 1 January 1990, taking into account, in particular, progress made in scientific knowledge and technology and in the light of the experience gained in applying this directive.

Proposal for a Council directive on the minimum safety and health requirements for activities in the transport sector

Activities in the transport sector often create dangerous working conditions, and transport-related maintenance, handling and loading work also expose workers to considerable risks. The proposal for a directive aims to set the minimum requirements for the prevention of dangerous situations and the protection of all the workers concerned.

Establishment of a safety, hygiene and health agency

The Commission's programme concerning safety, hygiene and health at work is high on the list of priorities of a significant social policy initiative.

In its Resolution of 21 December 1987, the Council welcomed the Commission communication on its programme concerning safety, hygiene and health at work. Among other things, it requested the Commission to examine the possible ways of improving the exchange of information and experience in the field concerned, in particular as regards the gathering and dissemination of data and the advisability of setting up Community machinery to study the repercussions at national level of Community measures in the field of health and safety at work.

Moreover, this Resolution called for an intensification of the co-operation with and between the bodies active in the field concerned.

The Council also stressed that it was fundamentally important for workers to be aware of the issues involved and to have access to information and, if necessary, to training in the measures recommended in the Commission's programme referred to above were to be successful.

Recognizing the dangers not only to health and safety, but also to the business environment and the labour markets of divergent health and safety conditions, employers and workers organizations have impressed upon the Commission the need to ensure that directives are implemented accurately, fully and equitably, and enforced correctly; and they demanded that appropriate advice and adequate assistance be provided to undertakings and organizations concerned in order to help them comply with the requirements imposed by Community directives.

In order to satisfy these demands and whilst retaining its right to supervise the implementation of Community law, the Commission will set up a safety, hygiene and health agency which will provide support for the implementation of programmes relating to the workplace, including technical and scientific assistance and co-ordination as well as assistance in the field of training. In so doing, it will bear in mind the existence and experience of the European Foundation for the Improvement of Living and Working Conditions (Dublin Foundation).

of 30 December 1989

for the implementation of Council Directive 89/656/EEC of 30 November 1989,
concerning the assessment of the safety aspects of personal protective
equipment with a view to the choice and use thereof
[89/c 328/02]

I.   Council Directive 89/656/EEC on the minimum requirements for the use by workers of personal protective equipment lays down in Article 6(1), that the general Regulations for use shall be established in each of the Member States. These Regulations must indicate in particular the circumstances or risk situations where the use of such equipment is necessary because collective means of protection cannot be used. Article 6(3) of the said Directive stipulates that the employers' and workers' organisations must be consulted in advance on establishing the above Regulations for use.

II.   The Annexes to the Directive, which are indicative and not exhaustive, contain useful information on establishing these Regulations.

The Commission considers that it could be useful to have other information available at the abovementioned consultation, in order to improve its effectiveness. The establishment of high-calibre Regulations for use should in fact be considered as an essential precondition for the optimum use of personal protective equipment. Of this supplementary information, the factors to be taken into account in choosing and using each of the main categories of personal protective equipment should be regarded as important data which would be of assistance to the employers' and workers' organisations at the consultation provided for in Article 6(3).

III.   Generally speaking, the Commission attaches considerable importance to the consultation and participation of workers and/or their representatives in all matters affecting the safety and health of workers (in accordance with the provisions of Article 11 of Council Directive 89/391/EEC of 12 June 1989).

Therefore, in the specific case of the use by workers of personal protective equipment, it is the opinion of the Commission that in implementing Article 8 of the Directive in question, the consultation and participation of the workers should be extended to cover all the data which might be of use, without prejudice to the provisions of the aforementioned Article 8.

IV.   With a view therefore to promoting the better implementation of the Council Directive on the minimum health and safety requirements for the use by workers of personal protective equipment, and whereas, taking into account the object of the Directive itself, the circulation of all supplementary relevant information and data should lead to the greater effectiveness of the provisions contained therein, and in particular those appearing in Article 6 (1) and (3) and Article 8, the Commission requests that the Member States ensure, by the method they judge the most appropriate, the widespread circulation of the data contained in the Annex to this communication, and in particular amongst the competent authorities and the employers' and workers' organisations, so that these data may serve as reference documents during implementation of Council Directive 89/656/EEC.

ANNEX

Non-exhaustive information for evaluating personal protective equipment

1. Industrial helmets.

2. Goggles and visors.

3. Ear protectors.

4. Respirators.

5. Gloves.

6. Boots and shoes.

7. Protective clothing.

8. Life jackets for industrial use.

9. Protection against falls.

1. INDUSTRIAL HELMETS

Risk	Origin and type of risk	Safety and performance criteria for selection of equipment

RISKS TO BE COVERED

Risk	Origin and type of risk	Safety and performance criteria for selection of equipment
Mechanical	— Objects falling, impact — Lateral crushing — Stud drivers	— Absorption of impact — Resistance to puncture — Lateral resistance — Resistance to shots
Electrical	— Low-voltage electricity	— Electrical insulation
Thermal	— Cold, heat — Splashes of molten metal	— Maintenance of characteristics at high and low temperatures — Resistance to splashes of molten metal
Non-visibility	— Not sufficiently noticeable	— Luminous/reflective colour

RISKS ARISING FROM THE EQUIPMENT

Risk	Origin and type of risk	Safety and performance criteria for selection of equipment
Discomfort, interference with work	— Inadequate comfort	— Ergonomic design: — weight — headroom — head fit — ventilation
Accidents and health hazards	— Poor compatibility — Poor hygiene — Poor stability, helmet falls off — Contact with flames	— Quality of materials — Ease of maintenance — Fit — Non-flammability and resistance to flame
Ageing	— Exposure to weather, ambient conditions, cleaning, use	— Resistance to industrial wear and tear — Maintenance of characteristics throughout useful life

RISKS ARISING FROM THE USE OF THE EQUIPMENT

Inadequate protection	— Wrong choice of equipment	— Select equipment in line with the nature and scale of risks and stress:
		  — follow manufacturer's instructions
		  — follow markings on equipment (e.g. level of protection, special uses)
		  — select equipment to suit user's individual requirements

	— Incorrect use of equipment	— Use equipment appropriataly, be aware of risk — Follow manufacturer's instructions
	— Equipment dirty, worn or deteriorated	— Maintain in good condition — Regular checks — Replace in good time — Follow manufacturer's instructions

2. GOGGLES AND VISORS

Risk	Origin and type of risk	Safety and performance criteria for selection of equipment

RISKS TO BE COVERED

Risk	Origin and type of risk	Safety and performance criteria
Specific	— Stress arising from use — Puncturing by low-power foreign bodies	— Eyepiece with adequate mechanical resistance and shatter-resistant — Imperviousness and resistance
Mechanical	— High-speed particles, splinters, splashing — Stud drivers	— Mechanical resistance
Thermal/Mechanical	— Burning particles at high speed	— Resistance to burning or molten materials
Cold	— Hypothermia of the eyes	— Close fit to face

Chemical	— Irritation from:
	  — gases
	  — aerosols
	  — dusts
	  — fumes
		— Imperviousness (lateral protection and chemical resistance)

Radiation	— Technical sources of infra-red, visible and ultraviolet radiation, ionizing radiation and laser rays — Natural radiation: daylight	— Filtering capacity of eyepiece — Imperviousness to radiation of frame — Frame opaque to radiation

RISKS ARISING FROM THE EQUIPMENT

Discomfort, interference with work	— Inadequate comfort:
	  — too bulky
	  — increased perspiration
	
	— inadequate grip, contact pressure too high
		— Ergonomic design:
		  — reduced bulk
		  — adequate ventilation, anti-misting eyepiece
		  — individual adaptability to the user

Accidents and health hazards	— Poor compatibility — Poor hygiene	— Quality of materials — Ease of maintenance
	— Risk of cuts from sharp edges	— Rounded edges and rims — Use of safety eyepieces

RISKS ARISING FROM THE EQUIPMENT

Accidents and health hazards	— Impairment of vision caused by poor optical quality, e.g. distortion, modification of colours, in particular signals, diffusion — Reduction of field of visibility — Reflection — Sudden severe changes in transparency (light-dark) — Misty eyepieces	— Check optical quality — Use abrasion-proof eyepieces — Eyepieces of adequate size — Anti-reflective eyepieces and frame — Eyepiece light reaction speed (photochromatic) — Anti-misting facility
Ageing	— Exposure to weather, ambient conditions, cleaning, use	— Resistance to industrial wear and tear — Maintenance of characteristics throughout useful life

RISKS ARISING FROM THE USE OF THE EQUIPMENT

Inadequate protection	— Wrong choice of equipment	— Select equipment in line with the type and scale of the risks and stress: — follow manufacturer's instructions — follow markings on equipment (e.g. level of protection, special uses) — Select equipment to suit user's individual requirements
	— Incorrect use of equipment	— Use equipment appropriately, be aware of risk — Follow manufacturer's instructions
	— Equipment dirty, worn or deteriorated	— Maintain in good condition — Regular checks — Replace in good time — Follow manufacturer's instructions

3. EAR PROTECTORS

Risk	Origin and type of risk	Safety and performance criteria for selection of equipment

RISKS TO BE COVERED

Noise	— Continuous noise — Impulse noise	— Sufficient noise reduction for all types of noise
Thermal	— Metal splashing, for example during welding	— Resistance to molten or burning materials

RISKS ARISING FROM THE EQUIPMENT

Discomfort, interference with work	— Inadequate comfort: — too bulky — too much pressure — increased perspiration — inadequate grip	— Ergonomic design: — bulk — pressure when worn and effort required to keep in place — adaptability to individual requirements
Restriction of hearing capacity	— Deterioration of ability to understand words, recognize signals and key sounds during work and to locate direction of noise	— Variation in noise reduction depending on frequency, reduction in hearing performance — Possibility of replacing shells with earplugs — Audio tests before selection — Use of appropriate electro-acoustic protection

RISKS ARISING FROM THE EQUIPMENT

Accidents and health hazards	— Poor compatibility — Poor hygiene — Unsuitable materials — Sharp edges — Pulls hair — Contact with burning objects — Contact with flame	— Quality of materials — Ease of maintenance — Possibility of replacing muffs with shells, use of disposable earplugs — Rounded edges and corners — Eliminate elements which pull hair — Resistance to combustion and melting — Non-flammability, resistance to flame
Ageing	— Exposure to weather, ambient conditions, cleaning, use	— Resistance to industrial wear and tear — Maintenance of characteristics throughout useful life

RISKS ARISING FROM THE USE OF THE EQUIPMENT

Inadequate protection	— Wrong choice of equipment	— Select equipment in line with nature and scale of risks and stress: — follow manufacturer's instructions — follow markings on equipment (e.g. level of protection, special uses) — Select equipment to suit user's individual requirements
	— Incorrect use of equipment	— Use equipment appropriate, be aware of risk — Follow manufacturer's instructions
	— Equipment dirty, worn or deteriorated	— Maintain in good condition — Regular checks — Replace in good time — Follow manufacturer's instructions

4. RESPIRATORS

Risk	Origin and type of risk	Safety and performance criteria for selection of equipment

RISKS TO BE COVERED

Effect of dangerous substances in inhaled air	— Particulate pollutants (dust, fumes, aerosols)	— Particle filter of the required efficiency (filter grade), depending on concentration, toxicity/health hazard and size range of particles — Particular attention should be given to liquid particles (droplets)
	— Gaseous and evaporative pollutants	— Selection of suitable gas filter type and appropriate filter grade, depending on concentration, toxicity/health hazard, length of time to be worn and nature of work
	— Particulate and gaseous aerosol pollutants	— Selection of suitable combined filter type according to same criteria as for particle and gas filters
Lack of oxygen in inhaled air	— Oxygen retention — Oxygen pressure	— Guaranteed oxygen supply through equipment — Respect oxygen capacity of equipment in relation to duration of use

RISKS ARISING FROM THE EQUIPMENT

Discomfort, interference with work	— Inadequate comfort:
	  — size
	  — bulk
	  — supply
	  — respiratory resistance
	
	  — microclimate in respirator
	  — use
		— Ergonomic design:
		  — adaptability
		  — small bulk, good weight distribution
		  — no interference with head movements
		  — respiratory resistance and high pressure in respiratory zone
		  — respirators with breathing valve, blower
		  — easy to handle/use

Accidents and health hazards	— Poor compatibility — Poor hygiene — Not airtight (leaks) — Accumulation of CO_2 in inhaled air — Contact with naked flames, sparks, spatters of molten metal — Reduction of field of vision — Contamination	— Quality of materials — Ease of maintenance and disinfection — Airtight fit to the face; imperviousness of equipment — Respirators with breathing valves, blower or CO_2 absorbers — Use of non-flammable materials — Adequate range of field of vision — Resistance, ease of decontamination
Ageing	— Exposure to weather, ambient conditions, cleaning, use	— Resistance to industrial wear and tear — Maintenance of characteristics throughout useful life

RISKS ARISING FROM THE USE OF THE EQUIPMENT

Inadequate protection	— Wrong choice of equipment	— Select equipment in line with nature and scale of risks and stress:
		  — follow manufacturer's instructions
		  — follow markings on equipment (e.g. level of protection, special uses)
		  — observe restrictions on use and duration of use; where oxygen concentration is too high or too low clean-air equipment should be used instead of filtered-air equipment
		— Select equipment to suit user's individual requirements (possibility of change)

	— Incorrect use of equipment	— Use equipment appropriately, be aware of risk — Follow information and instructions for use provided by manufacturer, safety organizations and test laboratories

	— Equipment dirty, worn or deteriorated	— Maintain in good condition
		  — Regular checks
		  — Respect maximum periods of use
		  — Replace in good time
		  — Follow manufacturer's instructions as safety rules

5. PROTECTIVE GLOVES

Risk	Origin and type of risk	Safety and performance criteria for selection of equipment

RISKS TO BE COVERED

Risk	Origin and type of risk	Safety and performance criteria for selection of equipment
General	— Contact — Use-related stress	— Area of hand covered — Resistance to tearing, stretching, abrasion
Mechanical	— By abrasive, sharp or pointed objects — Impact	— Resistance to penetration, puncture and cutting — Padding
Thermal	— Burning or cold materials, ambient temperature — Contact with naked flame — Effects of welding work	— Insulation against cold and heat — Non-flammability, resistance to flame — Protection from and resistance to radiation and splashes of molten metal
Electrical	— Electricity	— Electrical insulation
Chemical	— Action of chemicals	— Imperviousness, resistance
Vibration	— Mechanical vibration	— Vibration reduction
Contamination	— Contacts with radioactive materials	— Imperviousness, ease of decontamination, resistance

RISKS ARISING FROM THE EQUIPMENT

Risk	Origin and type of risk	Safety and performance criteria for selection of equipment
Discomfort, interference with work	— Inadequate comfort	— Ergonomic design: — bulk, grading of sizes, surface area, comfort, permeability to water vapour
Accidents and health hazards	— Poor compatibility — Poor hygiene — Gloves stick to the skin	— Quality of materials — Ease of maintenance — Good shaping design
Ageing	— Exposure to weather, ambient conditions, cleaning, use	— Resistance to industrial wear and tear — Maintenance of characteristics throughout useful life — Maintenance of size

RISKS ARISING FROM THE USE OF THE EQUIPMENT

Risk	Origin and type of risk	Safety and performance criteria for selection of equipment

Inadequate protection	— Wrong choice of equipment	— Select equipment in line with nature and scale of risks and stress:
		  — follow manufacturer's instructions
		  — follow markings on equipment (e.g. level of protection, special uses)
		— Select equipment to suit user's individual requirements

	— Incorrect use of equipment	— Use equipment appropriately, be aware of risk — Follow manufacturer's instructions
	— Equipment dirty, worn or deteriorated	— Maintain in good condition — Regular checks — Replace in good time — Follow manufacturer's instructions

150

6. SAFETY BOOTS AND SHOES

Risk	Origin and type of risk	Safety and performance criteria for selection of equipment

RISKS TO BE COVERED

Risk	Origin and type of risk	Safety and performance criteria for selection of equipment

Mechanical	— Objects falling on or crushing the front of the foot
	— Falls and impact on heel
	
	— Falls as a result of slipping
	— Treading on pointed or sharp objects
	— Damage to:
	  — the malleoli
	  — the metatarsus
	  — the leg
		— Resistance of the front of the boot or shoe
		
		— Energy-absorbing capacity of the heel
		— Reinforcement of instep
		— Resistance to slipping of sole
		— Puncture-proof sole
		
		— Protection for:
		  — the malleoli
		  — the metatarsus
		  — the leg

Electrical	— Low and medium voltage — High voltage	— Electrical insulation — Electrical conductibility
Thermal	— Cold, heat — Molten metal spatter	— Thermal insulation — Resistance, imperviousness
Chemical	— Harmful dusts or liquids	— Resistance and impermeability

RISKS ARISING FROM THE EQUIPMENT

Risk	Origin and type of risk	Safety and performance criteria for selection of equipment

Discomfort, interference with work	— Inadequate comfort:
	  — the shoe does not fit
	  — poor absorption of perspiration
	
	  — fatigue from using the equipment
	  — the shoe leaks
		— Ergonomic design:
		  — shape, padding, size
		  — vapour permeability and water absorption capacity
		
		  — flexibility, bulk
		
		  — waterproofing

Accidents and health hazards	— Poor compatibility — Poor hygiene — Risk of dislocation and sprains because of poor foot holding	— Quality of materials — Ease of maintenance — Stiffness across width of the shoe and arch support, fit
Ageing	— Exposure to weather, ambient conditions, cleaning, use	— Resistance to corrosion, abrasion and fatigue of the sole — Resistance to industrial wear and tear — Maintenance of characteristics throughout useful life
Static electricity	— Discharge of static electricity	— Electrical conductibility

RISKS ARISING FROM THE USE OF THE EQUIPMENT

Risk	Origin and type of risk	Safety and performance criteria for selection of equipment

Inadequate protection	— Wrong choice of equipment	— Select equipment in line with nature and scale of risks and stress:
		  — follow manufacturer's instructions
		  — follow markings on equipment (e.g. level of protection, special uses)
		— Select equipment to suit user's individual requirements

RISKS ARISING FROM THE EQUIPMENT

Inadequate protection	— Incorrect use of equipment	— Use equipment appropriately, be aware of risk — Follow manufacturer's instructions
	— Equipment dirty, worn or deteriorated	— Maintain in good condition — Regular checks — Replace in good time — Follow manufacturer's instructions

7. PROTECTIVE CLOTHING

Risk	Origin and type of risk	Safety and performance criteria for selection of equipment

RISKS TO BE COVERED

Risk	Origin and type of risk	Safety and performance criteria for selection of equipment
General	— Contact — Stress arising from use	— Coverage of torso — Resistance to tearing, stretching, prevention of spreading of tears
Mechanical	— Abrasive, pointed and sharp objects	— Resistance to penetration
Thermal	— Burning or cold materials, ambient temperature — Contact with naked flames — Welding work	— Insulation against cold and heat, maintenance of protective qualities — Non-flammability, resistance to flame — Protection from and resistance to radiation and splashes of molten metal
Electrical	— Electricity	— Electrical insulation
Chemical	— Chemical damage	— Impermeability and resistance to chemical damage
Humidity	— Clothing leaks	— Waterproofing
Non-visibility	— Clothing difficult to see	— Bright or reflective colour
Contamination	— Contact with radioactive materials	— Impermeability, ease of decontamination, resistance

RISKS ARISING FROM THE EQUIPMENT

Risk	Origin and type of risk	Safety and performance criteria for selection of equipment
Discomfort, interference with work	— Inadequate comfort	— Ergonomic design: — size, grading of sizes, surface area, comfort, permeability to water vapour
Accidents and health hazards	— Poor compatibility — Poor hygiene — Clothing sticks to the skin	— Quality of materials — Ease of maintenance — Good shaping design
Ageing	— Exposure to weather, ambient conditions, cleaning, use	— Resistance to industrial wear and tear — Maintenance of characteristics throughout useful life — Maintenance of size

RISKS ARISING FROM THE USE OF THE EQUIPMENT

Inadequate protection	— Wrong choice of equipment	— Select equipment in line with nature and scale of risks and stress: — follow manufacturer's instructions — follow markings on equipment (e.g. level of protection, special uses) — Select equipment to suit user's individual requirements
	— Incorrect use of equipment	— Use equipment appropriately, be aware of risk — Follow manufacturer's instructions
	— Equipment dirty, worn or deteriorated	— Maintain in good condition — Regular checks — Replace in good time — Follow manufacturer's instrucitons

8. LIFE JACKETS FOR INDUSTRIAL USE

Risk	Origin and type of risk	Safety and performance criteria for selection of equipment

RISKS TO BE COVERED

Drowning	— Fall into water of a person in work clothing, unconscious or deprived of physical faculties	— Buoyancy — Righting ability even if wearer is unconscious — Inflation time — Triggering of automatic inflation — Ability to keep mouth and nose out of the water

RISKS ARISING FROM THE EQUIPMENT

Discomfort, interference with work	— Constraint caused by size or poor design	— Ergonomic design which does not restrict vision, respiration or movement
Accidents and health hazards	— Jacket falls off if wearer falls into water — Damage to jacket during use — Function of inflation system affected — Inappropriate use	— Design (stays in position) — Resistance to mechanical damage (impact, crushing, perforation) — Maintenance of safety qualities under all conditions — Type of gas used for inflation (size of container, whether or not gas is harmful) — Efficiency of automatic inflation device (also after long storage) — Possibility of triggering inflation manually — Provision of a device for oral inflation even while jacket is worn — Outline instructions for use marked on jacket indelibly
Ageing	— Exposure to weather, ambient conditions, cleaning, use	— Resistance to chemical, biological and physical attack: seawater, detergents, hydrocarbons, micro-organisms (bacteria, mould) — Resistance to climatic factors: thermal stress, humidity, rain, splashing, solar radiation — Resistance of materials and protective covers: tearing, abrasion, non-flammability, spattering of molten metal (welding)

RISKS ARISING FROM THE USE OF THE EQUIPMENT

Inadequate protection	— Wrong choice of equipment	— Select equipment in line with nature and scale of risks and stress:
		  — follow manufacturer's instructions
		  — follow markings on equipment (e.g. level of protection, special uses)
		— Select equipment to suit user's individual requirements

	— Incorrect use of equipment	— Use equipment appropriately, be aware of risk — Follow manufacturer's instructions
	— Equipment dirty, worn or deteriorated	— Maintain in good condition — Regular checks — Replace in good time — Follow manufacturer's instructions

9. EQUIPMENT FOR PROTECTION AGAINST FALLS

Risk	Origin and type of risk	Safety and performance criteria for selection of equipment

RISKS TO BE COVERED

Impact	— Falls from a height — Loss of balance	— Resistance and suitability of equipment and anchorage point

RISKS ARISING FROM THE EQUIPMENT

Discomfort, interference with work	— Inadequate ergonomic design	— Ergonomic design:
		  — method of construction
		  — bulk
		  — flexibility
		  — ease of putting on
		  — gripping device with automatic length adjustment

	— Restriction of freedom of movement	

Accidents and health hazards	— Dynamic stress exerted on the user and equipment during braking	— Suitability of equipment:
		  — distribution of braking stress to parts of the body with absorption capacity
		  — reduction of braking force
		  — braking distance
		  — position of attaching device

	— Oscillation and lateral impact	— Anchorage point above the head, anchorage at other points
	— Static stress exerted on suspended body by straps	— Design of equipment (distribution of stress)
	— Slipping of link device	— Short link device (e.g. safety harness, espace line)
Ageing	— Change in mechanical resistance resulting from exposure to weather, ambient conditions, cleaning and use	— Resistance to corrosion — Resistance to industrial wear and tear — Maintenance of characteristics throughout useful life

RISKS ARISING FROM THE USE OF THE EQUIPMENT

Inadequate protection	— Wrong choice of equipment	— Select equipment in line with nature and scale of risks and stress:
		  — follow manufacturer's instructions
		  — follow markings on equipment (e.g. level of protection, special uses)
		— Select equipment to suit user's individual requirements

	— Incorrect use of equipment	— Use equipment appropriately, be aware of risk — Follow manufacturer's instructions
	— Equipment dirty, worn or deteriorated	— Maintain in good condition — Regular checks — Replace in good time — Follow manufacturer's instructions

DECISION OF THE COUNCIL AND THE REPRESENTATIVES OF THE GOVERNMENTS OF THE MEMBER STATES MEETING WITHIN THE COUNCIL

————— of 17 May 1990 —————

adopting a 1990 to 1994 action plan in the context of
the 'Europe against Cancer' programme
[90/238/Euratom, ECSC, EEC]

THE COUNCIL OF THE EUROPEAN COMMUNITIES AND THE REPRESENTATIVES OF THE GOVERNMENTS OF THE MEMBER STATES, MEETING WITHIN THE COUNCIL,

Having regard to the Treaties establishing the European Communities,

Having regard to the draft resolution submitted by the Commission,

Having regard to the opinion of the European Parliament,

Having regard to the opinion of the Economic and Social Committee,

Whereas, at its meetings in June 1985 in Milan and in December 1985 in Luxembourg, the European Council underlined the advantages of launching a European programme against cancer;

Whereas, at its meeting in December 1986 in London, the European Council decided that 1989 should be designated 'European Cancer Information Year' and specified that the aim would be to develop a sustained and concerted information campaign in all the Member States on the prevention, early screening and treatment of cancer;

Whereas the Council and the representatives of the Governments of the Member States, meeting within the Council, adopted on 7 July 1986 a resolution on a programme of action of the European Communities against cancer which is concerned principally with cancer prevention;

Whereas the Council and the representatives of the Governments of the Member States, meeting within the Council, adopted on 21 June 1988 Decision 88/351/EEC on a 1988 to 1989 action plan for an information and public awareness campaign in the context of the 'Europe against Cancer programme;

Whereas various Community measures to prevent cancers arising from exposure to ionizing radiation or to chemical carcinogens are already being implemented under the Treaties establishing the European Economic Community and the European Atomic Energy Community;

Whereas measures to reduce the risk of cancer from exposure to carcinogens are included in a number of existing Community programmes on the environment, worker protection, consumer protection, nutrition, agriculture and the internal market;

Whereas the right to health is a natural right and every European citizen has the right to the most appropriate treatment, regardless of social position;

Whereas the purpose of this action plan is to increase knowledge of the causes of cancer and the possible means of preventing and treating it;

Whereas occupational cancers account for 4% of all cancers and cause 30,000 deaths per year;

Whereas, by ensuring wider dissemination of knowledge of the causes, prevention, screening and treatment of cancer, as well as an improvement in the comparability of information about those matters, in particular concerning the nature and degree of risk of cancer arising from exposure to given substances or processes, the programme will contribute to the achievement of Community objectives while contributing to the overall reduction of risks of cancer;

Whereas it is advisable to promote the dissemination and the implementation of recommendations on the oncology content of training programmes which were approved in 1988 by the three advisory committees on the training of members of the health professions;

Whereas recognition must be given to the crucial role of members of the health professions and whereas theoretical and practical training for all professions and individuals involved in the prevention of cancer and the treatment of cancer sufferers must be encouraged in accordance with the conclusions of the European Organization for Research on Treatment of Cancer;

Whereas it is the view of the World Health Organization that palliative care can provide extremely valuable support both for patients for whom all treatment has failed and for their families; whereas such care should therefore be given recognition and assistance;

Whereas it is advisable to support training actions in respect of cancer for members of the health professions of one Member State in centres of excellence in another Member State;

Whereas encouragement should be given to information campaigns, in schools as well as elsewhere, on cancer and its prevention;

Whereas duplication of effort should be avoided by the promotion of exchanges of experience and by the joint development of basic information modules for the public, for health education and for training members of the health professions;

Whereas efforts should be made to achieve advances in treatment through controlled clinical tests;

Whereas public health policy as such, except in cases where the Treaties provide otherwise, is the responsibility of Member States but whereas promoting co-operation and the co-ordination of national activities as well as the stimulation of Community activities in this field makes a valuable contribution to the fight against cancer;

Whereas it is advisable to continue and strengthen, from 1990 to 1994, the action undertaken between 1987 and 1989 in the fields of prevention, information and health education, and training of members of the health professions,

HAVE DECIDED AS FOLLOWS:

Article 1

1. The Commission shall implement the 1990 to 1994 action plan set out in the Annex in close co-ordination with the competent authorities of the Member States.

2. The Commission shall be assisted by an advisory committee consisting of representatives of the Member States and chaired by the Commission representative.

    The duties of the committee shall be:

    —   to examine projects and measures involving co-financing from public funds,

    —   to co-ordinate at national level projects partly financed by non-governmental organizations.

3. The Commission representative shall submit to the committee a draft of the measures to be taken. The committee shall deliver its opinion on the draft, within a time limit which the chairman may lay down according to the urgency of the matter, if necessary by taking a vote.

    The opinion shall be recorded in the minutes; in addition, each Member State shall have the right to ask to have its position recorded in the minutes.

    The Commission shall take the utmost account of the opinion delivered by the committee. It shall inform the committee of the manner in which its opinion has been taken into account.

4. Furthermore, the Commission will involve the high-level Committee of Cancer Experts and the private bodies active in the fight against cancer closely in implementing the action plan. It will co-operate with both the World Health Organization and the International Agency for Research on Cancer.

5. The Commission will regularly publish technical information on the progress of the action plan and on potential Community financing in the various fields of action.

Article 2

1. The Community contribution estimated necessary for implementing the 1990 to 1994 action plan is ECU 50 million.

2. The Council and the Ministers for Health, meeting within the Council, will review this total amount in the light of the evaluation report referred to in Article 3 (2), with the possibility of increasing it to a total amount of ECU 55 million, if necessary, from 1 January 1993.

Article 3

1. The Commission will continuously assess the action undertaken.

2. The Council and the Ministers of Health will carry out a scientific evaluation of the effectiveness of the action undertaken. To this end the Commission will submit a report on the subject during the second half of 1992.

3. The Commission will keep the European Parliament, the Council and the Economic and Social Committee regularly informed of progress.

4. The Commission is also requested to encourage every kind of exchange with third countries in connection with the activities set out in the Annex.

Done at Brussels, 17 May 1990.

ANNEX

MEASURES TO BE IMPLEMENTED IN THE PERIOD 1990 TO 1994

I. CANCER PREVENTION (including screening)

In addition to the legislative action under way:

A. **Prevention of tobacco consumption**

— Stimulation of projects of European interest concerning the prevention of nicotine addiction, especially amongst such target groups as young people, women, teachers and members of the health professions.

— Stimulation of pilot projects to teach methods of breaking nicotine addiction to members of the health professions and to teachers.

— Stimulation of innovative information campaigns to prevent the use of tobacco among the general public and in the workplace.

— Financing of a study on the possibilities for putting tobacco-growing areas to other uses.

B. **Diet and cancer** (including alcohol)

— Stimulation of studies into eating habits and cancer in close conjunction with the Community medical research programme (meta-analyses, case studies, prospective studies, intervention studies on 'anti-promoting' agents).

— Drafting and publication of guidelines on nutrition aimed at improving cancer prevention.

C. **Campaign against carcinogenic agents**

— Continuation of all Community action concerning protection against ionizing radiation.

— Support for comparative studies of European interest aimed at improving protection against ultra-violet radiation.

— Support for European studies on the possible carcinogenic risks of certain chemicals.

— Continuation of the classification and labelling, at European level, of carcinogens and continuation of the information campaign by means of specialized annual publications.

D. **Systematic screening and early diagnosis**

— Continuation of comparative studies to improve the organization of cancer screening programmes.

— Extension and monitoring of the European network of breast cancer screening pilot programmes to help the Member States determine a general screening policy.

— Evaluation of existing cervical cancer screening programmes and setting up of a European network of regional or local pilot programmes.

— Continuation of evaluation studies on screening programmes for colorectal cancer and possible setting up of a European network.

— Promotion of studies of European interest on the effectiveness and feasibility of early screening for other cancers.

— Promotion of, and support for, screening programmes where the results of exploratory studies have proved positive, in close co-ordination with the AIM and RACE programmes.

E. **Cancer registers and similar measures**

— Support for exchanges of experience in establishing cancer registers in the Community and for setting up a European network in co-operation with the International Agency for Research on Cancer and in close co-ordination with the AIM and RACE programmes.

F. **Other aspects**

1. Evaluation of the operation of the various bone marrow banks.

2. Feasibility study on co-operation between such banks and, if appropriate, support for existing European co-operation.

3. Exchanging experience regarding the quality control of care given.

4. Establishing an up-to-date list of treatments recognized as worthwhile by the international scientific community.

II. HEALTH INFORMATION AND EDUCATION

A. **Information of the public**

— Possible updating of the European Code against Cancer.

— Repeat of European compaigns of cancer information, if possible during the second week of October. Encouraging, within this context, private and public television stations to run spots free of charge on the subject of the fight against cancer.

— Production of European information modules on the prevention, screening and treatment of cancers, adaptable to national requirements.

— Publicizing of the European Code among the general public by the partners in the action plan.

— Support for innovative information compaigns on cancer prevention among targeted groups.

— Informing workers, and migrant workers in particular, under existing Community Directives, of the fight against job-related cancers.

B.   **Health education and cancer**

—   Support for efforts to inform and increase the awareness of school teachers of the European Code against Cancer.

—   Dissemination of European teaching material for health education.

—   Promotion of pilot projects to promote awareness of the European Code among young people.

—   Encouragement at school of a change in dietary habits and, in particular, encouragement of the consumption of fruit and vegetables during break and at mealtimes.

III. TRAINING OF THE HEALTH PROFESSIONS

—   Support for the organization of national or regional meetings to promote the 1989 European recommendations on the cancerology content of basic training programmes for members of the health professions.

—   Support for setting up three European pilot networks of medical schools, nursing colleges and dental schools implementing the recommendations on training in cancer formulated in 1988 by the three European advisory committees on the training of the health professions.

—   Promotion of cancerology training projects.

—   Support for the mobility of the health professions between Member States in order to improve their specialized training in cancerology.

—   Collection and exchange of teaching material of European interest for the training of members of the health professions.

—   Exchange of experience and support for the organization of European seminars on the continuing education of members of the health professions.

—   Exchange of experience between Member States in the area of pain-relieving treatments, palliative and continuing care and the role of the health professions.

IV. RESEARCH AND CANCER

—   Contribution towards the preparation of a Fifth European Medical and Health Research Co-ordination Programme and a Sixth ECSC Medical Research Programme.

COMMISSION RECOMMENDATION

of 22 May 1990

to the Member States concerning the adoption of a European schedule of
occupational diseases
[90/326/EEC]

The Commission, under the terms of the Treaty establishing the European Economic
Community and in particular Article 155 thereof, and without prejudice to more
favourable national laws or regulations, recommends that the Member States:

1. introduce as soon as possible into their national laws, regulations or administrative
   provisions concerning scientifically recognized occupational diseases liable for
   compensation and subject to preventive measures, the European schedule in
   Annex I;

2. take steps to introduce into their national laws, regulations or administrative
   provisions the right of a worker to compensation in respect of occupational
   diseases if the worker is suffering from an ailment which is not listed in Annex I but
   which can be proved to be occupational in origin and nature, particularly if the
   ailment is listed in Annex II;

3. ensure as far as possible that all cases of occupational disease are reported and
   progressively make their statistics on occupational diseases compatible with the
   schedule in Annex I;

4. — develop and improve the various preventive measures for the diseases
     mentioned in the European schedule, turning, if necessary, to the Commis-
     sion for information on the experience acquired by Member States,

   — use for this purpose the European schedule as a reference document on the
     prevention of occupational diseases and certain work accidents;

5. — circulate notices on the occupational diseases in their national list, taking
     special account of the medical information notices on occupational diseases
     in the European schedule, drawn up by the Commission,

   — supply in particular all relevant information on diseases or agents recog-
     nized in their national legislation when requested to do so by another
     Member State through the Commission, and supply the Commission with
     statistical and epidemiological information on the incidence of occupa-
     tional diseases;

6. Provide the personnel responsible for implementing the national provisions
   resulting from this recommendation with adequate training;

7. — introduce a system for the collection of information on data concerning the
     epidemiology of the diseases listed in Annex II and any other disease of an
     occupational nature,

   — promote research in the field of ailments linked to an occupational activity,
     in particular the ailments listed in Annex II.

This recommendation shall not apply to diseases which are not recognized as being occupational in origin.

The Member States shall themselves determine the criteria for recognizing each occupational disease in accordance with their current national laws or practices.

The Commission requests the Member States to inform it, at the end of a three-year period, of the measures taken or envisaged in response to this recommendation. The Commission will then examine the extent to which this recommendation has been implemented in the Member States, in order to determine whether there is a need for binding legislation.

Done at Brussels, 22 May 1990.

ANNEX I

EUROPEAN SCHEDULE OF OCCUPATIONAL DISEASES

The diseases mentioned in this schedule must be linked directly to the occupation. The Commission will determine the criteria for recognizing each of the occupational diseases listed hereunder:

1. **Diseases caused by the following chemical agents:**

		EEC No
100	Acrylonitrile	608 003 004
101	Arsenic or compounds thereof	033 002 005
102	Beryllium (glucinium) or compounds thereof	
103.01	Carbon monoxide	006 001 002
103.02	Carbon oxychloride	
104.01	Hydrocyanic acid	
104.02	Cyanides and compounds thereof	006 007 005
104.03	Isocyanates	
105	Cadmium or compounds thereof	048 001 005
106	Chromium or compounds thereof	
107	Mercury or compounds thereof	080 001 000
108	Manganese or compounds thereof	
109.01	Nitric acid	007 004 001
109.02	Oxides of nitrogen	007 002 000
109.03	Ammonia	007 001 005
110	Nickel or compounds thereof	
111	Phosphorus or compounds thereof	015 001 001
112	Lead or compounds thereof	082 001 006
113.01	Oxides of sulphur	
113.02	Sulphuric acid	016 020 008
113.03	Carbon disulphide	006 003 003
114	Vanadium or compounds thereof	
115.01	Chlorine	017 001 007
115.02	Bromine	
115.04	Iodine	602 005 003
115.05	Fluorine or compounds thereof	009001 000
116	Aliphatic or alicyclic hydrocarbons derived from petroleum spirit or petrol	
117	Halogenated derivatives of the aliphathic or alicyclic hydrocarbons	
118	Butyl, methyl and isopropyl alcohol	
119	Ethylene glycol, diethylene glycol, I , 4-butanediol and the nitrated derivatives of the glycols and of glycerol	
120	Methyl ether, ethyl ether, isopropyl ether, vinyl ether, dichloroisopropyl ether, guaiacol, methyl ether and ethyl of ethylene glycol	

121	Acetone, chloroacetone, bromoacetone, hexafluoroacetone, methyl ethyl ketone, methyl n-butyl ketone, methyl isobutyl ketone, diacetone alcohol, mesityl oxide, 2-methylcyclohexanone	
122	Organophosphorus esters	
123	Organic acids	
124	Formaldehyde	
125	Aliphatic nitrated derivatives	
126.01	Benzene or counterparts thereof (the counterparts of benzene are defined by the formula: C_nH_{2n-6})	
126.02	Naphthalene or naphthalene counterparts (the counterpart of naphthalene is defined by the formula: C_nH_{2n-12})	
126.03	Vinylbenzene and divinylbenzene	
127	Halogenated derivatives of the aromatic hydrocarbons	
128.01	Phenols or counterparts or halogenated derivatives thereof	
128.02	Naphthols or counterparts or halogenated derivatives thereof	
128.03	Halogenated derivatives of the alkylaryl oxides	
128.04	Halogenated derivatives of the alkylaryl sulfonates	
128.05	Benzoquinones	
129.01	Aromatic amines or aromatic hydrazines or halogenated, phenolic, nitrified, nitrated or sulfonated derivatives thereof	
129.02	Aliphatic amines and halogenated derivatives thereof	
130.01	Nitrated derivatives of aromatic hydrocarbons	
130.02	Nitrated derivatives of phenols or their counterparts	
131	Antimony and derivatives thereof	OSI 003 009

2. Skin diseases caused by substances and agents not included under other headings

201	*Skin diseases and skin cancers caused by:*
201.01	Soot
201.02	Tar
20 1.03	Bitumen
201.04	Pitch
201.05	Anthracene or compounds thereof
201.06	Mineral and other oils
201.07	Crude paraffin
201.08	Carbazole or compounds thereof
201.09	By-products of the distillation of coal
202	Occupational skin ailments caused by scientifically recognized allergy provoking or irritative substances not included under other headings

3. **Diseases caused by the inhalation of substances and agents not included under other headings**

301	*Diseases of the respiratory system and cancers:*
301.11	Silicosis
301.12	Silicosis combined with pulmonary tuberculosis
301.21	Asbestosis
301.22	Mesothelioma following the inhalation of asbestos dust
301.31	Pneumoconioses caused by dusts of silicates
302	Complication of asbestos in the form of bronchial cancer
303	Broncho-pulmonary ailments caused by dusts from sintered metals
304.01	Extrinsic allergic alveolites
304.02	Lung diseases caused by the inhalation of dusts and fibres from cotton, flax, hemp, jute, sisal and bagasse
304.03	Respiratory ailments of an allergic nature caused by the inhalation of substances consistently recognized as causing allergies and inherent to the type of work
304.04	Respiratory ailments caused by the inhalation of dust from cobalt, tin, barium and graphite
304.05	Siderosis
305.01	Cancerous diseases of the upper respiratory tract caused by dust from wood

4. **Infectious and parasitic diseases:**

401	Infectious or parasitic diseases transmitted to man by animals or remains of animals
402	Tetanus
403	Brucellosis
404	Viral hepatitis
405	Tuberculosis
406	Amoebiasis

5. **Diseases caused by the following physical agents:**

502.01	Cataracts caused by heat radiation
502.02	Conjunctival ailments following exposure to ultra-violet radiation
503	Hypoacousis or deafness caused by noise
504	Diseases caused by atmospheric compression or decompression
505.01	Osteo-articular diseases of the hands and wrists caused by mechanical vibration
505.02	Angio-neurotic diseases caused by mechanical vibration
506.10	Diseases of the periarticular sacs due to pressure
506.21	Diseases due to overstraining of the tendon sheaths
506.22	Diseases due to overstraining of the peritendineum
506.23	Diseases due to overstraining of the muscular and tendonous insertions
506.30	Meniscus lesions following extended periods of work in a kneeling or squatting position
506.40	Paralysis of the nerves due to pressure
507	Miner's nystagmus
508	Diseases caused by ionizing radiation

ANNEX II

ADDITIONAL LIST OF DISEASES SUSPECTED OF BEING OCCUPATIONAL IN ORIGIN WHICH SHOULD BE SUBJECT TO NOTIFICATION AND WHICH MAY BE CONSIDERED AT A LATER STAGE FOR INCLUSION IN ANNEX I TO THE EUROPEAN SCHEDULE

2.1.	**Diseases caused by the following chemical agents:**	
2.101	Ozone	
2.102	Aliphatic hydrocarbons other than those referred to under heading 1.116 of Annex I	
2.103	Diphenyl	
2.104	Decalin	
2.105	Aromatic acids — aromatic anhydrides	
2.106	Diphenyl oxide	
2.107	Tetrahydrophurane	603 025 000
2.108	Thiopene	
2.109	Methacrylonitrile	608 001003
	Acetonitrile	
2.110	Hydrogen sulphide	016 001 004
2.111	Thioalcohols	
2.112	Mearcaptans and thioethers	
2.113	Thallium or compounds thereof	081 002 009
2.114	Alcohols or their halogenated derivatives not referred to under heading 1.118 of Annex I	
2.115	Glycols or their halogenated derivatives not referred to under heading 1.119 of Annex I	
2.116	Ethers or their halogenated derivatives not referred to under heading 1.120 of Annex I	
2.117	Ketones or their halogenated derivatives not referred to under heading 1.121 of Annex I	
2.118	Esters or their halogenated derivatives not referred to under heading 1.122 of Annex I	
2.119	Furfural	605 010 004
2.120	Thiophenols or counterparts or halogenated derivatives thereof	
2.121	Silver	
2.122	Selenium	034 002 008
2.123	Copper	
2.124	Zinc	
2.125	Magnesium	
2.126	Platinum	
2.127	Tantalum	
2.128	Titanium	
2.129	Terpenes	
2.130	Boranes	
2.140	Diseases caused by inhaling nacre dust	
2.141	Diseases caused by hormonal substances	

2.150	Dental caries associated with work in the chocolate, sugar and flour industries

2.2.   **Skin diseases caused by substances and agents not included under other headings:**

2.201	Allergic and orthoallergic skin ailments not recognized in Annex I

2.3.   **Diseases caused by inhaling substances not included under other headings:**

2.301	Pulmonary fibroses due to metals not included in the European schedule
2.302	Broncho-pulmonary ailments caused by dusts or fumes from aluminium or compounds thereof
2.303	Broncho-pulmonary ailments and cancers associated with exposure to the following:

— soot,
— tar,
— bitumen,
— pitch,
— anthracene or compounds thereof,
— mineral and other oils

2.304	Broncho-pulmonary ailments caused by man-made mineral fibres
2.305	Broncho-pulmonary ailments caused by synthetic fibres
2.306	Broncho-pulmonary ailments caused by dusts from basic slags

2.4.   **Infectious and parasitic diseases not described in Annex I:**

2.401	Parasitic diseases
2.402	Tropical diseases
2.403	Infectious diseases, not included in Annex I, of workers engaged in disease prevention, health care, domiciliary assistance or laboratory work and other activities where a risk of infection exists

2.5.   **Avulsion due to overstraining of the spinous processes**

ANNEX III

THE SITUATION IN THE MEMBER STATES

This Annex was adopted in 1989 and is for guidance only, as the situation is in constant development. It will be updated when the Commission presents its report on the impact of the present recommendation in accordance with item 4 of the explanatory memorandum.

1.   Belgium

Belgium has a list of occupational diseases carrying entitlement to compensation. The occupational diseases are broken down into the following categories:
1.   caused by chemical agents;
2.   caused by physical agents;
3.   caused by biological agents;
4.   of the skin due to various causes;
5.   of the respiratory tract due to various causes.

Furthermore, Belgium has lists of 'occupational' diseases not carrying entitlement to compensation, but which are now being studied with a view to possible inclusion in the list of occupational diseases carrying entitlement to compensation. The mixed system of compensation is not used in Belgium.

2.   Denmark

The list of occupational diseases contains seven categories:
1.   occupational diseases caused by chemical agents (Category A);
2.   occupational diseases of the skin caused by substances or agents which do not come under other headings (Category B);
3.   occupational diseases caused by the inhalation of substances or agents which do not come under other headings (Category C);
4.   infectious or parasitic occupational diseases (Category D);
5.   occupational diseases caused by physical agents (Category E);
6.   initial stages of malignant aliments caused by organic compounds (Category F);
7.   dental or periodontal diseases (Category G).

The mixed system of compensation is used.

3.   Federal Republic of Germany

The list of occupational diseases carrying entitlement to compensation contains six categories:
1.   diseases caused by chemical agents;
2.   diseases caused by physical agents;
3.   diseases caused by biological agents;
4.   respiratory tract and lung diseases;
5.   skin diseases;
6.   diseases not covered in the above.

Total: 59 occupational diseases carrying entitlement to compensation.
A mixed system is used on the basis of specific conditions governing compensation.

4. Greece

The list of occupational diseases carrying entitlement to compensation contains five categories:

1. (a) poisoning and allergies caused by 13 listed chemical substances;
   (b) skin diseases caused by chromium and cement;
2. parasitic and contagious diseases;
3. (a) diseases caused by physical agents;
   (b) miners' diseases;
4. skin diseases;
5. lung diseases.

Total: 52 occupational diseases carrying entitlement to compensation. The mixed system of compensation is not used.

5. Spain

The list of occupational diseases carrying entitlement to compensation contains six categories:

— diseases caused by chemical agents,
— diseases of the skin caused by agents which do not come under other headings:
— skin cancers,
— other skin diseases of occupational origin,
— pneumoconioses,
— infectious and parasitic diseases,
— diseases caused by physical agents,
— diseases not classifiable under other headings.

Total: 71 occupational diseases carrying entitlement to compensation. The mixed system is not used.

6. France

For the general scheme for employees, there are 91 occupational disease tables, which are not broken down by the agents responsible but by disease families and the products or agents responsible. Compensation for occupational diseases is on a flat-rate basis, but employees are given the benefit of the assumption that their disease is attributable to work if it meets the conditions set out in each table (symptoms of the disease, products or agents, period required for recognition, work involving exposure, and occasionally duration of exposure).

There is a mixed system for recognition and compensation in the case of pneumoconioses, with the procedure involving an approved doctor or a board of three specialists. A claim for compensation (not limited to a flat-rate) can be made in respect of any disease not covered by the tables by invoking the liability of the employer.

There is also a schedule for occupational diseases containing 47 tables for farmers and farm employees. It effectively corresponds largely to the first schedule, but with special features owing to the particular nature of the risks covered.

A total of 300 symptoms or groups of symptoms carry entitlement to compensation under the general scheme for employees and nearly the same number is covered by the farm scheme. New tables are established or the existing ones are revised when it is found, as a result of epidemiological study, that a new type of disease is, almost certainly, occupationally induced. Furthermore, consideration is being given to the extension of the mixed system.

7.   Ireland

The classification of occupational diseases is divided into four categories (A, B, C and D):

A: diseases caused by physical agents (14 diseases),
B: diseases caused by biological agents (10 diseases),
C: diseases caused by chemical agents (29 diseases),
D: diseases with various causes other than those above (three diseases).

Total: 56 occupational diseases carrying entitlement to compensation; seven further occupational diseases have carried entitlement to compensation since 1985. A mixed system of compensation exists only for certain respiratory diseases, including certain pneumoconioses, respiratory and skin diseases.

8.   Italy

There are two lists of occupational diseases:

—   one list covering occupational diseases in industry,
—   one list covering occupational diseases in agriculture.

The first list contains 49 headings carrying entitlement to compensation not classified by agents responsible. The second list contains 21 occupational diseases in agriculture carrying entitlement to compensation not classified by agents responsible.

Total: 70 occupational diseases carrying entitlement to compensation. The system of compensation is currently being amended.

9.   Luxembourg

The list of occupational diseases carrying entitlement to compensation contains six categories:

1.   diseases caused by chemical agents;
2.   diseases caused by physical agents;
3.   diseases caused by biological agents;
4.   respiratory tract and lung diseases (including pneumoconioses);
5.   skin diseases;
6.   diseases covered in the above.

Total: 55 occupational diseases carrying entitlement to compensation. The mixed system is used on the basis of specific conditions governing compensation.

10.   Netherlands

In the Netherlands the European Schedule of Occupational Diseases is used as a basic reference document for the diagnosis, reporting and registration of occupational diseases, provided, however, that there is a cause-and-effect link between the disease and the occupational activity. Under the Dutch social security system all cases of disease or incapacity for work give rise to compensation, regardless of the cause. The form this compensation takes does not depend on whether the disease is occupational in origin or not.

11. Portugal

There are two groups of occupational diseases:

(a) The diseases contained in a list published by the relevant Ministry, which is based on the French list and contains 89 tables of diseases giving the causal agent, the type of disease caused, the recognition period and a list of the main activities responsible. These occupational diseases are divided into seven categories:
    1. poisoning;
    2. lung ailments;
    3. dermatoses;
    4. diseases caused by physical agents;
    5. diseases caused by biological agents;
    6. tumours;
    7. mucous membrane allergies.

(b) Injuries, functional disorders or diseases not included in the above list for which no compensation will be obtained unless a link is established between the activity carried out by the worker and the ailment caused (mixed system).

12. United Kingdom

The list of occupational diseases contains four categories (A, B, C and D):
    A: diseases caused by physical agents (11 diseases);
    B: diseases caused by biological agents (9 diseases);
    C: diseases caused by chemical agents (29 diseases);
    D: diseases with various causes which do not come under the above categories (10 diseases).

Total: 59 occupational diseases carrying entitlement to compensation. There is no mixed system of compensation, except in the case of industrial accidents and certain specific diseases.

JOINT OPINION
10 January 1991
on new technologies, work organization and adaptability of the labour market
[Social Dialogue 1991]

Preamble

On 6 March 1987 the representatives of the employers' organizations affiliated to UNICE and CEEP and the representatives of the trade union organizations affiliated to the ETUC meeting within the Working Party on the Social Dialogue and the New Technologies, adopted a joint opinion on training and motivation and information and consultation of workers in connection with the introduction of new technologies into firms.

On the same occasion they confirmed their readiness to continue the dialogue on the consequences of the introduction of new technologies from the point of view of adaptability and flexibility with a view to contributing both to an increase in the competitiveness of European firms on world markets and to the improvement of working and employment conditions.

As a conclusion to the Working Party's exchanges of views on the draft concerning "New technologies, organization of work, adaptability of the labour market", the participants, noting that, according to the varying practices specific to each country, adaptability and flexibility are becoming more widespread throughout the Community by way of both legislation and agreements at various levels, adopted the following joint opinion:

1. A new approach

A. Adaptability and flexibility must make a positive contribution to greater competitiveness for firms, the safeguarding of existing levels of employment and the creation of new jobs and thereby contribute to a significant and lasting reduction in unemployment.

To this effect, it is important that, in the framework of "the co-operative strategy for growth and employment", productivity gains attributable to the introduction of new technologies and to the control of costs should, at the same time, help to strengthen the competitiveness and improve the profitability of firms, promote employment and encourage the improvement of working conditions and the quality of life.

To this end, investment, particularly of the type which creates jobs, should be increased and production should be organized more efficiently, the organization and content of work improved, working time organized and its duration determined in a manner which takes account not only of the specific problems of firms and of industrial sectors but also of the needs of workers as regards the organization of their working and private lives.

B. The participants agree on the need to seek new rules on a statutory or contractual basis, or new approaches which satisfy, at the same time, the specific requirements of firms, the concerns to improve the organization and conditions of work and the needs of the workers to take part in the social and cultural life of the community to which they belong. To this end, efforts must be made to identify in each country and at Community level provisions of any type which need to be adapted to the new situation.

2. The social dialogue

To this end, the participants underline the importance and value of a constructive dialogue.

This dialogue assumes that the social partners are able to manage change; that implies negotiations at appropriate levels to adapt, if necessary, existing agreements or, if necessary, to conclude new ones.

They recall the importance at company level of timely information and consultation of the workforce and/or its representatives, as stated in the joint opinion of 6 March 1987, particularly when defining the objectives of adaptability. These objectives will be all the easier to attain if a balance can be struck between the needs of the various parties concerned, both within the firm and outside.

3. Guidelines

With a view to the establishment of a large internal market, the social partners, while bearing in mind its social dimension, should adhere to the following guidelines as regards the adaptability of the labour market:

3.1 Qualifications and training

The changes resulting from technological development and/or from the introduction of new technologies as regards the organization and content of work should contribute to the mobilization and optimum use of human resources, to improved qualification levels by developing vocational and multi-skilled training throughout working life, to an improvement in working conditions and in the quality of work, to more rational use of equipment and machinery and to an increase in the competitiveness of firms.

In this context and in order to enable the potential offered by the restructuring of jobs and the necessary occupational mobility of the workforce to be fully exploited and serve the optimum adaptation of the firm:

— greater encouragement should be given to the implementation of appropriate continuing training programmes accessible to all so that the entire workforce can contribute to the effort needed to ensure rapid and continuous adaptation to the technological and structural changes affecting firms. In many cases a type of multi-skilled training will be needed to this effect;

— the necessary measures should be taken to provide all workers with good health and safety conditions at the workplace.

The motivation of the workforce at all levels of skills and responsibility, which, according to the joint opinion of 6 March 1987, is also based on good information and consultation practices, will be enhanced as a result and will enable firms more effectively to meet the need to increase their productivity.

3.2 Reduction and reorganization of working time

Reductions and reorganizations of working time must help to protect or create jobs while maintaining or improving the productivity of firms.

They should be achieved, at the levels concerned, by means of negotiations and/or agreements which can be of varied and innovatory form. Whilst encouraging the creation of jobs, these changes should help to promote a better use of production equipment compatible with changing market conditions and pursue the objective of improved working conditions, particularly from the point of view of the health of the workers and their ability to organize their time more effectively, both in work and outside.

3.3 Employment contracts other than open-ended, full-time contracts

Certain forms of employment contracts (for example, for fixed-duration and part-time work) can enable firms to adapt to changes in demand and also meet the needs of certain groups of wage-earners to organize their working time and leisure time with a greater degree of flexibility. These different forms of contract must, however, stipulate, in accordance with the statutory and contractual practices specific to each country and at Community level, the principal conditions applicable to their execution and benefit from the protection referred to in point 3.6 below.

3.4 Remuneration

The participants held an initial exchange of views on the problems entailed by the adaptation of wage and remunerations systems to the new forms of work organization, the new production and management systems of firms and the need for competitiveness. In order to give greater consideration to the developments and discussions taking place in this complex field in the different countries, they agreed to continue their joint deliberations on the basis of an analysis of the current trends in these areas in the Member States of the Community .

3.5 Employment

As regards the adaptability of the workforce, the social partners consider that priority should in the first instance be given to adaptability measures within firms and operating units so as to increase competitiveness and avoid redundancies, in accordance with the joint opinion on the creation of a European geographical and occupational mobility area and improving the operation of the labour market in Europe, adopted on 13 February 1990.

Should redundancies prove inevitable, the end result of national procedures and of the procedures laid down in the directives on collective redundancies will be all the more satisfactory for the parties concerned if it can be applied in a dynamic labour market able to make full use of the periods fixed for identifying the most appropriate measures for the retraining of redundant workers with a view to their re-employment. This entails sustained efforts of co-operation between the firms concerned, the workers' representatives in these firms, the employers' and workers' organizations and the institutions that regulate the labour market, with a view to the implementation of appropriate methods for identifying job opportunities and creating alternative jobs.

3.6 Worker protection

Furthermore, the social partners consider that, in accordance with arrangements specific to each country and in accordance with Community legislation, all workers in the firm, whatever their status and whatever the size of the firm for which they work, should be entitled to:

— the same health and safety protection at the workplace;

— the same social security cover on the understanding that such cover must observe certain minimum limits in line with the systems specific to each country or with Community agreements and, where appropriate, must be proportional to the hours worked and to pay;

— adequate protection in law or on the basis of agreements as regards other appropriate aspects of employment relationships.

PART THREE
Directives

COUNCIL DIRECTIVE

of 25 July 1977

on the approximation of the laws, regulations and administrative provisions
of the Member States relating to the provision of safety signs at
places of work
[77/576/EEC]

THE COUNCIL OF THE EUROPEAN COMMUNITIES,

Having regard to the Treaty establishing the European Economic Community, and in particular Article 100 thereof,

Having regard to the proposal from the Commission,

Having regard to the opinion of the European Parliament,

Having regard to the opinion of the Economic and Social Committee,

Whereas, in its resolution of 21 January 1974 concerning a social action programme, the Council affirmed the need to improve safety and protection of health in places of work, as part of the improvement of living and working conditions;

Whereas the freedom of movement of persons and services has considerably increased the risk of accidents at work and occupational diseases, in particular because of the differences in the organization of work within the Member States, the different languages and the resulting misunderstandings and errors; whereas these difficulties, which constitute an obstacle to the functioning of the common market, can be reduced by the introduction of a Community system of safety signs;

Whereas the use of uniform safety signs has positive effects both for workers at places of work, inside or outside undertakings, and for other persons having access to such places;

Whereas a Community system of safety signs can be effective only if it is ensured by means of unified provisions, if the presentation of the signs is as simple and striking as possible, if it makes the minimum use of explanatory texts and, furthermore, if those concerned receive full and repeated information thereon;

Whereas technical progress and the future development of international methods of signposting require that safety signs be brought up to date; whereas, in order to facilitate the carrying out of the necessary measures with regard to Community signs, close collaboration should be instituted between the Member States and the Commission; whereas a Special Committee should be set up for the purpose,

HAS ADOPTED THIS DIRECTIVE:

Article 1

1. This Directive shall apply to safety signs at places of work.

2. This Directive shall not apply to:

   (a) signs used in rail, road, inland waterway, marine or air transport;

   (b) signs laid down for the placing of dangerous substances and preparations on the market;

   (c) coal mines.

Article 2

1. For the purposes of this Directive:

   (a) *system of safety signs*

   means a system of signs referring to a specific object or situation and providing safety information by means of a safety colour or sign;

   (b) *safety colour*

   means a colour to which a specific safety meaning is assigned;

   (c) *contrasting colour*

   means a colour contrasting with the safety colour and providing additional information;

   (d) *safety sign*

   means a sign combining geometrical shape, colour and symbol to provide specific safety information;

   (e) *prohibition sign*

   means a safety sign prohibiting behaviour likely to cause danger;

   (f) *warning sign*

   means a safety sign giving warning of a hazard;

   (g) *mandatory sign*

   means a safety sign prescribing a specific obligation;

   (h) *emergency sign*

   means a safety sign indicating, in the event of danger, an emergency exit, the way to an emergency installation or the location of a rescue appliance;

   (i) *information sign*

   means a safety sign providing safety information other than that referred to in points (e) to (h);

   (j) *additional sign*

   means a safety sign used only in conjunction with one of the safety signs referred to in points (e) to (h) and providing additional information;

(k) *symbol*

means a pictural representation, describing a specific situation, used on one of the safety signs referred to in points (e) to (h).

2. The meaning and use of safety and contrast colours and the shape, design and meaning of safety signs shall be as defined in Annex I.

Article 3

Member States shall take all necessary measures to ensure that:

— safety signs at all places of work conform to the principles laid down in Annex I,

— only those safety signs defined in Annex II are used to indicate the dangerous situations and to provide the information specified in that Annex;

— road traffic signs in force are used to regulate internal works traffic.

Article 4

Any amendments required to adapt Annex I, points 2 to 6, and Annex II to technical progress and to future developments in international methods regarding signs shall be adopted in accordance with the procedure laid down in Article 6.

Article 5

1. A Committee of Representatives of the Member States, with a Commission representative as chairman, is hereby set up.

2. The Committee shall establish its rules of procedure.

Article 6

1. Where the procedure laid down in this Article is invoked, the matter shall be referred to the Committee by its chairman, either on his own initiative or at the request of a representative of a Member State.

2. The Commission representative shall submit to the Committee a draft of the measures to be taken. The Committee shall give its opinion on the draft within the time laid down by the chairman, having regard to the urgency of the matter. Decisions shall be taken by a majority of 41 votes, the votes of the Member States being weighted as laid down in Article 148 (2) of the Treaty. The chairman shall not vote.

3. (a) Proposed measures which are in accordance with the opinion of the Committee shall be taken by the Commission.

(b) Where the proposed measures are not in accordance with the opinion of the Committee, or if no opinion is delivered, the Commission shall forthwith submit to the Council a proposal on the measures to be taken. The Council shall act by a qualified majority.

(c) If the Council has not acted within three months of receiving the proposal, the proposed measures shall be adopted by the Commission.

Article 7

1.   Member States shall adopt and publish by 1 January 1979 the measures necessary to comply with this Directive and shall inform the Commission immediately thereof. They shall apply these measures from 1 January 1981 at the latest.

2.   Member States shall communicate to the Commission the text of any national provisions which they adopt in the field covered by this Directive.

Article 8

This Directive is addressed to the Member States.

Done at Brussels, 25 July 1977.

ANNEX I

Basic principles of the system of safety signs

1. GENERAL

1.1. The objective of the system of safety signs is to draw attention rapidly and unambiguously to objects and situations capable of causing specific hazards.

1.2. Under no circumstances is the system of safety signs a substitute for the requisite protective measures.

1.3. The system of safety signs may be used only to give information related to safety.

1.4. The effectiveness of the system of safety signs is dependent in particular on the provision of full and constantly repeated information to all persons likely to benefit therefrom.

2. SAFETY COLOURS AND CONTRASTING COLOURS

2.1. Meaning of safety colours

Table 1

Safety colour	Meaning or purpose	Examples of use
Red	Stop Prohibition	Stop signs Emergency shutdown devices Prohibition signs
	This colour is also used to identify fire-fighting equipment.	
Yellow	Caution! Possible danger	Identification of dangers (fire, explosion, radiation, chemical hazards, etc.) Identification of steps, dangerous passages, obstacles
Green	No danger First aid	Identification of emergency routes and emergency exits Safety showers First aid stations and rescue points
Blue (¹)	Mandatory signs Information	Obligation to wear individual safety equipment Location of telephone

(¹) Counts as a safety colour only when used in conjunction with a symbol or words on a mandatory sign or information sign bearing instructions relating to technical prevention.

2.2. Contrasting colours and symbol colours

Table 2

Safety colour	Contrasting colour	Symbol colour
Red	White	Black
Yellow	Black	Black
Green	White	White
Blue	White	White

3. GEOMETRICAL FORM AND MEANING OF SAFETY SIGNS

Table 3

Geometrical form	Meaning
◯	Mandatory and prohibition signs
△	Warning signs
▢ ▭	Emergency, information and additional signs

4. COMBINATIONS OF SHAPES AND COLOURS AND THEIR MEANINGS FOR SIGNS

Table 4

Shape / Colour	◯	△	▭
Red	Prohibition	· · · ·	Fire-fighting equipment
Yellow	· · · ·	Caution, possible danger	· · · ·
Green	· · · ·	· · · ·	No danger Rescue equipment
Blue	Mandatory	· · · ·	Information or instruction

5. DESIGN OF SAFETY SIGNS

5.1. Prohibition signs

Background: white; symbol or wording: black.

The safety colour red must appear around the edge and in a transverse bar and must cover at least 35 % of the surface of the sign.

5.2. Warning, mandatory, emergency and information signs

Background: safety colour; symbol or wording: contrasting colour.

A yellow triangle must have a black edge. The safety colour must cover at least 50 % of the surface of the sign.

5.3. Additional signs

Background: white; wording: black;

or

background: safety colour; wording: contrasting colour.

5.4. Symbols

The design must be as simple as possible and details not essential to comprehension must be left out.

6. YELLOW/BLACK DANGER IDENTIFICATION

(Proportion of safety colour at least 50 %)

Identification of permanent risk locations such as:

— locations where there is a risk of collision, falling, stumbling or of falling loads,

— steps, holes in floors, etc.

ANNEX II

SÆRLIG SIKKERHEDSSKILTNING — BESONDERE SICHERHEITSKENNZEICHNUNG — SPECIAL SYSTEM OF SAFETY SIGNS — SIGNALISATION PARTICULIÈRE DE SÉCURITÉ — SEGNALETICA PARTICOLARE DI SICUREZZA — BIJZONDERE VEILIGHEIDSSIGNALERING

1. Forbudstavler — Verbotszeichen — Prohibition signs — Signaux d'interdiction — Segnali di divieto — Verbodssignalen

a)

b)

c)

Rygning forbudt
Rauchen verboten
No smoking
Défense de fumer
Vietato fumare
Verboden te roken

Rygning og åben ild forbudt
Feuer, offenes Licht und Rauchen verboten
Smoking and naked flames forbidden
Flamme nue interdite et défense de fumer
Vietato fumare o usare fiamme libere
Vuur, open vlam en roken verboden

Ingen adgang for fodgængere
Für Fußgänger verboten
Pedestrians forbidden
Interdit aux piétons
Vietato ai pedoni
Verboden voor voetgangers

d)

e)

Sluk ikke med vand
Verbot, mit Wasser zu löschen
Do not extinguish with water
Défense d'éteindre avec de l'eau
Divieto di spegnere con acqua
Verboden met water te blussen

Ikke drikkevand
Kein Trinkwasser
Not drinkable
Eau non potable
Acqua non potabile
Geen drinkwater

2. Advarselstavler — Warnzeichen — Warning signs — Signaux d'avertissement — Segnali di avvertimento — Waarschuwingssignalen

a)

Brandfarlige stoffer
Warnung vor feuergefährlichen
Stoffen
Flammable matter
Matières inflammables
Materiale infiammabile
Ontvlambare stoffen

b)

Eksplosionsfarlige stoffer
Warnung vor explosionsgefährlichen
Stoffen
Explosive matter
Matières explosives
Materiale esplosivo
Explosieve stoffen

c)

Giftige stoffer
Warnung vor giftigen Stoffen
Toxic matter
Matières toxiques
Sostanze velenose
Giftige stoffen

d)

Ætsende stoffer
Warnung vor ätzenden Stoffen
Corrosive matter
Matières corrosives
Sostanze corrosive
Bijtende stoffen

e)

Ioniserende stråling
Radioaktivitet/Røntgenstråling
Warnung vor radioaktiven Stoffen oder
ionsisierenden Strahlen
Radioactive matter
Matières radioactives
Radiazioni pericolose
Radioactieve stoffen

f)

Kran i arbejde
Warnung vor schwebender Last
Beware, overhead load
Charges suspendues
Attenzione ai carichi sospesi
Hangende lasten

g)

Pas på kørende transport
Warnung vor Flurförderzeugen
Beware, industrial trucks
Chariots de manutention
Carrelli di movimentazione
Transportvoertuigen

h)

Farlig elektrisk spænding
Warnung vor gefährlicher elektrischer
Spannung
Danger: electricity
Danger électrique
Tensione elettrica pericolosa
Gevaar voor elektrische spanning

i)

Giv agt
Warnung vor einer Gefahrenstelle
General danger
Danger général
Pericolo generico
Gevaar

3. Påbudstavler — Gebotszeichen — Mandatory signs — Signaux d'obligation — Segnali di prescrizione — Gebods-signalen

a)

Øjenværn påbudt
Augenschutz tragen
Eye protection must be worn
Protection obligatoire de la vue
Protezione degli occhi
Oogbescherming verplicht

b)

Hovedværn påbudt
Schutzhelm tragen
Safety helmet must be worn
Protection obligatoire de la tête
Casco di protezione
Veiligheidshelm verplicht

c)

Høreværn påbudt
Gehörschutz tragen
Ear protection must be worn
Protection obligatoire de l'ouïe
Protezione dell'udito
Gehoorbescherming verplicht

d)

Åndedrætsværn påbudt
Atemschutz tragen
Respiratory equipment must be used
Protection obligatoire des voies respira-toires
Protezione vie respiratorie
Adembescherming verplicht

e)

Fodværn påbudt
Schutzschuhe tragen
Safety boots must be worn
Protection obligatoire des pieds
Calzature di sicurezza
Veiligheidsschoenen verplicht

f)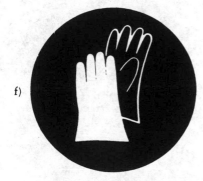

Beskyttelseshandsker påbudt
Schutzhandschuhe tragen
Safety gloves must be worn
Protection obligatoire des mains
Guanti di protezione
Veiligheidshandschoenen verplicht

4. Redningstavler — Rettungszeichen — Emergency signs — Signaux de sauvetage — Segnali di salvataggio — Reddings-signalen

a)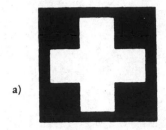

 b)

Førstehjælp
Hinweis auf „Erste Hilfe"
First aid post
Poste premiers secours
Pronto soccorso
Eerste hulp-post

c)

eller/oder/or/ou/o/of

e)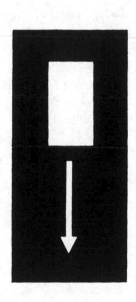

d)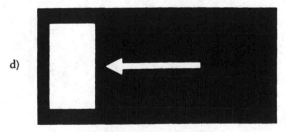

Retningsangivelse til nødudgang
Fluchtweg (Richtungsangabe für Flucht-weg)
Emergency exit to the left
Issue de secours vers la gauche
Uscita d'emergenza a sinistra
Nooduitgang naar links

Nødudgang
(anbringes over udgangen)
Fluchtweg
(über dem Fluchtausgang anzubringen)
Emergency exit
(to be placed above the exit)
Sortie de secours
(à placer au-dessus de la sortie)
Uscita d'emergenza
(da collocare sopra l'uscita)
Nooduitgang
(te plaatsen boven de uitgang)

COMMISSION DIRECTIVE
of 21 June 1979

amending the Annexes to Council Directive 77/576/EEC on the approximation
of the laws, regulations and administrative provisions of the Member States
relating to the provision of safety signs at places of work
[79/640/EEC]

THE COMMISSION OF THE EUROPEAN COMMUNITIES,

Having regard to the Treaty establishing the European Economic Community,

Having regard to Council Directive 77/576/EEC of 25 July 1977 on the approximation of the laws, regulations and administrative provisions of the Member States relating to the provisions of safety signs at places of work, and in particular Articles 4, 5 and 6 thereof,

Whereas the provisions in the Annexes to the abovementioned Directive relating to a uniform system of safety signs at places of work need to be regularly adapted to take account of technical progress and the future development of international methods of signposting;

Whereas Annex I contains no regulations concerning the relationship between dimensions of safety signs and distance of observation and no precise definition of the colorimetric and photometric properties of the materials used for such signs; whereas, when approving the Directive, the Council asked that these omissions be promptly rectified; whereas the addition which has accordingly been made to Annex I is in line with the current international standards in this field;

Whereas it seems necessary to include in Annex II a new sign warning of the presence of laser beams; whereas here also the sign on which there is unanimous international agreement can serve as a model;

Whereas the provisions of this Directive are in accordance with the opinion of the Committee for the Adjustment to Technical Progress and to Future Development in International Methods of Directive 77/576/EEC on the Approximation of the Laws, Regulations and Administrative Provisions of the Member States Relating to the Provision of Safety Signs at Places of Work,

HAS ADOPTED THIS DIRECTIVE:

Article 1

The Annexes to Council Directive 77/576/EEC are amended as provided in the following Articles.

Article 2

In Annex I:

1. the following paragraph shall be inserted after paragraph 5.4 of section 5, 'Design of safety signs':

    5.5. Dimensions of safety signs

    The dimensions of safety signs may be determined in accordance with the formula:

    $$\frac{A}{2000} > I$$

    Where A is the area of the sign in m2 and I the greatest distance in m from which the sign must be understood.

    Note: This formula is applicable for distances up to about 50 m.'

2. after section 5, 'Design of safety signs', the following new section 6 shall be inserted:

    '6. COLORIMETRIC AND PHOTOMETRIC PROPERTIES OF MATERIALS

    As regards the colour and photometric properties of working substances the 150 standards and the standards of the International Lighting Commission (CIE — Commission internationale de l'eclairage) are recommended .'

3. The existing section 6 'Yellow/black danger identification' shall become section 7.

Article 3

In Annex II, No 2, 'Warning signs' the following sign is added:

Laserstråler
Warnung vor Laserstrahl
Laser beam
Rayonnements laser
Raggio laser
Lasertraat'

Article 4

Member States shall bring into force the laws, regulations or administrative provisions necessary to comply with the provisions of this Directive by January 1981 at the latest. They shall forthwith inform the Commission thereof.

Article 5

This Directive is addressed to the Member States.

Done at Brussels, 21 June 1979.

COUNCIL DIRECTIVE

of 29 June 1978

on the approximation of the laws, regulations and administrative provisions
of the Member States on the protection of the health of workers
exposed to vinyl chloride monomer
[78/610/EEC]

THE COUNCIL OF THE EUROPEAN COMMUNITIES,

Having regard to the Treaty establishing the European Economic Community, and in particular Article 100 thereof,

Having regard to the proposal from the Commission,

Having regard to the opinion of the European Parliament,

Having regard to the opinion of the Economic and Social Committee,

Whereas, in the past it was recognized that vinyl chloride monomer was capable of giving rise only to the generally reversible disease known as 'occupational acro-osteolysis'; whereas more recent evidence from epidemiological studies and animal experimentation indicates that prolonged and/or repeated exposure to high concentrations of vinyl chloride monomer in the atmosphere may give rise to a vinyl chloride monomer syndrome encompassing, in addition to occupational acro-osteolysis, the skin disease scleroderma and liver disorders;

Whereas vinyl chloride monomer should also be regarded as a carcinogen which may cause angiosarcoma, a rare malignant tumour which can also occur without any known cause;

Whereas, although working conditions are considerably better than those under which the above syndrome formerly occurred, a comparison of protective measures taken by each Member State reveals certain differences; whereas, therefore, in the interests of balanced economic and social development, these national laws, which directly affect the functioning of the common market, should be harmonized and improved;

Whereas the first step should be to take technical preventive and protective measures based on the latest scientific knowledge so that the values of concentrations of vinyl chloride monomer in the atmosphere in the works can be reduced to an extremely low figure;

Whereas medical surveillance of workers in the vinyl chloride monomer and vinyl chloride polymer industry should take account of the latest medical knowledge, in order that the health of workers in this important sector of the chemical industry may be protected;

Whereas the urgent need to harmonize laws in this field is recognized by both sides of industry which took part in the discussion on this specific problem; whereas effect must therefore be made towards the approximation, while the improvement is being maintained, of the laws, regulations and administrative provisions of the Member States as envisaged in Article 117 of the Treaty;

Whereas the provisions of this Directive constitute minimal requirements which may be re-examined in the light of the experience gained and of progress in medical techniques and knowledge in this field, the final objective being to achieve optimum protection of workers,

HAS ADOPTED THIS DIRECTIVE:

Article 1

1. The object of this Directive is the protection of workers:

    — employed in works in which vinyl chloride monomer is produced, re-claimed, stored, discharged into containers, transported or used in any way whatsoever, or in which vinyl chloride monomer is converted into vinyl chloride polymers, and

    — exposed to the effects of vinyl chloride monomer in a working area.

2. This protection shall comprise:

    — technical preventive measures,

    — the establishment of limit values for the atmospheric concentration of vinyl chloride monomer in the working area,

    — the definition of measuring methods and the fixing of provisions for monitoring the atmospheric concentration of vinyl chloride monomer in the working area,

    — if necessary, personal protection measures,

    — adequate information for workers on the risks to which they are exposed and the precautions to be taken,

    — the keeping of a register of workers with particulars of the type and duration of their work and the exposure to which they have been subjected,

    — medical surveillance provisions.

Article 2

For the purpose of this Directive:

(a) *'working area'* means a section of a works with defined boundaries which may comprise one or more workplaces. It is characterized by the fact that the individual worker spends irregular periods of time there at various workplaces in the course of his duty or duties, that the length of time spent at these individual workplaces cannot be more closely defined and that further sub-division of the working area into smaller units is not possible;

(b) *'technical long-term limit value'* means the value which shall not be exceeded by the mean concentration, integrated with respect to time, of vinyl chloride monomer in the atmosphere of a working area, the reference period being the year, with account being taken only of the concentrations measured during the periods in which the plant is in operation and of the duration of such periods.

For guidance and for practical reasons, Annex I contains a table of the corresponding limit values obtained from statistics with a view to being able to detect, over shorter periods, the risk of the technical long-term limit value's being exceeded.

The concentration values recorded during the alarm periods referred to in Article 6 shall not be taken into account in the calculation of the mean concentration.

(c)   *'competent doctor'* means the doctor responsible for the medical surveillance of the workers referred to in Article 1(1).

Article 3

1.   The fundamental aim of the technical measures adopted to meet the requirements of this Directive shall be to reduce to the lowest possible levels the concentrations of vinyl chloride monomer to which workers are exposed. All working areas in works referred to in Article 1(1) shall therefore be monitored for the atmospheric concentration of vinyl chloride monomer.

2.   For the works referred to in Article 1(1), the technical long-term limit value shall be three parts per million.

An adjustment period not exceeding one year in which to comply with the technical long-term limit value of three parts per million shall be provided for in the case of existing plant at such works.

Article 4

1.   The concentration of vinyl chloride monomer in the working area may be monitored by continuous or discontinuous methods. The permanent sequential method shall be regarded as being a continuous method.

However, the use of a continuous or permanent sequential method shall be obligatory in enclosed vinyl chloride monomer polymerization plant.

2.   In the case of continuous or permanent sequential measurements over a period of one year, the technical long-term limit value shall be considered as having been complied with if the arithmetic mean concentration is found not to exceed this value.

In the case of discontinuous measurements, the number of values measured shall be such that it is possible to predict with a confidence co-efficient of at least 95% — accepting the relevant assumptions made in Annex I — that the actual mean annual concentration will not exceed the technical long-term limit value.

3.   Any measurement system which records accurately for the purposes of analysis at least one third of the technical long-term limit value concentration shall be regarded as suitable.

4.   If non-selective systems of measurement are used for measuring vinyl chloride monomer, the measurement recorded shall be taken as the total vinyl chloride monomer concentration value.

5.   Measuring instruments shall be calibrated at regular intervals. Calibration shall be carried out by suitable methods based on the latest state of the art.

Article 5

1. Measurements of the atmospheric concentration of vinyl chloride monomer in a working area for the purpose of verifying compliance with the technical long-term limit value shall be carried out using measuring points chosen so that the results obtained are as representative as possible of the individual vinyl chloride monomer exposure level of workers in that area.

2. Depending on the size of the working area, there may be one or more measuring points. If there is more than one measuring point, the mean value for the various measuring points shall be considered in principle as the representative value for the whole working area.

   If the results obtained are not representative of the vinyl chloride monomer concentration in the working area the measuring point for checking compliance with the technical long-term limit value shall be that point in the working area where the worker is exposed to the highest mean concentration.

3. Measurements carried out as described in this Article may be combined with measurements based on individual sampling, *i.e.* using devices worn by exposed persons for the purpose of verifying the suitability of the measuring points chosen and of obtaining any other information relevant to technical prevention and medical surveillance.

Article 6

1. In order that abnormal increases in vinyl chloride monomer concentration levels may be detected, a monitoring system capable of detecting such increases shall be provided in places where they may occur.

   In cases involving such an increase in the concentration level, technical measures shall be taken without delay to determine and to remedy the causes thereof.

2. The value corresponding to the alarm threshold shall not exceed, at a measuring point, 15 parts per million for mean values measured over a period of one hour, 20 parts per million for mean values measured over 20 minutes or 30 parts per million for mean values measured over two minutes. If the alarm threshold is exceeded, personal protection measures shall be taken without delay.

Article 7

Appropriate personal protection measures shall be provided for certain operations (*eg.* cleaning of autoclaves, servicing and repairs) during which it cannot be guaranteed that concentrations will be kept below the limit values through operational or ventilation measures.

Article 8

Employers shall inform the workers referred to in Article 1(1), both upon recruitment or prior to their taking up their activities and at regular intervals thereafter, of the health hazards associated with vinyl chloride monomer and of the precautions to be taken when this substance is being handled.

Article 9

1. Employers shall keep a register of the workers referred to in Article 1(1), with particulars of the type and duration of work and the exposure to which they have been subjected. This register shall be given to the competent doctor.

2. A worker shall, at his request, be given the opportunity to note the paticulars in the register concerning him.

3. Employers shall make available to workers' representatives at the undertaking, at their request, the results of the measurements taken at the places of work.

Article 10

1. Employers shall be required to ensure that the workers referred to in Article 1(1) are examined by the competent doctor, both upon recruitment or prior to their taking up their activities and subsequently.

2. Without prejudice to national provisions, the competent doctor shall determine in each individual case the frequency and type of the examination provided for in paragraph 1. The necessary guidelines are given in Annex II.

3. Member States shall take the necessary steps to ensure that the registers referred to in Article 9 and the medical records are kept for at least 30 years from the date on which the activity of the workers referred to in Article 1(1) was taken up.

   For workers already engaged in such activity on the date of entry into force of the provisions adopted pursuant to this Directive, the 30-year period shall commence on that date.

   Member States shall determine how the registers and the medical records are to be used for study and research purposes.

Article 11

1. Member States shall bring into force the laws, regulations and administrative provisions necessary to comply with this Directive within 18 months of its notification and shall forthwith inform the Commission thereof.

2. Member States shall communicate to the Commission the texts of the provisions of national law which they adopt in the field covered by this Directive.

Article 12

This Directive is addressed to the Member States.

Done at Luxembourg, 29 June 1978.

ANNEX I

STATISTICAL BASIS FOR THE TECHNICAL LONG-TERM LIMIT VALUE
(Article 2 (b))

1. Owing to differences in definition, the recommended values for the permissible atmospheric concentration substances injurious to health at the workplace currently vary from country to country.

    This Directive is concerned with a new, statistically-defined reference value — the technical long-term limit value — which should be regarded as a mean annual value.

2. The limit values for shorter reference periods are based on data obtained by extensive measurement of vinyl chloride monomer concentrations in the vinyl chloride polymer industry. These measurements accord with the data resulting from observations both on other substances injurious to health and for other sectors of industry.

    The data can be summarized as follows:

(a) the distributions of concentrations of substances injurious to health can be represented log normally;

(b) the logarithmic variance $\sigma2$ (τ, T) is a function of the reference period τ from which the mean of the individual values is calculated and of the assessment period T over which all individual values extend.

    This relationship can, with a degree of approximation, be expressed by the following equation:

$$\sigma2 \ (\tau, T) = 2.5 \times 10^{-2} \log (T/\tau).$$

Assuming these data, a mean ratio of the limit values for shorter reference periods to the technical long-term limit value can be established:

Reference period	Limit value in parts per million (rounded off)	Ratio of short-term value to technical limit value
One year	3	
One month	5	17
One week	6	195
Eight hours	7	23
One hour	8	255

The above limit values for reference periods shorter than one year must have a maximum 5 % probability of being exceeded when the annual arithmetic mean of atmospheric vinyl chloride monomer concentrations is three parts per million.

ANNEX II

GUIDELINES FOR THE MEDICAL SURVEILLANCE OF WORKERS
(Article 10 (2))

Current knowledge indicates that over-exposure to vinyl chloride monomer can give rise to the following disorders and diseases:

— sclerodermatous skin disorders,

— circulatory disorders in the hands and feet (similar to Raynaud's syndrome),

— acro-osteolysis (affecting certain bone structures, particularly the phalanges in the hand),

— liver and spleen fibroses (similar to perilobular fibrosis, known as Banti's syndrome),

— lung function disorders,

— thrombocytopenia,

— hepatic angiosarcoma.

Medical surveillance of the workers should take account of all symptoms and syndromes, with particular emphasis on the area of greatest risk. As far as is known at present, no symptoms occurring separately or in combination have been identified as precursors or transitional stages of hepatic sarcoma. As no specific methods of preventive analysis are known for this disease, medical action shall include at least the following measures as minimum requirements:

(a) records of the workers' medical and occupational history,

(b) clinical examination of the extremities, the skin and the abdomen,

(c) X-ray of the hand bones (every two years).

Further tests, particularly laboratory tests, are desirable. These should be decided by the competent doctor in the light of the most recent developments in industrial medicine.

The following laboratory tests are suggested at present for prognostic epidemiological surveys:

— urinalysis (glucose, proteins, salts, bile pigments, urobilinogen),

— erythrocyte sedimentation rate,

— blood platelet count,

— determination of total bilirubin level,

— determination of transaminase levels (SGOT, SGPT),

— determination of gamma glutamyl transferase (GT) level,

— thymol turbidity test,

— alkaline phosphatase level,

— determination of cryoglobulin.

3. As in the case of all biological examinations, the results of the tests shall be interpreted in the light of the laboratory techniques used and their normal values. Generally speaking, the significance of a functional disorder is assessed after joint consideration of the results obtained from various examinations and by developments in the anomalies observed. As a general rule, abnormal results shall be investigated and, if necessary, additional specialist examinations carried out.

4. The competent doctor shall decide in each case whether a worker is suitable for a working area.

    The competent doctor shall also decide what contra-indications apply. The most important of these are:

    —   typical vascular and neurovascular lesions,

    —   lung function disorders,

    —   clinical or biological hepatic insufficiency,

    —   diabetes,

    —   chronic renal insufficiency,

    —   thrombocytopenia and hemorrhagic disorders,

    —   certain chronic skin diseases such as scleroderma,

    —   abuse of alcohol and/or addiction to drugs.

    This list, which is intended merely for guidance, has been drawn up using pathological data obtained from previous retrospective studies.

COUNCIL DIRECTIVE

of 27 November 1980

on the protection of workers from the risks related to exposure to chemical, physical and biological agents at work
[80/1107/EEC]

THE COUNCIL OF THE EUROPEAN COMMUNITIES,

Having regard to the Treaty establishing the European Economic Community, and in particular Article 100 thereof,

Having regard to the proposal from the Commission, drafted following consultation with the Advisory Committee on Safety, Hygiene and Health Protection at work,

Having regard to the opinion of the European Parliament,

Having regard to the opinion of the Economic and Social Committee,

Whereas the Council resolution of 29 June 1978 on an action programme of the European Communities on safety and health at work, provides for the harmonization of provisions and measures regarding the protection of workers with respect to chemical, physical and biological agents; whereas efforts must therefore be made towards approximation, while the improvement is being maintained, of the laws, regulations and administrative provisions of the Member States in accordance with Article 117 of the Treaty;

Whereas certain differences are revealed by an examination of the measures taken by Member States to protect workers from the risks related to exposure to chemical, physical and biological agents at work; whereas, therefore, in the interests of balanced development, these measures, which directly affect the functioning of the common market, should be approximated and improved; whereas this approximation and improvement should be based on common principles;

Whereas the said protection should as far as possible be ensured by measures to prevent exposure or keep it at as low a level as is reasonably practicable;

Whereas to this end it is appropriate that the Member States should, when they adopt provisions in this field, comply with a set of requirements, including in particular the laying down of limit values; whereas an initial list of agents may be adopted in this Directive for the application of further more specific requirements; whereas the Member States will determine whether and to what extent each of these requirements is applicable to the agent concerned;

Whereas provision should be made, within the time limits set by this Directive, for the implementation, in respect of a limited number of agents, of provisions to ensure, for the workers concerned, appropriate surveillance of their state of health during exposure and the provision of appropriate information;

Whereas the Council will lay down the limit values and other specific requirements for certain agents in individual Directives;

Whereas certain technical aspects concerning the specific requirements established in the individual directives can be reviewed in the light of experience and progress made in the technical and scientific fields;

Whereas representatives of employers and workers have a role to play in the protection of workers;

Whereas, since the Hellenic Republic is to become a member of the European Economic Community on the 1 January 1981 in accordance with the 1979 Act of Accession, it should be granted a longer period in which to implement this Directive so as to enable it to set up the necessary legislative, social and technical structures, in particular those concerning consultation of both sides of industry, the setting up of a system for monitoring the health of workers as well as the supervision of such implementation,

HAS ADOPTED THIS DIRECTIVE:

Article 1

1. The aim of this Directive is the protection of workers against risks to their health and safety, including the prevention of such risks, arising or likely to arise at work from exposure to chemical, physical and biological agents considered harmful.

2. This Directive shall not apply to:

   — workers exposed to radiation covered by the Treaty establishing the European Atomic Energy Community,

   — sea transport,

   — air transport.

Article 2

For the purposes of this Directive:

(a) 'agent' means any chemical, physical or biological agent present at work and likely to be harmful to health;

(b) 'worker' means any employed person exposed or likely to be exposed to such agents at work;

(c) 'limit value' means the exposure limit or biological indicator limit in the appropriate medium, depending on the agent.

Article 3

1. In order that the exposure of workers to agents be avoided or kept at as low a level as is reasonably practicable Member States shall, when they adopt Provisions for the protection of workers, concerning an agent, take:

   — the measures set out in Article 4,

   — the additional measures set out in Article 5, where the agent appears in the initial list in Annex I.

2. For the purposes of paragraph 1, the Member States shall determine the extent, if any, to which each of the measures provided for in Articles 4 and 5 is to apply, taking into account the nature of the agent, the extent and duration of the exposure, the gravity of the risk and the available knowledge concerning it, together with the degree of urgency of the measures to be adopted.

3. Member States shall adopt the measures necessary to ensure:

    —   in the case of the agents listed in Annex II, Part A, appropriate surveillance of the state of health of workers during the period of exposure,

    —   in the case of the agents listed in Annex II, Part B, access for workers and/ or their representatives at the place of work to appropriate information on the dangers which these agents present.

4. The adoption of the measures referred to in paragraph 3 by the Member States shall not oblige them to apply paragraphs 1 and 2.

Article 4

The measures referred to in the first indent of Article 3(1) shall be:

1. limitation of the use of the agent at the place of work;

2. limitation of the number of workers exposed or likely to be exposed;

3. prevention by engineering control;

4. establishment of limit values and of sampling procedures, measuring procedures and procedures for evaluating results;

5. protection measures involving the application of suitable working procedures and methods;

6. collective protection measures;

7. individual protection measures, where exposure cannot reasonably be avoided by other means;

8. hygiene measures;

9. information for workers on the potential risks connected with their exposure, on the technical preventive measures to be observed by workers, and on the precautions taken by the employer;

10. use of warning and safety signs;

11. surveillance of the health of workers;

12. keeping updated records of exposure levels, lists of workers exposed and medical records;

13. emergency measures for abnormal exposures;

14. if necessary, general or limited ban on the agent, in cases where use of the other means available does not make it possible to ensure adequate protection.

Article 5

The additional measures referred to in the second indent of Article 3(1) shall be:

1. providing medical surveillance of workers prior to exposure and thereafter at regular intervals. In special cases, it shall be ensured that a suitable form of health surveillance is available to workers who have been exposed to the agent, after exposure has ceased;

2. access by workers and/or their representatives at the place of work to the results of exposure measurements and to the anonymous collective results of the biological tests indicating exposure when such tests are provided for;

3. access by each worker concerned to the results of his own biological tests indicating exposure;

4. informing workers and/or their representatives at the place of work where the limit values referred to in Article 4 are exceeded, of the causes thereof and of the measures taken or to be taken in order to rectify the situation;

5. access by workers and/or their representatives at the place of work to appropriate information to improve their knowledge of the dangers to which they are exposed.

Article 6

Member States shall see to it that:

— workers' and employers' organizations are consulted before the provisions for the implementation of the measures referred to in Article 3 are adopted and that workers' representatives in the undertakings or establishments, where they exist, can check that such provisions are applied or can be involved in their application,

— any worker temporarily suspended on medical grounds in accordance with national laws or practices from exposure to the action of an agent is, where possible, provided with another job,

— the measures adopted in implementation of this Directive are consistent with the need to protect public health and the environment.

Article 7

This Directive and the individual Directives referred to in Article 8 shall not prejudice the right of Member States to apply or introduce laws, regulations or administrative provisions ensuring greater protection for workers.

Article 8

1. In the individual Directives which it adopts on the agents listed in Annex 1, the Council shall, acting on a proposal from the Commission, lay down the limit value or values and the other specific requirements applicable.

2. The titles of the individual Directives shall include serial numbers.

3. Adaptation to technical progress in accordance with the procedure in Article 10 shall be restricted to the technical aspects listed in Annex III under the conditions laid down in the individual Directives.

Article 9

1. With a view to the adaptation to technical progress referred to in Article 8(3) a committee is hereby established consisting of representatives of the Member States and presided over by a representative of the Commission.

2. The Committee shall draw up its own rules of procedure.

Article 10

1. Where the procedure laid down in this Article is invoked, matters shall be referred to the Committee by the chairman, either on his own initiative or at the request of the representative of a Member State.

2. The representative of the Commission shall submit to the Committee a draft of the measures to be taken. The Committee shall deliver its opinion on this draft within a time limit which the chairman may set according to the urgency of the matter. Decisions shall be taken by a majority of 41 votes, the votes of Member States being weighted as provided for in Article 148(2) of the Treaty. The chairman shall not vote.

3. (a) The Commission shall take the proposed measures where they are in accordance with the opinion of the Committee.

    (b) Where the proposed measures are not in accordance with the opinion of the Committee, or if no opinion is delivered the Commission shall without delay propose to the Council the measures to be taken. The Council shall act by a qualified majority.

    (c) If the Council has not acted within three months of receiving the proposal, the proposed measures shall be adopted by the Commission.

Article 11

1. Member States shall bring into force the laws, regulations and administrative provisions necessary to comply with this Directive within a period of three years of its notification and shall forthwith inform the Commission thereof.

    However, in the case of Article 3(3), first indent, this period shall be four years.

    In derogation from the above provisions, the time limits laid down in the first and second sub-paragraphs shall be four and five years respectively in the case of the Hellenic Republic.

2. Member States shall communicate to the Commission the provisions of the national law which they adopt in the field governed by this Directive.

Article 12

This Directive is addressed to the Member States.

Done at Brussels, 27 November 1980

ANNEX I

List of agents referred to in Article 3(1), second indent, and Article 8(1)

Acrylonitrile
Asbestos
Arsenic and compounds
Benzene
Cadmium and compounds
Mercury and compounds
Nickel and compounds
Lead and compounds
Chlorinated hydrocarbons: — chloroform
    — paradichlorobenzene
    — carbon tetrachloride

ANNEX II

List of agents referred to in Article 3(3), first indent

1. Asbestos
2. Lead and compounds

List of agents referred to in Article 3(3), second indent

1. Asbestos
2. Arsenic and compounds
3. Cadmium and compounds
4. Mercury and compounds
5. Lead and compounds

ANNEX III

Technical aspects referred to in Article 8(3).

1. Sampling procedures and measuring methods (including quality control) with respect to the limit values in so far as such procedures and methods have no effect on the quantitative significance of those limit values.

2. Practical recommendations on medical surveillance before and during exposure and after such exposure has ceased and keeping of records on the results of such medical surveillance.

3. Practical procedures regarding the establishment and keeping of records concerning ambient measurement results and lists of exposed workers.

4. Practical recommendations for alarm systems to be installed at workplaces where abnormal exposures are likely to occur.

5. Practical recommendations for emergency measures to be taken in the event of abnormal emissions.

6. Collective and individual protection measures for certain operations (*eg.* servicing and repairs) during which it cannot be guaranteed that concentrations or intensities of the agents will be kept below the limit values.

7. Procedures regarding general hygiene requirements, and means of ensuring personal hygiene.

8. Signs to identify areas where significant exposure is likely to occur and to indicate the precautions which have to be taken.

COUNCIL DIRECTIVE

of 16 December 1988

amending Directive 80/1107/EEC on the protection of workers from the risks
related to exposure to chemical, physical and biological agents at work
[88/642/EEC]

THE COUNCIL OF THE EUROPEAN COMMUNITIES,

Having regard to the European Economic Community, and in particular Article
118a thereof,

Having regard to the proposal from the Commission,

In co-operation with the European Parliament,

Having regard to the opinion of the Economic and Social Committee,

Whereas, for improved protection of workers with respect to chemical, physical
and biological agents at work, it is necessary to strengthen the provisions contained
in Council Directive 80/1107/EEC, as last amended by the Act of Accession of
Spain and Portugal;

Whereas the Council resolution of 27 February 1984 on a second programme of
action of the European Communities on safety and health at work provides for the
harmonization of provisions and measures regarding the protection of workers
with respect to certain chemical, physical and biological agents; whereas, in the
interests of balanced development, it is therefore necessary to harmonize and
improve those measures, while adapting them to take account of technical
progress; whereas this harmonization and improvement should be based on
common principles;

Whereas the Council resolution of 21 December 1987 on safety, hygiene and
health at work stresses the importance of improving the safety and health of
workers at the place of work;

Whereas, in accordance with Decision 74/375/EEC, as amended by the Act of
Accession of Spain and Portugal, the Advisory Committee on Safety, Hygiene and
Health at Work is to be consulted by the Commission with a view to drawing up
proposals in this field;

Whereas, for certain agents, the Council will lay down, in individual Directives, the
limit values of a binding nature for occupational exposure and, where appropriate,
other specific requirements;

Whereas provision should be made at Community level for drawing up for the
other agents indicative limit values which the Member States would, *inter alia*, take
into account when establishing national limit values;

Whereas representatives of employers and workers have a role to play in the
protection of workers;

Whereas the provisions of this Directive are minimum requirements and in no way prevent Member States from maintaining or taking other measures so as to protect workers further,

HAS ADOPTED THIS DIRECTIVE:

Article 1

Directive 80/1107/EEC is hereby amended as follows:

1. The following sub-paragraph is added to Article 3(1):

    'The Council, in accordance with the procedure laid down in Article 118a of the Treaty, may amend Annex I with a view, *inter alia,* to inserting in it agents in respect of which a binding limit value or binding limit values and/or other specific requirements appear necessary.'

2. Article 4 is amended as follows:

    (a) point 4 is replaced by the following:

    '4. (a) in the case of any activity likely to involve a risk of exposure of workers, determination of the nature and degree of the workers' exposure so that any risk to their safety or health can be assessed and the measures to be taken can be defined;

    (b) establishment of limit values and of sampling procedures, measuring procedures, and procedures for evaluating results; in the case of chemical agents, the establishment of sampling procedures, measuring procedures and procedures for evaluating results, in accordance with the reference method described in Annex IIa or a method yielding equivalent results;

    (c) when a limit value is exceeded, identification without delay of the reasons for the limit being exceeded and implementation as soon as possible of appropriate measures to remedy the situation';

    (b) point 9 is replaced by the following:

    '9. appropriate measures shall be taken by the employer to ensure that workers and/or their representatives in undertakings or establishments receive full information on, and instruction in:

    (a) the potential risks connected with their exposure, the technical preventive measures to be observed by workers and the precautions taken by the employer and to be taken by workers;

    (b) the risk assessment methods used, the existence of a limit value as referred to in point 4 (b) and the need to carry out measurements, and the action to be taken as laid down in point 4(c), in the event of a limit value being exceeded.'

3.   Article 8(1) is replaced by the following:

'1.   The Council shall, in accordance with the procedure laid down in Article 118a of the Treaty, fix in the individual directives that it adopts with regard to the agents listed in Annex I a binding limit value or binding limit values and/or other specific requirements.'

4.   The following paragraph is added to Article 8:

'4.   Without prejudice to paragraph 1, for agents other than those listed in Annex 1, indicative limit values shall be drawn up in accordance with the procedure laid down in Article 10.

The Member States shall take account, *inter alia,* of those indicative limit values when establishing the limit values referred to in Article 4(4)(b).

Indicative limit values shall reflect expert evaluations based on scientific data.'

5.   Article 9(1) is replaced by the following:

'1.   With a view to the adaptation to technical progress referred to in Article 8(3) and to the establishment of indicative limit values as referred to in Article 8(4), a committee is hereby established consisting of representatives of the Member States and chaired by a representative of the Commission.'

6.   Annex IIa, which appears in the Annex to this Directive, is inserted.

Article 2

1.   This Directive shall be without prejudice to the right of Member States to apply or adopt other laws, regulations and administrative provisions laying down more stringent standards.

2.   Member States shall adopt the laws, regulations and administrative provisions necessary to comply with this Directive not later than two years after its notification. They shall forthwith inform the Commission thereof.

3.   Member States shall communicate to the Commission the provisions of national law which they adopt in the field covered by this Directive.

Article 3

This Directive is addressed to the Member States.

Done at Brussels, 16 December 1988.

ANNEX

'ANNEX II a

REFERENCE METHOD REFERRED TO IN ARTICLE 4(4)(b)

A.   DEFINITIONS

I.   Suspended matter

1.   *Physico-chemical definitions*

(a)   "Dust" means a disperse distribution of solids in air, brought about by mechanical processes or stirred up.

(b)   "Fume" means a disperse distribution of solids in air, brought about by thermal and/or chemical processes.

(c)   "Mist" means a disperse distribution of liquids in air, brought about by condensation or dispersion.

2.   *Occupational medicine and toxicological definitions of particle populations*

(a)   Dusts, like fumes and mists, fall into the category of suspended matter.

In assessing the health risks of suspended matter, account must be taken of particle size as well as specific dangerous effect, concentration and exposure time.

(b)   Only part of the total suspended matter within a worker's breathing area is inhaled. This is termed the inspirable fraction.

Important factors here are the inspiration rate around the nose and mouth and flow conditions about the head.

(c)   Depending on its size, the inspirable fraction may be deposited in various areas of the respiratory tract.

Deposition has, *inter alia* a considerable effect on the point and nature of noxious effect.

The fraction of the inspirable fraction reaching the alveoli is called the respirable fraction.

The respirable fraction is of particular interest in occupational medicine.

II.   Limit value

(a)   The limit value is stated as the eight-hour time-weighted average concentration of exposure of a substance in gaseous, vaporous or suspended form in the air at the workplace.

Exposure means the presence of a chemical agent in the air within the breathing area of a worker.

It is described in terms of concentration over a reference period.

This section does not concern limit values for biological indicators.

(b)   In addition it may be necessary to limit, for certain substances, permissible upward excursions from the average eight-hour time-weighted exposure to substances for shorter terms.

Monitoring then relates to the average concentration of the substance for the shorter term in question.

(c)   The limit value for gases and vapours is stated in terms independent of temperature and air pressure variables in ml/m^3 (ppm) and in terms dependent on those variables in mg/m^3 for a temperature of 20°C and a pressure of 101,3 kPa.

The limit value for suspended matter is given in mg/m^3 for operating conditions at the workplace.

B.   ASSESSMENT OF EXPOSURE AND MEASURING STRATEGY

1.   Basics

(a)   If the presence of one or more agents in gaseous, vaporous or suspended form in the air at the workplace cannot for certain be ruled out, an assessment must be made to see whether the limit values are complied with.

(b)   In this assessment, all points which might be relevant to exposure must be carefully looked into, for example:

— agents used or produced,

— operations, technical installations and processes,

— temporal and spatial distribution of concentrations of agents.

(c)   A limit value is complied with if the assessment shows that exposure does not exceed it.

If the information obtained is insufficient to establish reliably whether the limit values are complied with, it must be supplemented by workplace measurements.

(d)   If the assessment shows that a limit value is not complied with:

— the reasons for the limit being exceeded must be identified and appropriate measures to remedy the situation must be implemented as soon as possible.

— the assessment must be repeated.

(e)   If the assessment shows that the limit values are complied with, subsequent measurements at appropriate intervals must, if necessary, be taken to ensure that the situation continues to prevail.

The nearer the concentration recorded comes to the limit value, the more frequently measurements must be taken.

(f)   If the assessment shows that, on a long-term basis, owing to the arrangement of the work process, the limit values are complied with and there is no substantial change in conditions at the workplace likely to lead to a change of workers' exposure, the frequency of checks on compliance through measurements may be curtailed.

In such cases, however, it must regularly be checked whether the assessment leading to that conclusion is still applicable.

(g) If workers are exposed simultaneously or consecutively to more than one agent, this fact must be taken into consideration in evaluating the health risk to which they are exposed.

2. **Requirements for persons who carry out measurements**

Those carrying out measurements must possess the necessary expertise and facilities.

3. **Requirements for measuring procedures**

(a) The measuring procedure must give results representative of worker exposure.

(b) To ascertain the exposure of the worker at the workplace, where possible personal sampling devices should be used, attached to workers' bodies.

Where a group of workers is performing identical or similar tasks at the same place and has similar exposure, sampling such as to be representative of the group may be carried out within that group.

Fixed-point measuring systems may be used if the results make it possible to assess exposure of the worker at the workplace.

Samples should as far as possible be taken at breathing height and in the immediate vicinity of workers.

If in doubt, the point of greatest risk is to be taken as the measuring point.

(c) The measuring procedure used must be appropriate to the agent to be measured, its limit value and the workplace atmosphere.

The result must show the concentration of the agent exactly and in the same terms as the limit value.

(d) If the measuring procedure is not specific to the agent to be measured, the full value recorded must be counted as applying to the agent to be measured.

(e) The limits of detection, sensitivity and precision of the measuring procedure must be appropriate to the limit value.

(f) The accuracy of the measuring procedure should be ensured.

(g) The measuring procedure must have been tested under practical conditions of use.

(h) If the European Committee for Standardization (CEN) publishes general requirements for the performance of measuring procedures and devices for workplace measurements together with provisions on testing, they should be referred to when selecting appropriate measuring procedures.

4. **Measurement specifications for detecting representative particle populations in the air at the workplace**

   (a) Suspended matter concentration should be measured in relation to effect; therefore, when sampling, either the inspirable fraction or the respirable fraction should be measured.

   This requires particle separation according to aerodynamic diameter equivalent to the deposition occurring in breathing.

   Since appropriate equipment for workplace sampling is not yet available, practical specifications for uniform measurement are needed.

   (b) The fraction of suspended matter which can be breathed in by a worker through the mouth and/or the nose is deemed to be inspirable.

   By way of example, in measurement practice, devices with an inspiration rate of 1,25 m/s +/- 10% or devices in conformity with ISO/TR 7708 1983 (E) are used for sampling.

   In the first of these two cases, cited by way of example:

   — with sampling devices attached to the person, the inlet should be directed parallel to the worker's face throughout sampling,

   — with fixed-point sampling, the position and shape of the inlet should enable samples representative of workers' exposure covering various directions of flow to be taken,

   — the position of the sampling device inlet is of little significance where there are very low flow rates for the surrounding air,

   — with surrounding flow rate of 1 m/s and above, omnidirectional sampling in the horizontal plane is recommended.

   (c) The respirable fraction of suspended matter comprises a population passed through a separation system equivalent in its effect to the theoretical separation function of a sedimentation separator giving 50% separation of particles with an aerodynamic diameter of 5 μm (Johannesburg Convention, 1979).

   (d) If the CEN establishes specifications for the collection of suspended material at the workplace, they should be applied, by way of preference.

   Other methods may be used provided that they yield the same conclusion or a stricter conclusion in relation to compliance with the limit values.'

——————— **of 24 June 1982** ———————

on the major accident hazards of certain industrial activities
[82/501/EEC]

THE COUNCIL OF THE EUROPEAN COMMUNITIES,

Having regard to the Treaty establishing the European Economic Community, and in particular Articles 100 and 235 thereof,

Having regard to the proposal from the Commission,

Having regard to the opinion of the European Parliament,

Having regard to the opinion of the Economic and Social Committee,

Whereas the objectives and principles of the Community environment policy were fixed by the action programmes of the European Communities on the environment of 22 November 1973 and 17 May 1977, and having regard in particular to the principle that the best policy consists in preventing the creation of pollution or nuisances at source; whereas to this end technical progress should be conceived and directed so as to meet the concern for the protection of the environment;

Whereas the objectives of the Community policy of health and safety at work were fixed by the Council resolution of 29 June 1978 on an action programme of the European Communities on safety and health at work, and having regard in particular to the principle that the best policy consists in obviating possible accidents at source by the integration of safety at the various stages of design, construction and operation;

Whereas the Advisory Committee on Safety, Hygiene and Health Protection at Work, set up by Decision 74/325/EEC, has been consulted;

Whereas the protection of the public and the environment and safety and health protection at work call for particular attention to be given to certain industrial activities capable of causing major accidents; whereas such accidents have already occurred in the Community and have had serious consequences for workers and, more generally, for the public and the environment;

Whereas, for every industrial activity which involves, or may involve, dangerous substances and which, in the event of a major accident, may have serious consequences for man and the environment, the manufacturer must take all necessary measures to prevent such accident and to limit the consequences thereof;

Whereas the training and information of persons working on an industrial site can play a particularly important part in preventing major accidents and bringing the situation under control in the event of such accidents;

Whereas, in the case of industrial activities which involve or may involve substances that are particularly dangerous in certain quantities, it is necessary for the manufacturer to provide the competent authorities with information including details of the substances in question and high-risk installations and situations, with a view to reducing the hazards of major accidents and enabling the necessary steps to be taken to reduce their consequences;

Whereas it is necessary to lay down that any person outside the establishment liable to be affected by a major accident should be appropriately informed of the safety measures to be taken and of the correct behaviour to be adopted in the event of an accident;

Whereas, if a major accident occurs, the manufacturer must immediately inform the competent authorities and communicate the information necessary for assessing the impact of that accident;

Whereas Member States should forward information to the Commission regarding major accidents occurring on their territory, so that the Commission can analyze the hazards from major accidents;

Whereas this Directive does not preclude the conclusion by a Member State of agreements with third countries concerning the exchange of information to which it is privy at internal level other than that obtained through the Community arrangements for the exchange of information set up by this Directive;

Whereas disparity between provisions already applicable or being prepared in the various Member States on measures to prevent major accidents and limit their consequences for man and the environment may create unequal conditions of competition and hence directly affect the functioning of the common market; whereas the approximation of laws provided for in Article 100 of the Treaty should therefore be carried out in this field,

Whereas it seems necessary to combine this approximation of laws with action by the Community aimed at attaining one of the Community objectives in the field of environmental protection and health and safety at work; whereas, in pursuance of this aim, certain specific provisions should therefore be laid down; whereas, since the necessary powers have not been provided by the Treaty, Article 235 of the Treaty should be invoked,

HAS ADOPTED THIS DIRECTIVE:

Article 1

1. This Directive is concerned with the prevention of major accidents which might result from certain industrial activities and with the limitation of their consequences for man and the environment. It is directed in particular towards the approximation of the measures taken by Member States in this field.

2. For the purposes of this Directive:

   (a) *Industrial activity* means:

   — any operation carried out in an industrial installation referred to in Annex I involving, or possibly involving, one or more dangerous substances and capable of presenting major accident hazards, and also transport carried out within the establishment for internal reasons and the storage associated with this operation within the establishment,

   — any other storage in accordance with the conditions specified in Annex II;

(b) *Manufacturer* means:

— any person in charge of an industrial activity;

(c) *Major accident* means:

— an occurrence such as a major emission, fire or explosion resulting from uncontrolled developments in the course of an industrial activity, leading to a serious danger to man, immediate or delayed, inside or outside the establishment, and/or to the environment, and involving one or more dangerous substances;

(d) *Dangerous substances* means:

— for the purposes of Articles 3 and 4, substances generally considered to fulfil the criteria laid down in Annex IV,

— for the purposes of Article 5, substances in the lists in Annex III and Annex II in the quantities referred to in the second column.

Article 2

This Directive does not apply to the following:

1. nuclear installations and plant for the processing of radioactive substances and material;

2. military installations;

3. the manufacture and separate storage of explosives, gunpowder and munitions;

4. extraction and other mining operations;

5. installations for the disposal of toxic and dangerous waste which are covered by Community Acts in so far as the purpose of those Acts is the prevention of major accidents.

Article 3

Member States shall adopt the provisions necessary to ensure that, in the case of any of the industrial activities specified in Article 1, the manufacturer is obliged to take all the measures necessary to prevent major accidents and to limit their consequences for man and the environment.

Article 4

Member States shall take the measures necessary to ensure that all manufacturers are required to prove to the competent authority at any time, for the purposes of the controls referred to in Article 7(2), that they have identified existing major-accident hazards, adopted the appropriate safety measures, and provided the persons working on the site with information, training and equipment in order to ensure their safety.

Article 5

1. Without prejudice to Article 4, Member States shall introduce the necessary measures to require the manufacturer to notify the competent authorities specified in Article 7:

    — if, in an industrial activity as defined in Article 1 (2) (a), first indent, one or more of the dangerous substances listed in Annex III are involved, or it is recognized that they may be involved, in the quantities laid down in the said Annex, such as:

    — substances stored or used in connection with the industrial activity concerned,

    — products of manufacture,

    — by-products, or

    — residues,

    — or if, in an industrial activity as defined in Article 1 (2) (a), second indent, one or more of the dangerous substances listed in Annex II are stored in the quantities laid down in the second column of the same Annex.

    The notification shall contain the following:

    (a) information relating to substances listed, respectively, in Annex II and Annex III, that is to say:

    — the data and information listed in Annex V,

    — the stage of the activity in which the substances are involved or may be involved,

    — the quantity (order of magnitude),

    — the chemical and/or physical behaviour under normal conditions of use during the process,

    — the forms in which the substances may occur or into which they may be transformed in the case of abnormal conditions which can be foreseen,

    — if necessary, other dangerous substances whose presence could have an effect on the potential hazard presented by the relevant industrial activity;

    (b) information relating to the installations, that is to say:

    — the geographical location of the installations and predominant meteorological conditions and sources of danger arising from the location of the site,

    — the maximum number of persons working on the site of the establishment and particularly of those persons exposed to the hazard,

    — a general description of the technological processes,

    — a description of the sections of the establishment which are important from the safety point of view, the sources of hazard and the conditions under which a major accident could occur, together with a description of the preventive measures planned,

— the arrangements made to ensure that the technical means necessary for the safe operation of plant and to deal with any malfunctions that arise are available at all times;

(c) information relating to possible major-accident situations, that is to say:

— emergency plans, including safety equipment, alarm systems and resources available for use inside the establishments in dealing with a major accident,

— any information necessary to the competent authorities to enable them to prepare emergency plans for use outside the establishment in accordance with Article 7(1),

— the names of the person and his deputies or the qualified body responsible for safety and authorized to set the emergency plans in motion and to alert the competent authorities specified in Article 7.

2. In the case of new installations, the notification referred to in paragraph 1 must reach the competent authorities a reasonable length of time before the industrial activity commences.

3. The notification specified in paragraph 1 shall be updated periodically to take account of new technical knowledge relative to safety and of developments in knowledge concerning the assessment of hazards.

4. In the case of industrial activities for which the quantities, by substance, laid down in Annex II or III, as appropriate, are exceeded in a group of installations belonging to the same manufacturer which are less than 500 metres apart, the Member States shall take the necessary steps to ensure that the manufacturer supplies the amount of information required for the notification referred to in paragraph 1, without prejudice to Article 7, having regard to the fact that the installations are a short distance apart and that any major-accident hazards may therefore be aggravated.

Article 6

In the event of modification of an industrial activity which could have significant consequences as regards major-accident hazards, the Member States shall take appropriate measures to ensure that the manufacturer:

— revises the measures specified in Articles 3 and 4,

— informs the competent authorities referred to in Article 7 in advance, if necessary, of such modification in so far as it affects the information contained in the notification specified in Article 5.

Article 7

1. The Member States shall set up or appoint the competent authority or authorities who, account being taken of the responsibility of the manufacturer, are responsible for:

— receiving the notification referred to in Article 5 and the information referred to in the second indent of Article 6,

— examining the information provided,

— ensuring that an emergency plan is drawn up for action outside the establishment in respect of whose industrial activity notification has been given,

and, if necessary,

— requesting supplementary information,

— ascertaining that the manufacturer takes the most appropriate measures, in connection with the various operations involved in the industrial activity for which notification has been given, to prevent major accidents and to provide the means for limiting the consequences thereof.

2. The competent authorities shall organize inspections or other measures of control proper to the type of activity concerned, in accordance with national regulations.

Article 8

1. Member States shall ensure that persons liable to be affected by a major accident originating in a notified industrial activity within the meaning of Article 5 are informed in an appropriate manner of the safety measures and of the correct behaviour to adopt in the event of an accident.

2. The Member States concerned shall at the same time make available to the other Member States concerned, as a basis for all necessary consultation within the framework of their bilateral relations, the same information as that which is disseminated to their own nationals.

Article 9

1. This Directive shall apply to both new and existing industrial activities.

2. 'New industrial activity' shall also include any modification to an existing industrial activity likely to have important implications for major-accident hazards.

3. In the case of existing industrial activities, this Directive shall apply at the latest on 8 January 1985.

However, as regards the application of Article 5 to an existing industrial activity, the Member States shall ensure that the manufacturer shall submit to the competent authority, at the latest on 8 January 1985, a declaration comprising:

— name or trade name and complete address,

— registered place of business of the establishment and complete address,

— name of the director in charge,

— type of activity,

— type of production or storage,

— an indication of the substances or category of substances involved, as listed in Annexes II or III.

4. Moreover, Member States shall ensure that the manufacturer shall, at the latest on 8 July 1989, supplement the declaration provided for in paragraph 3, second subparagraph, with the data and information specified in Article 5. Manufacturers shall normally be obliged to forward such supplementary declaration to the competent authority; however, Member States may waive the obligation on manufacturers to submit the supplementary declaration; in that event such declaration shall be submitted to the competent authority at the explicit request of the latter.

Article 10

1. Member States shall take the necessary measures to ensure that, as soon as a major accident occurs, the manufacturer shall be required:

    (a) to inform the competent authorities specified in Article 7 immediately;

    (b) to provide them with the following information as soon as it becomes available:

    — the circumstances of the accident,

    — the dangerous substances involved within the meaning of Article 1(2)(d),

    — the data available for assessing the effects of the accident on man and the environment,

    — the emergency measures taken;

    (c) to inform them of the steps envisaged:

    — to alleviate the medium and long-term effects of the accident,

    — to prevent any recurrence of such an accident.

2. The Member States shall require the competent authorities:

    (a) to ensure that any emergency and medium and long-term measures which may prove necessary are taken;

    (b) to collect, where possible, the information necessary for a full analysis of the major accident and possibly to make recommendations.

Article 11

1. Member States shall inform the Commission as soon as possible of major accidents which have occurred within their territory and shall provide it with the information specified in Annex VI as soon as it becomes available.

2. Member States shall inform the Commission of the name of the organization which might have relevant information on major accidents and which is able to advise the competent authorities of the other Member States which have to intervene in the event of such an accident.

3. Member States may notify the Commission of any substance which in their view should be added to Annexes II and III and of any measures they may have taken concerning such substances. The Commission shall forward this information to the other Member States.

Article 12

The Commission shall set up and keep at the disposal of the Member States a register containing a summary of the major accidents which have occurred within the territory of the Member States, including an analysis of the causes of such accidents, experience gained and measures taken, to enable the Member States to use this information for prevention purposes.

Article 13

1. Information obtained by the competent authorities in pursuance of Articles 5, 6, 7, 9, 10 and 12 and by the Commission in pursuance of Article 11 may not be used for any purpose other than that for which it was requested.

2. However this Directive shall not preclude the conclusion by a Member State of agreements with third countries concerning the exchange of information to which it is privy at internal level other than that obtained through the Community machinery for the exchange of information set up by the Directive.

3. The Commission and its officials and employees shall not divulge the information obtained in pursuance of this Directive. The same requirement shall apply to officials and employees of the competent authorities of the Member States as regards any information they obtain from the Commission.

    Nevertheless, such information may be supplied:

    — in the case of Articles 12 and 18,

    — when a Member State carries out or authorizes the publication of information concerning that Member State itself.

4. Paragraphs 1, 2 and 3 shall not preclude the publication by the Commission of general statistical data or information on makers of safety containing no specific details regarding particular undertakings or groups of undertakings and not jeopardizing industrial secrecy.

Article 14

The amendments necessary for adapting Annex V to technical progress shall be adopted in accordance with the procedure specified in Article 16.

Article 15

1. For the purposes of applying Article 14, a Committee responsible for adapting this Directive to technical progress (hereinafter referred to as 'the Committee') is hereby set up. It shall consist of representatives of the Member States and be chaired by a representative of the Commission.

2. The Committee shall draw up its own rules of procedure.

Article 16

1. Where the procedure laid down in this Article is to be followed, matters shall be referred to the Committee by the chairman, either on his own initiative or at the request of the representative of a Member State.

2. The representative of the Commission shall submit to the Committee a draft of the measures to be adopted. The Committee shall deliver its opinion on the draft within a time limit which may be determined by the chairman according to the urgency of the matter. It shall decide by a majority of 45 votes, the votes of the Member States being weighted as provided for in Article 148 (2) of the Treaty. The chairman shall not vote.

3. (a) The Commission shall adopt the measures envisaged where these are in accordance with the opinion of the Committee.

    (b) Where the measures envisaged are not in accordance with the opinion of the Committee, or in the absence of an opinion, the Commission shall forthwith submit a proposal to the Council on the measures to be adopted. The Council shall act by a qualified majority.

    (c) If the Council does not act within three months of the proposal being submitted to it, the measures proposed shall be adopted by the Commission.

Article 17

This Directive shall not restrict the right of the Member States to apply or to adopt administrative or legislative measures ensuring greater protection of man and the environment than that which derives from the provisions of this Directive.

Article 18

Member States and the Commission shall exchange information on the experience acquired with regard to the prevention of major accidents and the limitation of their consequences; this information shall concern, in particular, the functioning of the measures provided for in this Directive. Five years after notification of this Directive, the Commission shall forward to the Council and the European Parliament a report on its application which it shall draw up on the basis of this exchange of information.

Article 19

At the latest on 8 January 1986 the Council shall, on a proposal from the Commission, review Annexes I, II and III.

Article 20

1. Member States shall take the measures necessary to comply with this Directive at the latest on 8 January 1984. They shall forthwith inform the Commission thereof.

2. Member States shall communicate to the Commission the provisions of national law which they adopt in the field covered by this Directive.

ANNEX I

INDUSTRIAL INSTALLATIONS WITHIN THE MEANING OF ARTICLE 1

1. Installations for the production or processing of organic or inorganic chemicals using for this purpose, in particular:

   — alkylation
   — amination by ammonolysis
   — carbonylation
   — condensation
   — dehydrogenation
   — esterification
   — halogenation and manufacture of halogens
   — hydrogenation
   — hydrolysis
   — oxidation
   — polymerization
   — sulphonation
   — desulphurization, manufacture and transformation of sulphur-containing compounds
   — nitration and manufacture of nitrogen-containing compounds
   — manufacture of phosphorus-containing compounds
   — formulation of pesticides and of pharmaceutical products.
   — installations for the processing of organic and inorganic chemical substances, using for this purpose, in particular:
     — distillation
     — extraction
     — solvation
     — mixing.

2. Installations for distillation, refining or other processing of petroleum or petroleum products.

3. Installations for the total or partial disposal of solid or liquid substances or incineration or chemical decomposition.

4. Installations for the production or processing of energy gases, for example, LPG, ENG, SNG.

5. Installations for the dry distillation of coal or lignite.

6. Installations for the production of metals or non-metals by the wet process or by means of electrical energy.

ANNEX II
STORAGE AT INSTALLATIONS OTHER THAN THOSE COVERED BY ANNEX I
('ISOLATED STORAGE')

The quantities set out below relate to each installation or group of installations belonging to the same manufacturer where the distance between the installations is not sufficient to avoid, in foreseeable circumstances, any aggravation of major-accident hazards. These quantities apply in any case to each group of installations belonging to the same manufacturer where the distance between the installations is less than approximately 500 m.

Substances or groups of substances	Quantities (tonnes)	
	For application of Articles 3 and 4	For application of Article 5
1. Flammable gases as defined in Annex IV (c) (i)	50	300
2. Highly flammable liquids as defined in Annex IV (c) (ii)	10 000	100 000
3. Acrylonitrile	350	5 000
4. Ammonia	60	600
5. Chlorine	10	200
6. Sulphur dioxide	20	500
7. Ammonium nitrate	500([2])	5 000([2])
8. Sodium chlorate	25	250([2])
9. Liquid oxygen	200	2 000([2])

([1]) Member States may provisionally apply Article 5 to quantities of at least 500 tonnes until the revision of Annex II mentioned in Article 19.

([2]) Where this substance is in a state which gives it properties capable of creating a major-accident hazard.

ANNEX III

LIST OF SUBSTANCES FOR THE APPLICATION OF ARTICLE 5

The quantities set out below relate to each installation or group of installations belonging to the same manufacturer where the distance between the installations is not sufficient to avoid, in foreseeable circumstances, any aggravation of major-accident hazards. These quantities apply in any case to each group of installations belonging to the same manufacturer where the distance between the installations is less than approximately 500m.

	Name	Quantity[(2)]	CAS No	EEC No
1.	4-Aminodiphenyl	1 kg	92-67-1	
2.	Benzidine	1 kg	92-87-5	612-042-00-2
3.	Benzidine salts	1 kg		
4.	Dimethylnitrosamine	1 kg	62-75-9	
5.	2-Naphthylamine	1 kg	91-59-8	612-022-00-3
6.	Beryllium (powders, compounds)	10 kg		
7.	Bis(chloromethyl)ether	1 kg	542-88-1	603-046-00-5
8.	1,3-Propanesulfone	1 kg	1120-71-4	
9.	2,3,7,8-Tetrachlorodibenzo-p-dioxin (TCDD)	1 kg	1746-01-6	
10.	Arsenic pentoxide, Arsenic (V) acid and salts	500 kg		
11.	Arsenic trioxide, Arsenious (III) acid and salts	100 kg		
12.	Arsenic hydride (Arsine)	10 kg	7784-42-1	
13.	Dimethylcarbamoyl chloride	1 kg	79-44-7	
14.	4-(Chloroformyl) morpholine	1 kg	15159-40-7	
15.	Carbonyl chloride (Phosgene)	20 t	75-44-5	006-002-00-8
16.	Chlorine	50 t	7782-50-5	017-001-00-7
17.	Hydrogen sulphide	50 t	7783-06-04	016-001-00-4
18.	Acrylonitrile	200 t	107-13-1	608-003-00-4
19.	Hydrogen cyanide	20 t	74-90-8	006-006-00-X
20.	Carbon disulphide	200 t	75-15-0	006-003-00-3
21.	Bromine	500 t	7726-95-6	035-001-00-5
22.	Ammonia	500 t	7664-41-7	007-001-00-5
23.	Acetylene (Ethyne)	50 t	74-86-2	601-015-00-0
24.	Hydrogen	50 t	1333-74-0	001-001-00-9
25.	Ethylene oxide	50 t	75-21-8	603-023-00-X
26.	Propylene oxide	50 t	75-56-9	603-055-00-4
27.	2-Cyanopropan-2-ol (Acetone cyanohydrin)	200 t	75-86-5	608-004-00-X
28.	2-Propenal (Acrolein)	200 t	107-02-8	605-008-00-3
29.	2-Propen-1-ol (Allyl alcohol)	200 t	107-18-6	603-015-00-6
30.	Allylamine	200 t	107-11-9	612-046-00-4

31.	Antimony hydride (Stibine)	100 kg	7803-52-3	
32.	Ethyleneimine	50 t	151-56-4	613-001-00-1
33.	Formaldehyde (concentration ≥ 90 %)	50 t	50-00-0	605-001-01-2
34.	Hydrogen phosphide (Phosphine)	100 kg	7803-51-2	
35.	Bromomethane (Methyl bromide)	200 t	74-83-9	602-002-00-3
36.	Methyl isocyanate	1 t	62483-9	615-001-00-7
37.	Nitrogen oxides	50 t	11104-93-1	
38.	Sodium selenite	100 kg	10102-18-8	
39.	Bis(2-chloroethyl) sulphide	1 kg	505-60-2	
40.	Phosacetim	100 kg	4104-14-7	015-092-00-8
41.	Tetraethyl lead	50 t	78-00-2	
42.	Tetramethyl lead	50 t	75-74-1	
43.	Promurit (1-(3,4Dichlorophenyl)-3-triazenethio-carboxamide)	100 kg		5836-73-7
44.	Chlorfenvinphos	100 kg	470-90-6	015-071-00-3
45.	Crimidine	100 kg	535-89-7	613-004-00-8
46.	Chloromethyl methyl ether	1 kg		107-30-2
47.	Dimethyl phosphoramidocyanidic acid	1 t		63917-41-9
48.	Carbophenothion	100 kg	786-19-6	015-044-00-6
49.	Dialifos	100 kg	10311-84-9	015-088-00-6
50.	Cyanthoate	100 kg	3734-95-0	015-070-00-8
51.	Amiton	1 kg	78-53-5	
52.	Oxydisulfoton	100 kg	2497-07-6	015-096-00-X
53.	OO-Diethyl S-ethylsulphinylmethyl phosphorothioate	100 kg	2588-05-8	
54.	OO-Diethyl S-ethylsulphonylmethyl phosphorothioate	100 kg	2588-06-9	
55.	Disulfoton	100 kg	298-04-4	015-060-00-3
56.	Demeton	100 kg	8065-48-3	
57.	Phorate	100 kg	298-02-2	015-033-00-6
58.	OO-Diethyl S-ethylthiomethyl phosphorothioate	100 kg	2600-69-3	
59.	OO-Diethyl S-isopropylthiomethyl phosphorodithioate	100 kg	78-52-4	
60.	Pyrazoxon	100 kg	108-34-9	015-023-00-1
61.	Pensulfothion	100 kg	115-90-2	015-090-00-7
62.	Paraoxon (Diethyl 4-nitrophenyl phosphate)	100 kg	311-45-5	
63.	Parathion	100 kg	56-38-2	015-034-00-1
64.	Azinphos-ethyl	100 kg	2642-71-9	015-056-00-1
65.	OO-Diethyl S-propylthiomethyl phosphorodithioate	100 kg	3309-68-0	
66.	Thionazin	100 kg	297-97-2	
67.	Carbofuran	100 kg	1563-66-2	006-026-00-9
68.	Phosphamidon	100 kg	13171-21-6	015-022-00-6

69.	Tirpate (2,4-Dimethyl-1,3-dithiolane 2-carboxaldehyde O-methylcarbamoyloxime)	100 kg	26419-73-8	
70.	Mevinphos	100 kg	7786-34-7	015-020-00-5
71.	Parathion-methyl	100 kg	298-00-0	015-035-00-7
72.	Azinphos-methyl	100 kg	86-50-0	015-039-00-9
73.	Cycloheximide	100 kg	66-81-9	
74.	Diphacinone	100 kg	82-66-6	
75.	Tetramethylenedisulphotetramine	1 kg	80-12-6	
76.	EPN	100 kg	2104-64-5	015-036-00-2
77.	4-Fluorobutyric acid	1 kg	462-23-7	
78.	4-Fluorobutyric acid, salts	1 kg		
79.	4-Fluorobutyric acid, esters	1 kg		
80.	4-Fluorobutyric acid, amides	1 kg		
81.	4-Fluorocrotonic acid	1 kg	37759-72-1	-
82.	4-Fluorocrotonic acid, salts	1 kg		
83.	4-Fluorocrotonic acid, esters	1 kg		
84.	4-Fluorocrotonic acid, amides	1 kg		
85.	Fluoroacetic acid	1 kg	14449-0	607-081-00-7
86.	Fluoroacetic acid, salts	1 kg		
87.	Fluoroacetic acid, esters	1 kg		
88.	Fluoroacetic acid, amides	1 kg		
89.	Fluenetil	100 kg	4301-50-2	607-078-00-0
90.	4-Fluoro-2-hydroxybutyric acid	1 kg		
91.	4-Fluoro-2-hydroxybutyric acid, salts	1 kg		
92.	4-Fluoro-2-hydroxybutyric acid, esters	1 kg		
93.	4-Fluoro-2-hydroxybutyric acid, amides	1 kg		
94.	Hydrogen fluoride	50 t	7664-39-3	009-002-00-6
95.	Hydroxyacetonitrile (Glycolonitrile)	100 kg	107-164	
96.	1,2,3,7,8,9-Hexachlorodibenzo-p-dioxin	100 kg	19408-74-3	
97.	Isodrin	100 kg	465-73-6	602-050-00-4
98.	Hexamethylphosphoramide	1 kg	680-31-9	
99.	Juglone (S-Hydroxynaphthalene-1,4-dione)	100 kg	481-39-0	
100.	Warfarin	100 kg	81-81-2	607-056-00-0
101.	4,4'-Methylenebis (2-chloroaniline)	10 kg	101-14-4	
102.	Ethion	100 kg	563-12-2	015-047-00-2
103.	Aldicarb	100 kg	116-06-3	006-017-00-X
104.	Nickel tetracarbonyl	10 kg	13463-39-3	028-001-00-1
105.	Isobenzan	100 kg	297-78-9	602-053-00-0
106.	Pentaborane	100 kg	19624-22-7	
107.	1-Propen-2-chloro-1,3-diol-diacetate	10 kg	10118-72-6	
108.	Propyleneimine	50 t		75-55-8

109.	Oxygen difluoride	10 kg	7783-41-7	
110.	Sulphur dichloride	1 t	10545-99-0	016-013-00-X
111.	Selenium hexafluoride	10 kg	7783-79-1	
112.	Hydropen selenide	10 kg	7783-07-5	
113.	TEPP	100 kg	10749-3	015-025-00-2
114.	Sulfotep	100 kg	3689-245	015-027-00-3
115.	Dimefox	100 kg	11-5-264	015-061-00-9
116.	l-Tri(cyclohexyl) stannyl-lH-1, 2,4-triazole	100 g	41083-11-8	
117.	Triethylenemelamine	10 kg	51-18-3	
118.	Cobalt (powders, compounds)	100 kg		
119.	Nickel (powders, compounds)	100 kg		
120.	Anabasine	100 kg	494-52-0	
121.	Tellurium hexafluoride	100 kg	7783-80-4	
122.	Trichloromethanesulphenyl chloride	100 kg	59442-3	
123.	1,2-Dibromoethane (Ethylene dibromide)	50 t	106-934	602-010-00-6
124.	Flammable substances as defined in Annex IV (c) (i)	200 t		
125.	Flammable substances as defined in Annex IV (c) (ii)	50 000 t		
126.	Diazodinitrophenol	10 t	7008-81-3	
127.	Diethylene glycol dinitrate	10 t	693-21-0	603-033-00-4
128.	Dinitrophenol, salts	50 t		609-017-00-3
129.	l-Guanyl-4-nitrosaminoguanyl-l-tetrazene	10 t	109-27-3	
130.	Bis (2,4,6-trinitrophenyl)amine	50 t	131-73-7	612-018-00-1
131.	Hydrazine nitrate	50 t	13464-97-6	
132.	Nitroglycerine	10 t	55-63-0	603-034-00-X
133.	Pentaerythritol tetranitrate	50 t	78-5	603-035-00-5
134.	Cyclotrimethylene trinitramine	50 t	121-82-4	
135.	Trinitroaniline	50 t	26952-42-1	
136.	2,4,6-Trinitroanisole	50 t	606-35-9	609-011-00-0
137.	Trinitrobenzene	50 t	25377-32-6	609-005-00-8
138.	Trinitrobenzoic acid	50 t	129-66-8	
139.	Chlorotrinitrobenzene	50 t	28260-61-9	610-004-00-X
140.	N-Methyl-N,2,4,6-N-tetranitroaniline	50 t	479-45-8	612-017-00-6
141.	2,4,6-Trinitrophenol (Picric acid)	50 t	88-89-1	609-009-00-X
142.	Trinitrocresol	50 t	28905-71-7	609-012-00-6
143.	2,4,6-Trinitrophenetole	50 t	4732-14-3	
144.	2,4,6-Trinitroresorcinol (Styphnic acid)	50 t	82-71-3	609-018-00-9
145.	2,4,6-Trinitrotoluene	50 t	118-96-7	609-008-00-4
146.	Ammonium nitrate ([1])	5 000 t	6484-52-2	
147.	Cellulose nitrate (containing >12·6 % nitrogen)	100 t	9004-70-0	603-037-00-6

148.	Sulphur dioxide	1000 t	7446-09-05	016-011-00-9
149.	Hydrogen chloride (liquefied gas)	250 t	7647-01-0	017-002-00-2
150.	Flammable substances as defined in Annex IV (c) (iii)	200 t		
151.	Sodium chlorate (¹)	250 t	7775-09-9	017-005-00-9
152.	tert-Butyl peroxyacetate (concentration ≥ 70 %)	50 t	107-71-1	
153.	tert-Butyl peroxyisobutyrate (concentration ≥ 80 %)	50 t	109-13-7	
154.	tert-Butyl peroxymaleate (concentration ≥ 80 %)	50 t	1931-62-0	
155.	tert-Butyl peroxy isopropyl carbonate (concentration ≥ 80 %)	50 t	2372-21-6	
156.	Dibenzyl peroxydicarbonate (concentration ≥ 90 %)	50 t	2144-45-8	
157.	2,2-Bis (tert-butylperoxy) butane (concentration ≥ 70 %)	50 t	2167-23-9	
158.	1,1-Bis (tert-butylperoxy) cyclohexane (concentration ≥ 80 %)	50 t	3006-86-8	
159.	Di-sec-butyl peroxydicarbonate (concentration ≥ 80 %)	50 t	19910-65-7	
160.	2,2-Dihydroperoxypropane (concentration ≥ 30 %)	50 t	2614-76-8	
161.	Di-n-propyl peroxydicarbonate (concentration ≥ 80 %)	50 t	16066-38-9	
162.	3,3,6,6,9,9-Hexamethyl-1,2,4, 5-tetroxacyclononane (concentration ≥ 75%)	50 t	22397 33 7	
163.	Methyl ethyl ketone peroxide (concentration ≥ 60 %)	50 t	1338-23-4	
164.	Methyl isobutyl ketone peroxide (concentration ≥ 60 %)	50 t	37206-20-5	
165.	Peracetic acid (concentration ≥ 60 %)	50 t	79-21-0	607-094-00-8
166.	Lead azide	50 t	13424-46-9	082-003-00-7
167.	Lead 2,4,6-trinitroresorcinoxide (Lead sryphnate)	50 t	15245-44-0	609-019-00-4
168.	Mercury fulminate	10 t	628 86 45	080-005-00-2
169.	Cyclotetramethylenetetranitramine	50 t	2691-41-0	
170.	2,2',4,4',6,6'-Hexanitrostilbene	50 t	20062-22-0	
171.	1,3,5-Triamino-2,4,6-trinitrobenzene	50 t	3058-38-6	
172.	Ethylene glycol dinitrate	10 t	628-96-6	603-032-00-9
173.	Ethyl nitrate	50 t	625-58-1	007-007-00-8

174.	Sodium picramate	50 t	831-52-7
175.	Barium azide	50 t	18810-58-7
176.	Di-isobutyryl peroxide (concentration ≥ 50 %)	50 t	3437-84l
177.	Diethyl peroxydicarbonate (concentration ≥ 30 %)	50 t	14666-78-5
178.	tert-Butyl peroxypivalate (concentration ≥ 77 %)	50 t	927-07-1

1. Where this substance is in a state which gives it properties capable of creating a major-accident hazard. *NB:* The EEC numbers correspond to those in Directive 67/548/EEC and its amendments.

ANNEX IV

INDICATIVE CRITERIA

(a) Very toxic substances

– substances which correspond to the first line of the table below,

– substances which correspond to the second line of the table below and which, owing to their physical and chemical properties, are capable of entailing major-accident hazards similar to those caused by the substance mentioned in the first line:

	LD 50 (oral) ([1]) mg/kg body weight	LD 50 (cutaneous) ([2]) mg/kg body weight	LC 50 ([3]) mg/l (inhalation)
1	LD 50 $\leqslant$ 5	LD 50 $\leqslant$ 10	LC 50 $\leqslant$ 0·1
2	5 < LD 50 $\leqslant$ 25	10 < LD 50 $\leqslant$ 50	0·1 < LC 50 $\leqslant$ 0·5

([1]) LD 50 oral in rats.
([2]) LD 50 cutaneous in rats or rabbits.
([3]) LC 50 by inhalation (four hours) in rats.

(b) Other toxic substances

The substances showing the following values of acute toxicity and having physical and chemical properties capable of entailing major-accident hazards:

LD 50 (oral) ([1]) mg/kg body weight	LD 50 (cutaneous) ([2]) mg/kg body weight	LC 50 ([3]) mg/l (inhalation)
25 < LD 50 $\leqslant$ 200	50 < LD 50 $\leqslant$ 400	0·5 < LC 50 $\leqslant$ 2

([1]) LD 50 oral in rats.
([2]) LD 50 cutaneous in rats or rabbits.
([3]) LC 50 by inhalation (four hours) in rats.

(c) Flammable substances

(i) *Flammable gases:*
substances which in the gaseous state at normal pressure and mixed with air become flammable and the boiling point of which at normal pressure is 20°C or below;

(ii) *Highly flammable liquids:*
substances which have a flash point lower than 21°C and the boiling point of which at normal pressure is above 20°C;

(iii) *Flammable liquids:*
substances which have a flash point lower than 55°C and which remain liquid under pressure, where particular processing conditions, such as high pressure and high temperature, may create major-accident hazards.

(d) Explosive substances

Substances which may explode under the effect of flame or which are more sensitive to shocks or friction than dinitrobenzene.

ANNEX V

DATA AND INFORMATION TO BE SUPPLIED IN CONNECTION WITH
THE NOTIFICATION PROVIDED FOR IN ARTICLE 5

If it is not possible or if it seems unnecessary to provide the following information, reasons must be given.

1. IDENTITY OF THE SUBSTANCE

   Chemical name

   CAS number

   Name according to the IUFAC nomenclature

   Other names

   Empirical formula

   Composition of the substance

   Degree of purity

   Main impurities and relative percentages

   Detection and determination methods available to the installation

   Description of the methods used or references to scientific literature

   Methods and precautions laid down by the manufacturer in connection with handling, storage and fire

   Emergency measures laid down by the manufacturer in the event of accidental dispersion

   Methods available to the manufacturer for rendering the substance harmless

2. BRIEF INDICATION OF HAZARDS

   — For man:

       — immediate ....

       — delayed ....

   — For the environment:

       — immediate ....

       — delayed ....

ANNEX VI

INFORMATION TO BE SUPPLIED TO THE COMMISSION BY THE MEMBER STATES PURSUANT TO ARTICLE 11

REPORT OF MAJOR ACCIDENT

Member State:
Authority responsible for report:
Address:

1. **General data**
   Date and time of the major accident:
   Country, administrative region, etc.:
   Address:
   Type of industrial activity:

2. **Type of major accident**

   Explosion ☐   Fire ☐   Emission of dangerous substances ☐
   Substance(s) emitted:

3. **Description of the circumstances of the major accident**

4. **Emergency measures taken**

5. **Cause(s) of major accident**

   Known: ☐
   (to be specified)

   Not known: ☐

   Information will be supplied as soon as possible   ☐

6. **Nature and extent of damage**
   (a) *Within the establishment*
      – casualties   . . . . . killed
                                            . . . . . injured
                                            . . . . . poisoned

      – persons exposed to the major accident   . . . . .

      – material damage   ☐

      – the danger is still present   ☐

      – the danger no longer exists   ☐

   (b) *Outside the establishment*
      – casualties   . . . . . killed
                                            . . . . . injured
                                            . . . . . poisoned

      – persons exposed to the major accident   . . . . .

      – material damage   ☐

      – damage to the environment   ☐

      – the danger is still present   ☐

      – the danger no longer exists   ☐

7. **Medium and long-term measures**, particularly those aimed at preventing the recurrence of similar major accidents (to be submitted as the information becomes available).

ANNEX VII

STATEMENT RE ARTICLE 8

The Member States shall consult one another in the framework of their bilateral relations on the measures required to avert major accidents originating in a notified industrial activity within the meaning of Article 5 and to limit the consequences for man and the environment. In the case of new installations, this consultation shall take place within the time limits laid down in Article 5 (2).

COUNCIL DIRECTIVE

of 28 July 1982

on the protection of workers from the risks related to exposure to metallic lead
and its ionic compounds at work

(first individual Directive within the meaning of Article 8 of Directive 80/1107/EEC)

[82/605/EEC]

THE COUNCIL OF THE EUROPEAN COMMUNITIES,

Having regard to the Treaty establishing the European Economic Community, and in particular Article 100 thereof,

Having regard to the proposal from the Commission,

Having regard to the opinion of the European Parliament,

Having regard to the opinion of the Economic and Social Committee,

Whereas the Council resolution of 29 June 1978 on an action programme of the European Communities on safety and health at work, provides for the establishment of specific harmonized procedures regarding the protection of workers with respect to lead:

Whereas Council Directive 80/1107/EEC of 27 November 1980 on the protection of workers from the risks related to exposure to chemical, physical and biological agents at work, lays down certain provisions which have to be taken into account for this protection; whereas that Directive provides for the laying down in individual Directives of limit values and specific requirements for those agents listed in Annex I, which include lead;

Whereas metallic lead and its ionic compounds are toxic agents found in a large number of circumstances at work; whereas many workers are therefore exposed to a potential health risk;

Whereas, therefore, preventive measures for the protection of the health of workers exposed to lead and the commitment envisaged for Member States with regard to the surveillance of their health are important;

Whereas workers exposed to lead in the extractive industries must enjoy a level of health protection similar to that laid down in this Directive; whereas, given the specific nature of such activities, the implementation of such protection will need to be covered by special provisions embodied in a subsequent Directive;

Whereas this Directive includes minimum requirements which will be reviewed on the basis of experience acquired and of developments in technology and medical knowledge in this area, the objective being to attain greater protection of workers,

HAS ADOPTED THIS DIRECTIVE:

Article 1

1. This Directive, which is the first individual Directive within the meaning of Article 8 of Directive 80/1107/EEC has as its aim the protection of workers against risks to their health, including the prevention of such risks, arising or likely to arise at work from exposure to metallic lead and its ionic compounds; it shall not apply to alkylated lead compounds. It shall lay down limit values and other specific requirements.

2. This Directive shall not apply to:

    — sea transport,
    — air transport,
    — mining and quarrying of lead-containing ores and the preparation of lead-ore concentrate at the site of the mine or quarry.

3. This Directive shall not prejudice the right of Member States to apply or introduce laws, regulations or administrative provisions ensuring greater protection for workers or for a particular category of workers.

Article 2

1. Any work likely to involve a risk of absorbing lead shall be assessed in such a way as to determine the nature and degree of the exposure to lead of the workers.

    Annex I contains an indicative, non-exhaustive list of activities where there is reason to consider that there may be a risk of absorbing lead.

2. If the assessment provided for in paragraph 1 reveals the presence of at least one of the following conditions:

    — exposure to a concentration of lead in air greater that $40 \, \mu g/m^3$, calculated as a time-weighted average over 40 hours per week,

    — a blood-lead level greater than $40 \, \mu g \, Pb/100 \, ml$ blood in individual workers,

    the provisions regarding information set out in Article 11(1) shall apply and appropriate measures shall be taken to minimize the risk of absorbing lead which arises through smoking, eating and drinking at the place of work.

3. If the assessment provided for in paragraph 1 reveals that the blood-lead level of workers due to lead absorption is between $40 \, \mu g$ and $50 \, \mu g \, Pb/100 \, ml$ blood, Member States shall endeavour to carry out biological monitoring of the workers concerned in accordance with the procedures laid down by the Member States.

4. If the assessment provided for in paragraph 1 reveals the presence of at least one of the following conditions:

    — exposure to a concentration of lead in air greater than $75 \, \mu g/m^3$, calculated as a time-weighted average over 40 hours per week,

    — a blood-lead level greater than $50 \, \mu g \, Pb/100 \, ml$ blood in individual workers,

    the protection provided for in this Directive, in particular the lead-in-air monitoring and the medical surveillance set out in Articles 3 and 4, is to be given to the workers concerned.

5.   The assessment provided for in paragraph 1 shall be the subject of consultation with the workers and/or their representatives within the undertaking or establishment and shall be revised where there is reason to believe that it is incorrect or there is a material change in the work.

Article 3

1.   All lead-in-air measurements shall be representative of worker exposure to particles containing lead.

Particles containing lead within the meaning of this Directive shall be those particles captured by equipment having the sampling characteristics specified in Annex II, point 1, and analyzed in accordance with the methods indicated in Annex II, point 2.

2.   Monitoring of the concentration of lead in air shall take place at least every three months.

This frequency may, however, be reduced in the cases listed in paragraph 3.

3.   Frequency of monitoring may be reduced to once a year, provided that there is no material change in the work and conditions of exposure, where:

(i)   the results of the measurements for individual workers or for groups of workers have shown that on the previous two consecutive occasions on which monitoring was carried out:

—   the lead-in-air concentration did not exceed 100 $\mu g/m^3$, or

—   the conditions or exposure did not fluctuate appreciably, or

(ii)   the blood-lead level of any worker does not exceed 60 μg Pb/ 100 ml blood.

4.   The monitoring for a worker or group of workers, as stipulated in paragraph 2, shall entail taking one or more air samples.

Without prejudice to the second indent of Article 7(b), sampling shall be carried out in such a way as to permit assessment of the probable maximum risk to which the individual worker or workers are exposed, account being taken of the work done, the working conditions and the length of exposure during the course of the work. The workers concerned and/or their representatives within the undertaking or establishment shall be consulted to this end.

For the initial monitoring, after it has been established that the values laid down in Article 2(4) have been exceeded, the duration of the sampling period shall not be less than four hours.

Subsequently this duration shall not be less than four hours if the results obtained on the occasion of the preceding monitoring have shown higher lead-in-air concentration values than those obtained before that monitoring.

Where groups of workers are performing identical or similar tasks in the same location and are thus being exposed to the same health risk, sampling may be carried out on a group basis. In such a case, sampling shall be carried out for at least one worker out of 10.

5. The specifications referred to in paragraph 1 and Annex II, with the exception of the specification concerning the air intake velocity given at point 1 (a) of the Annex, and the technical aspects of this Article shall be adapted in the light of technical progress in accordance with the procedure set out in Article 10 of Directive 80/1107/EEC, within the limits laid down in Annex III to that Directive.

Article 4

1. Workers shall be subject to medical (clinical and biological) surveillance. This surveillance must start prior to or at the beginning of the exposure. The frequency of clinical assessment shall be at least once a year during the period of employment. Biological monitoring shall be carried out, in accordance with paragraph 2, at least every six months.

    This surveillance shall take account not only of the magnitude of the exposure but also of the individual worker's susceptibility to lead.

2. The biological monitoring shall, apart from the exception mentioned in paragraph 3, include measuring the blood-lead level (PbB).

    This monitoring may also include measuring one or more of the following biological indicators:

    — delta aminolæ vulinic acid in urine (ALAU),

    — zinc protoporphyrin (ZPP),

    — delta aminolæ vulinic acid dehydratase in blood (ALAD).

    The methods of measuring the biological indicators referred to above are listed in Annex III and may be adapted in accordance with the procedure specified in Article 10 of Directive 80/1107/EEC.

3. The PbB measurement referred to in paragraph 2 may be replaced by that of ALAU when dealing with workers who have been subjected for a period of less than one month to risks of high exposure.

4. The frequency of biological monitoring may be reduced to once a year where at the same time:

    — the results of the measurements for individuals or for groups of workers have shown, on the previous two consecutive occasions on which monitoring was carried out, a lead-in-air concentration higher than the value laid down in the first indent of Article 2(4) and lower than 100 $\mu g/m^3$,

    — the PbB level of any individual worker does not exceed the value laid down in the second indent of Article 2(4).

5. Practical recommendations to which Member States may refer for clinical assessment are set out in Annex IV and may be adapted in accordance with the procedure set out in Article 10 of Directive 80/1107/EEC.

Article 5

1. Where the biological monitoring carried out in accordance with Article 4(2) reveals an individual PbB level higher than 60µg Pb/100ml blood but lower than the limit value set out in Article 6(1)(b), a clinical examination shall be carried out as soon as possible. However, this clinical examination may be deferred until a repeat determination of the PbB level, undertaken within one month, shows that the value of 60 µg Pb/100 ml blood continues to be exceeded.

   Thereafter, biological monitoring and clinical assessment shall be carried out at shorter intervals than those laid down in Article 4(1) at least until the PbB level is below 60 µg Pb/100 ml blood.

2. Following the clinical examination referred to in paragraph 1, the doctor or authority responsible for the medical surveillance of the workers should advise on any protective or preventive measures to be taken on an individual basis; these may include, where appropriate, the withdrawal of the worker concerned from exposure to lead or a reduction in the period of his exposure.

Article 6

1. The following limit values shall be applied:

   (a) lead-in-air concentration:

   150 µg/m³, calculated as a time-weighted average over 40 hours per week;

   (b) value of the biological parameters:

   PbB level in individual workers: 70 µg/ Pb/100 ml blood. However, a PbB level of between 70 and 80 µg Pb/100 ml blood shall be allowed if the ALAU level remains lower than 20 mg/g creatinine or the ZPP level remains lower than 20 µg/g haemoglobin or the ALAD level remains greater than six European units.

2. Where biological monitoring is based solely on ALAU measurement in accordance with Article 4(3), the following limit value shall be applied for ALAU: 20 mg/g creatinine.

3. The Council, acting on a proposal from the Commission, and taking into account in particular progress made in scientific knowledge and technology as well as experience gained in the application of this Directive, shall re-examine the limit values for the biological parameters within five years of adoption of this Directive, with a view to setting a maximum blood-lead limit value of 70 µg Pb/100 ml blood.

Article 7

For the purpose of establishing whether or not the lead-in-air limit value fixed in Article 6(1)(a) has been exceeded, it is appropriate to proceed as follows:

   (a) If the total sampling period is of 40 hours in one week then the lead-in-air concentrations obtained can be compared directly with the limit value laid down in Article 6(1)(a);

(b)   If the total sampling period is less than 40 hours in one week then:

— the limit value laid down in Article 6(1)(a) shall not be considered as having been exceeded if the concentration obtained by sampling in accordance with Article 3(4) is below the numerical level of the limit value,

— if the concentration referred to in the first indent exceeds the numerical level of the limit value then at least three additional lead-in-air samples shall be taken which are representative of average exposure to lead; the total period over which each of these three samples is taken shall be at least four hours.

If, from four samples taken over a period of one week, it is found that three levels of concentration are below the numerical level of the limit value, then it shall be deemed that this limit value has not been exceeded.

Article 8

1.   Where the lead-in-air limit value laid down in Article 6(1)(a) is exceeded the reasons for the limit being exceeded shall be identified and appropriate measures to remedy the situation shall be taken as soon as possible.

The doctor or authority responsible for the medical surveillance of the workers shall judge whether an immediate determination of the biological parameters of the workers concerned should be carried out.

In order to check the effectiveness of the measures mentioned in the first sub-paragraph, a further determination of the lead-in-air concentrations on the basis of the procedures laid down in Articles 3 and 7 shall be carried out.

2.   Where the measures referred to in the first sub-paragraph of paragraph 1 cannot, owing to their nature or magnitude. be taken within one month and a further determination of lead-in-air concentrations shows that the lead-in-air limit values continue to be exceeded, work may not be continued in the affected area until adequate measures have been taken for the protection of the workers concerned, in the light of the opinion of the doctor or authority responsible for medical surveillance.

Where the exposure cannot reasonably be reduced by other means and where the wearing of individual respiratory protective equipment proves necessary, this may not be permanent and shall be kept to the strict minimum necessary for each worker.

3.   In the case of incidents likely to lead to significant increases in exposure to lead, workers shall be immediately evacuated from the affected area. Only workers whose presence is required to carry out the necessary repairs may enter the affected area on condition that they use suitable protective apparatus.

4.   In the case of certain operations in respect of which it is foreseen that the limit value set out in paragraph 1 will be exceeded and in respect of which technical preventive measures for limiting concentrations in the air are not reasonably practicable, the employer shall define the measures intended to ensure protection

of the workers during operations of this kind. The workers and/or their representatives in the undertaking or establishment shall be consulted on these measures before such operations are effected.

Article 9

1. Where the biological limit value laid down in Article 6(1)(b) has been exceeded:

   — the necessary steps shall be taken immediately to ascertain the reasons for this excess and to remedy the situation. Such measures may, depending on the magnitude of the excess, and where it is considered desirable by the doctor or authority responsible for the medical surveillance of the workers include the immediate withdrawal of the worker concerned from all exposure to lead,

   — a further determination of the PbB level shall be made within three months. Following this determination, the worker concerned must not continue at his work or at any other work involving an equal or greater risk of exposure to lead if the biological limit value continues to be exceeded. The worker concerned may be assigned, following an opinion from the doctor or authority responsible for medical surveillance, to other work involving a lesser risk of exposure. In this case, he shall be subject to more frequent medical assessments.

   However, Member States may take different measures for workers who, having been exposed to lead over a number of years, have a very high body burden of lead when this Directive becomes applicable.

2. The worker concerned or the employer may ask for a review of the assessments referred to in paragraph 1.

Article 10

1. For all work carried out under the conditions set out in Article 2(4), appropriate measures shall be taken to ensure that:

   (a) (i) the risk of absorbing lead through smoking, eating or drinking is avoided,

   (ii) areas are set aside where workers can eat and drink without risking contamination by lead,

   (iii) in very hot workplaces where workers should be encouraged to drink, workers are provided with drinking water or other drinks not contaminated by the lead present in the workplace;

   (b) (i) workers are provided with appropriate working or protective clothing, taking into account the physico-chemical properties of the lead compounds to which they are exposed.

   (ii) this working or protective clothing remains within the undertaking. It may, however, be laundered in establishments outside the undertaking which are equipped for this sort of work, if the undertaking itself does not carry out cleaning; where this is the case, the clothing shall be transported in closed containers,

(iii) working or protective clothing and street clothes are stored separately,

(iv) workers are provided with adequate and appropriate washing facilities, including showers in the case of dusty operations.

2. The cost of the measures taken pursuant to paragraph 1 shall not be borne by the workers.

Article 11

1. For all work carried out under the conditions set out in Article 2(2), appropriate measures shall be taken so that workers and their representatives in the undertaking or establishment are provided with adequate information on:

— the potential risks to health from lead exposure, including the potential risks for the foetus and infants being breast-fed,

— the existence of statutory limit values and the need for biological and atmospheric monitoring,

— hygiene requirements, including the need to refrain from smoking, eating or drinking at the workplace,

— the precautions to be taken as regards the wearing and use of protective equipment and clothing,

— the special precautions to be taken to minimize exposure to lead.

2. In addition to the measures referred to in paragraph 1, for all work carried out under the conditions set out in Article 2(4), appropriate measures shall be taken so that;

(a) workers and/or their representatives within the undertaking or establishment have access to:

— the results of lead-in-air measurements,

— the statistical (non-personalized) results of biological monitoring,

and explanations of the significance of these results are available to them;

(b) if the results exceed the lead-in-air limit value laid down in Article 6(1)(a) the workers concerned and their representatives in the undertaking or establishment are informed as quickly as possible of the excess and the reason for it and the workers and/or their representatives in the undertaking or establishment are consulted on the measures to be taken or, in an emergency, are informed of the measures which have been taken;

(c) each time PbB tests, ALAU tests or any other biological measurements for assessing lead exposure are carried out, the workers concerned are informed, on the authority of the doctor responsible, of the results of those measurements and the interpretation placed on the results.

Article 12

The doctor or authority responsible for medical surveillance of the workers shall have access to all information necessary for determining the extent of workers' exposure to lead, including the results of the lead-in-air monitoring.

Article 13

Steps shall be taken to ensure that individual data relating to the exposure of workers and their clinical and biological examinations are recorded and stored in an appropriate form, in accordance with national laws and practices.

Article 14

1. Member States shall bring into force the laws, regulations and administrative provisions necessary to comply with this Directive by 1 January 1986 at the latest and shall forthwith inform the Commission thereof.

2. Member States shall communicate to the Commission the texts of the provisions of national law which they adopt in the field covered by this Directive.

Article 15

This Directive is addressed to the Member States.

Done at Brussels, 28 July 1982

ANNEX I

List of activities referred to in the second sub-paragraph of Article 2(1)

1. Handling of lead concentrate.
2. Lead and zinc smelting and refining (primary and secondary).
3. Lead arsenate spray manufacture and handling.
4. Manufacture of lead oxides.
5. Production of other lead compounds (including that part of the production of alkyl lead compounds, where it includes exposure to metallic lead and its ionic compounds).
6. Manufacture of paints, enamels, mastics and colours containing lead.
7. Battery manufacture and recycling.
8. Craftwork in tin and lead.
9. Manufacture of lead solder.
10. Lead ammunition manufacture.
11. Manufacture of lead-based or lead-alloy objects.
12. Use of paints, enamels, mastics and colours containing lead.
13. Ceramic and craft pottery industries.
14. Crystal glass industries.
15. Plastic industries using lead additives.
16. Frequent use of lead solder in an enclosed space.
17. Printing work involving the use of lead.
18. Demolition work, especially the processes of scraping off, burning off and name-cutting executed on materials coated with paint containing lead, as well as the breaking up of plant (*eg.* lead furnaces).
19. Use of lead ammunition in an enclosed space.
20. Automobile construction and repair work.
21. Manufacture of leaded steel.
22. Lead tempering of steel.
23. Lead coating.
24. Recovery of lead and metallic residues containing lead.

ANNEX II

Technical specifications referred to in the second sub-paragraph of Article 3 (1)

1. The equipment is that which complies with the technical specifications listed below:

    (a) *Air intake velocity at the orifice*: 1.25 m/s ± 10 %;

    (b) *Air flow rate*: at least 1 I/min.;

    (c) *Sampling head characteristics*: a closed face sampling head should be used, to avoid filter contamination;

    (d) *Intake orifice diameter*: at least 4 mm diameter in order to avoid wall effects;

    (e) *Filter or intake orifice position*: as far as possible kept parallel to the face of the worker during the whole sampling period;

    (f) *Filter efficiency*: a minimum of 95 % efficiency for all particles sampled down to an aerodynamic diameter of 0.3 μm;

    (g) *Filter homogeneity*: maximum homogeneity of the lead content in the filter to allow for comparison between two halves of the same filter.

2. The lead-in-air sample collected in accordance with the procedures in point 1 is to be analyzed by atomic absorption spectroscopy or any other method which gives equivalent results.

ANNEX III

Methods of measuring biological indicators referred to in Article 4(2)

PbB:	Atomic absorption spectroscopy,
ALAU:	Davis or equivalent method,
ZPP:	Haematofluorimetry or equivalent method,
ALAD:	European standardized method or equivalent method.

Appropriate quality control programmes will be established by the Commission.

ANNEX IV

Practical recommendations for the clinical assessment of workers referred to in Article 4(5)

1. Current knowledge indicates that large-scale absorption may produce adverse effects in the following systems:

   — hematopoietic,
   — gastro-intestinal,
   — central and peripheral nervous,
   — renal.

2. The doctor in charge of the medical surveillance of the worker exposed to lead should be familiar with the exposure conditions or circumstances of each worker.

3. Clinical assessment of the workers should be carried out in accordance with sound practice; it should include the following measures:

   — records of the worker's medical and occupational history,
   — physical examination and a personal interview with special attention to the associated symptoms of early lead poisoning,
   — evaluation of the pulonary status (for possible use of respiratory protective equipment).

   Blood analyses (and, in particular, establishment of the hematocrit level) and urine analysis should be carried out during the first medical examination and then regularly according to the doctor's judgement.

4. In addition to the decisions based on the results of biological monitoring, the examining doctor will establish the cases where exposure or continued exposure to lead is contra-indicated. The most important of these contra-indications are:

   (i) — congenital abnormalities:

      — thalassemia,
      — G — 6 — PD deficiency:

   (ii) — acquired conditions:

      — anaemia,
      — renal deficiencies,
      — hepatic deficiencies.

5. Use of chelating agents:

   The prohphylactic use of chelating agents, sometimes called 'preventive therapy' is medically and ethically unacceptable. Many chelating agents may be considered nephrotoxic when administered for long periods.

6. Intoxication therapy:

   To be carried out by specialists.

COUNCIL DIRECTIVE

of 19 September 1983

on the protection of workers from the risks related to exposure to asbestos at work
(second individual Directive within the meaning of Article 8 of Directive 80/1107/EEC)
[83/477/EEC]

THE COUNCIL OF THE EUROPEAN COMMUNITIES,

Having regard to the Treaty establishing the European Economic Community, and in particular Article 100 thereof,

Having regard to the proposal from the Commission,

Having regard to the opinion of the European Parliament,

Having regard to the opinion of the Economic and Social Committee,

Whereas the Council resolution of 29 June 1978 on an action programme of the European Communities on safety and health at work provides for the establishment of specific harmonized procedures regarding the protection of workers with respect to asbestos;

Whereas Council Directive 80/1107/EEC of 27 November 1980 on the protection of workers from the risks related to exposure to chemical, physical and biological agents at work laid down certain provisions which have to be taken into account for this protection; whereas that Directive provides for the laying down in individual Directives of limit values and specific requirements for those agents listed in Annex I, which include asbestos;

Whereas asbestos is a harmful agent found in a large number of circumstances at work; whereas many workers are therefore exposed to a potential health risk; whereas crocidolite is considered to be a particularly dangerous type of asbestos;

Whereas, although current scientific knowledge is not such that a level can be established below which risks to health cease to exist, a reduction in exposure to asbestos will nonetheless reduce the risk of developing asbestos-related disease; whereas this Directive includes minimum requirements which will be reviewed on the basis of experience acquired and of developments in technology in this area;

Whereas optical microscopy, although it does not allow a counting of the smallest fibres detrimental to health, is the most currently used method for the regular measuring of asbestos;

Whereas, therefore, preventive measures for the protection of the health of workers exposed to asbestos and the commitment envisaged for Member States with regard to the surveillance of their health are important,

HAS ADOPTED THIS DIRECTIVE:

Article 1

1. This Directive, which is the second individual Directive within the meaning of Article 8 of Directive 80/1107/EEC, has as its aim the protection of workers against risks to their health, including the prevention of such risks, arising or likely to arise from exposure to asbestos at work. It lays down limit values and other specific requirements.

2. This Decision shall not apply to:

    — sea transport,

    — air transport.

3. This Directive shall not prejudice the right of Member States to apply or introduce laws, regulations or administrative provisions ensuring greater protection for workers, in particular as regards the replacement of asbestos by less-dangerous substitutes.

Article 2

For the purposes of this Directive, 'asbestos' means the following fibrous silicates:

   — Actinolite, CAS No 77536-66-4,

   — Asbestos grunerite (amosite) CAS No I 217273-5,

   — Anthophyllite, CAS No 77536-67-5,

   — Chrysotile, CAS No 12001-29-5,

   — Crocidolite, CAS No 12001-28-4,

   — Tremolite, CAS No 77536-68-6.

Article 3

1. This Directive shall apply to activities in which workers are or may be exposed in the course of their work to dust arising from asbestos or materials containing asbestos.

2. In the case of any activity likely to involve a risk of exposure to dust arising from asbestos or materials containing asbestos, this risk must be assessed in such a way as to determine the nature and degree of the workers' exposure to dust arising from asbestos or materials containing asbestos.

3. If the assessment referred to in paragraph 2 shows that the concentration of asbestos fibres in the air at the place of work in the absence of any individual protective equipment is, at the option of the Member States, at a level as measured or calculated in relation to an eight-hour reference period,

    — lower than 0·25 fibre per cm^3 and/or

    — lower than a cumulative dose of 15·00 fibre-days per cm^3 over three months,

Articles 4, 7, 13, 14(2), 15 and 16 shall not apply.

4. The assessment provided for in paragraph 2 shall be the subject of consultation with the workers and/or their representatives within the undertaking or establishment and shall be revised where there is reason to believe that it is incorrect or there is a material change in the work.

Article 4

Subject to Article 3(3), the following measures shall be taken:

1. The activities referred to in Article 3(1) must be covered by a notification system administered by the responsible authority of the Member State.

2. The notification must be submitted by the employer to the responsible authority of the Member State, in accordance with national laws, regulations and administrative provisions. This notification must include at least a brief description of:

    —   the types and quantities of asbestos used,

    —   the activities and processes involved,

    —   the products manufactured.

3. Workers and/or their representatives in undertakings or establishments shall have access to the documents which are the subject of notification concerning their own undertaking or establishment in accordance with national laws.

4. Each time an important change occurs in the use of asbestos or of materials containing asbestos, a new notification must be submitted.

Article 5

The application of asbestos by means of the spraying process must be prohibited.

Article 6

For all activities referred to in Article 3(1), the exposure of workers to dust arising from asbestos or materials containing asbestos at the place of work must be reduced to as low a level as is reasonably practicable and in any case below the limit values laid down in Article 8, in particular through the following measures if appropriate:

1. The quantity of asbestos used in each case must be limited to the minimum quantity which is reasonably practicable.

2. The number of workers exposed or likely to be exposed to dust arising from asbestos or materials containing asbestos must be limited to the lowest possible figure.

3. Work processes must, in principle, be so designed as to avoid the release of asbestos dust into the air.

    If this is not reasonably practicable, the dust should be eliminated as near as possible to the point where it is released.

4. All buildings and/or plant and equipment involved in the processing or treatment of asbestos must be capable of being regularly and effectively cleaned and maintained.

5. Asbestos as a raw material must be stored and transported in suitable sealed packing.

6. Waste must be collected and removed from the place of work as soon as possible in suitable sealed packing with labels indicating that it contains asbestos. This measure shall not apply to mining activities.

    The waste referred to in the preceding paragraph shall then be dealt with in accordance with Council Directive 78/319/EEC of 20 March 1978 on toxic and dangerous waste.

Article 7

Subject to Article 3(3), the following measures shall be taken:

1. In order to ensure compliance with the limit values laid down in Article 8, the measurement of asbestos in the air at the place of work shall be carried out in accordance with the reference method described in Annex I or any other method giving equivalent results. Such measurement must be planned and carried out regularly, with sampling being representative of the personal exposure of the worker to dust arising from asbestos or materials containing asbestos.

    For the purposes of measuring asbestos in the air, as referred to in the preceding paragraph, only fibres with a length of more than five micrometres and a length/breadth ratio greater than 3:1 shall be taken into consideration.

    The Council, acting on a proposal from the Commission, and taking account in particular of progress made in scientific knowledge and technology and of experience gained in the application of this Directive, shall re-examine the provisions of the first sentence of paragraph 1 within five years following the adoption of this Directive, with a view to establishing a single method for measurement of asbestos-in-air concentrations at Community level.

2. Sampling shall be carried out after consulting the workers and/or their representatives in undertakings or establishments.

3. Sampling shall be carried out by suitably qualified personnel. The samples taken shall be subsequently analyzed in laboratories equipped to analyze them and qualified to apply the necessary identification techniques.

4. The amount of asbestos in the air shall be measured as a general rule at least every three months and, in any case, whenever a technical change is introduced. The frequency of measurements may, however, be reduced in the circumstances specified in paragraph 5.

5. The frequency of measurements may be reduced to once a year where:

    —   there is no substantial change in conditions at the place of work, and

    —   the results of the two preceding measurements have not exceeded half the limit values fixed in Article 8.

Where groups of workers are performing identical or similar tasks at the same place and are thus being exposed to the same health risk, sampling may be carried out on a group basis.

6. The duration of sampling must be such that representative exposure can be established for an eight hour reference period (one shift) by means of measurements or time-weighted calculations. The duration of the various sampling processes shall be determined also on the basis of point 6 of Annex I.

Article 8

The following limit values shall be applied:

(a) concentration of asbestos fibres other than crocidolite in the air at the place of work:

1·00 fibres per cm^3 measured or calculated in relation to an eight-hour reference period;

(b) concentration of crocidolite fibres in the air at the place of work:

0·50 fibres per cm^3 measured or calculated in relation to an eight-hour reference period;

(c) concentration of asbestos fibres in the air at the place of work in the case of mixtures of crocidolite and other asbestos fibres:

the limit value is at a level calculated on the basis of the limit values laid down in (a) and (b), taking into account the proportions of crocidolite and other asbestos types in the mixture.

Article 9

The Council, acting on a proposal from the Commission, shall, taking into account, in particular, progress made in scientific knowledge and technology and in the light of experience gained in applying this Directive, review the provisions laid down in Article 3(3) and in Article 8 before 1 January 1990.

Article 10

1. Where the limit values laid down in Article 8 are exceeded, the reasons for the limits being exceeded must be identified and appropriate measures to remedy the situation must be taken as soon as possible.

Work may not be continued in the affected area until adequate measures have been taken for the protection of the workers concerned.

2. In order to check the effectiveness of the measures mentioned in the first subparagraph of paragraph 1, a further determination of the asbestos-in-air concentrations shall be carried out immediately.

3. Where exposure cannot reasonably be reduced by other means and where the wearing of individual respiratory protective equipment proves necessary, this may not be permanent and shall be kept to the strict minimum necessary for each worker.

Article 11

1. In the case of certain activities in respect of which it is foreseeable that the limit values laid down in Article 8 will be exceeded and in respect of which technical preventive measures for limiting asbestos-in-air concentrations are not reasonably practicable, the employer shall determine the measures intended to ensure protection of the workers while they are engaged in such activities, in particular the following:

    (a) workers shall be issued with suitable respiratory equipment and other personal protective equipment, which must be worn; and

    (b) warning signs shall be put up indicating that it is foreseeable that the limit values laid down in Article 8 will be exceeded.

2. The workers and/or their representatives in the undertaking or establishment shall be consulted on these measures before the activities concerned are carried out.

Article 12

1. A plan of work shall be drawn up before demolition work or work on removing asbestos and/or asbestos-containing products from buildings, structures, plant or installations or from ships is started.

2. The plan referred to in paragraph 1 must prescribe the measures necessary to ensure the safety and health of workers at the place of work.

    The plan must in particular specify that:

    — as far as is reasonably practicable, asbestos and/or asbestos-containing products are removed before demolition techniques are applied,

    — the personal protective equipment referred to in Article 11(1)(a) is provided, where necessary.

Article 13

1. In the case of all activities referred to in Article 3(1), and subject to Article 3 (3), appropriate measures shall be taken to ensure that:

    (a) the places in which the above activities take place shall:

        (i) be clearly demarcated and indicated by warning signs;

        (ii) not be accessible to workers other than those who by reason of their work or duties are required to enter them;

        (iii) constitute areas where there should be no smoking;

(b) areas are set aside where workers can eat and drink without risking contamination by asbestos dust;

(c) (i) workers are provided with appropriate working or protective clothing;

(ii) this working or protective clothing remains within the undertaking. It may, however, be laundered in establishments outside the undertaking which are equipped for this sort of work if the undertaking does not carry out the cleaning itself; in that event the clothing shall be transported in closed containers;

(iii) separate storage places are provided for working or protective clothing and for street clothes;

(iv) workers are provided with appropriate and adequate washing and toilet facilities, including showers in the case of dusty operations;

(v) protective equipment shall be placed in a well defined place and shall be checked and cleaned after each use; appropriate measures shall be taken to repair or replace defective equipment before further use.

2. Workers may not be charged with the cost of measures taken pursuant to paragraph 1.

Article 14

1. In the case of all activities referred to in Article 3 appropriate measures shall be taken to ensure that workers and their representatives in the undertaking or establishment receive adequate information concerning:

— the potential risks to health from exposure to dust arising from asbestos or materials containing asbestos,

— the existence of statutory limit values and the need for the atmosphere to be monitored,

— hygiene requirement, including the need to refrain from smoking,

— the precautions to be taken as regards the wearing and use of protective equipment and clothing,

— special precautions designed to minimize exposure to asbestos.

2. In addition to the measures referred to in paragraph 1, and subject to Article 3(3), appropriate measures shall be taken to ensure that:

(a) workers and/or their representatives in the undertaking or establishment have access to the results of asbestos-in-air concentration measurements and can be given explanations of the significance of those results;

(b) if the results exceed the limit values laid down in Article 8 the workers concerned and their representatives in the undertaking or establishment are informed as quickly as possible of the fact and the reason for it and the workers and/or their representatives in the undertaking or establishment are consulted on the measures to be taken or, in an emergency, are informed of the measures which have been taken.

Article 15

Subject to Article 3(3) the following measures shall be taken:

1. An assessment of each worker's state of health must be available prior to the beginning of exposure to dust arising from asbestos or materials containing asbestos at the place of work.

    This assessment must include a specific examination of the chest. Annex II gives practical recommendations to which the Member States may refer for the clinical surveillance of workers; these recommendations shall be adapted to technical progress in accordance with the procedure set out in Article 10 of Directive 80/1107/EEC.

    A new assessment must be available at least once every three years for as long as exposure continues.

    An individual health record shall be established in accordance with national laws and practices for each worker referred to in the first sub-paragraph.

2. Following the clinical surveillance referred to in point 1, the doctor or authority responsible for the medical surveillance of the workers should, in accordance with national laws, advise on or determine any individual protective or preventive measures to be taken; these may include, where appropriate, the withdrawal of the worker concerned from all exposure to asbestos.

3. Information and advice must be given to workers regarding any assessment of their health which they may undergo following the end of exposure.

4. The worker concerned or the employer may request a review of the assessments referred to in point 1 in accordance with national laws.

Article 16

Subject to Article 3(3) the following measures shall be taken:

1. The employer must enter the workers responsible for carrying out the activities referred to in Article 3(1) in a register, indicating the nature and duration of the activity and the exposure to which they have been subjected. The doctor and/or the authority responsible for medical surveillance shall have access to this register. Each worker shall have access to the results in the register which relate to him personally. The workers and/or their representatives shall have access to anonymous, collective information in the register.

2. The register referred to in point 1 and the medical records referred to in point 1 of Article 15 shall be kept for at least 30 years following the end of exposure, in accordance with national laws.

Article 17

Member States shall keep a register of recognized cases of asbestosis and mesothelioma.

Article 18

1. Member States shall adopt the laws, regulations and administrative provisions necessary to comply with this Directive before 1 January 1987. They shall forthwith inform the Commission thereof. The date 1 January 1987 is, however, postponed until 1 January 1990 in the case of asbestos-mining activities.

2. Member States shall communicate to the Commission the provisions of national law which they adopt in the field covered by this Directive.

Article 19

This Directive is addressed to the Member States.

Done at Brussels, 19 September 1983.

ANNEX I

Reference method referred to in Article 7(1) for the measurement of asbestos in air at the place of work

1. Samples shall be taken within the individual worker's breathing zone: *ie* within a hemisphere of 300 mm radius extending in front of the face and measured from the mid-point of a line joining the ears.

2. Membrane filters (mixed esters of cellulose or cellulose nitrate) of pore size 0·8 to 1.2 micrometres with printed squares and a diameter of 25 mm shall be used.

3. An open-faced filter holder fitted with a cylindrical cowl extending between 33 and 44 mm in front of the filter exposing a circular area of at least 20 mm in diameter shall be used. In use, the cowl shall point downwards.

4. A portable battery-operated pump carried on the worker's belt or in a pocket shall be used. The flow shall be smooth and the rate initially set at 1·0 litres per minute ±5 %. The flow rate shall be maintained within ±10 % of the initial rate during the sampling period.

5. The sampling time shall be measured to within a tolerance of 2%.

6. The optimal fibre-loading on filters shall be within the range 100 to 400 fibres/ mm².

7. In order of preference the whole filter, or a section of the filter, shall be placed on a microscope slide, made transparent using the acetone-triacetin method, and covered with a glass coverslip.

8. A binocular microscope shall be used for counting and shall have the following features:

   — Koehler illumination,

   — its substage assembly shall incorporate an Abbe or achromatic phase-contrast condenser in a centering focusing mount. The phase-contrast centering adjustment shall be independent of the condenser centering mechanism,

   — a 40 times bar-focal positive phase-contrast achromatic objective with a numerical aperture of 0·65 to 0·70 and phase ring absorption within the range 65 to 85 %,

   — 12.5 times compensating eyepieces; at least one eyepiece must permit the insertion of a graticule and be of the focusing type,

   — a Walton-Beckett circular eyepiece graticule with an apparent diameter in the object plane of 100 micrometres + 2 micrometres, when using the specified objective and eyepiece, checked against a stage micrometer.

9. The microscope shall be set up according to the manufacturer's instructions, and the detection limit checked using a 'phase-contrast test slide'. Up to code 5 on the AIA test slides or up to block 5 on the HSE/NPL mark 2 test slide must be visible when used in the way specified by the manufacturer. This procedure shall be carried out at the beginning of the day of use.

10. Samples shall be counted in accordance with the following rules:

— a countable fibre is any fibre referred to in the second sub-paragraph of point 1 of Article 7 which does not touch a particle with a maximum diameter greater than three micrometers,

— any countable fibre with both ends within the graticule area shall be counted as one fibre; any fibre with only one end within the area shall count as half,

— graticule areas for counting shall be chosen at random within the exposed area of the filter,

— an agglomerate of fibres which at one or more points on its length appears solid and undivided but at other points is divided into separate strands (a split fibre) is counted as a single fibre if it conforms with the description in the second sub-paragraph of point 1 of Article 7 and indent 1 of this paragraph, the diameter measured being that of the undivided part, not that of the split part,

— in any other agglomerate of fibres in which individual fibres touch or cross each other (a bundle), the fibres shall be counted individually if they can be distinguished sufficiently to determine that they conform with the description in the second sub-paragraph of point 1 of Article 7 and indent 1 of this paragraph. If no individual fibres meeting the definition can be distinguished, the bundle is considered to be a counuble fibre if, taken as a whole, it conforms with the description in the second sub-paragraph of point 1 of Article 7 and indent 1 of this paragraph,

— if more than one-eighth of a graticule area is covered by an agglomerate of fibres and/or particles, the graticule area must be rejected and another counted,

— 100 fibres shall be counted, which will enable a minimum of 20 graticule areas to be examined, or 100 graticule areas shall be examined.

11. The mean number of fibres per graticule is calculated by dividing the number of fibres counted by the number of graticule areas examined. The effect on the count of marks on the filter and contamination shall be kept below three fibres/100 graticule areas and shall be assessed using blank filters.

Concentration in air, (number per graticule area x exposed area of filter)/ graticule area x volume of air collected).

ANNEX II

Practical recommendations for the clinical assessment of workers, as referred to in Article 15(1)

1. Current knowledge indicates that exposure to free asbestos fibres can give rise to the following diseases:

   — asbestosis,

   — mesothelioma,

   — bronchial carcinoma,

   — gastro-intestinal carcinoma.

2. The doctor and/or authority responsible for the medical surveillance of workers exposed to asbestos must be familiar with the exposure conditions or circumstances of each worker.

3. Clinical surveillance of workers should be carried out in accordance with the principles and practices of occupational medicine; it should include at least the following measures:

   — keeping records of a worker's medical and occupational history,

   — a personal interview,

   — a clinical examination of the chest,

   — a respiratory function examination.

   Further examinations, including a standard format radiograph of the chest and laboratory tests such as a sputum cytology test, are desirable. These examinations should be decided upon for each worker when he is the subject of medical surveillance, in the light of the most recent knowledge available to occupational medicine.

COUNCIL DIRECTIVE

of 12 May 1986

on the protection of workers from the risks related to exposure to noise at work
[86/188/EEC]

THE COUNCIL OF THE EUROPEAN COMMUNITIES,

Having regard to the Treaty establishing the European Economic Community, and in particular Article 100 thereof,

Having regard to the proposal from the Commission, drawn up after consulting the Advisory Committee on Safety, Hygiene and Health Protection at Work;

Having regard to the opinion of the European Parliament,

Having regard to the opinion of the Economic and Social Committee,

Whereas the Council resolutions of 29 June 1978 and 27 February 1984 on action programmes of the European Communities on safety and health at work provide for the implementation of specific harmonized procedures for the protection of workers exposed to noise; whereas the measures adopted in this field vary from State to State and it is recognized that they urgently need to be approximated and improved;

Whereas exposure to high noise levels is encountered in a large number of situations and therefore many workers are exposed to a potential safety and health hazard;

Whereas a reduction of exposure to noise reduces the risk of hearing impairment caused by noise;

Whereas, where the noise level at the workplace involves a risk for the health and safety of workers, limiting exposure to noise reduces that risk without prejudice to the applicable provisions on the limitation of noise emission;

Whereas the most effective way of reducing noise levels at work is to incorporate noise prevention measures into the design of installations and to choose materials, procedures and working methods which produce less noise; whereas the priority aim must be to achieve the said reduction at source;

Whereas the provision and use of personal ear protectors is a necessary complementary measure to the reduction of noise at source, where exposure cannot reasonably be avoided by other means;

Whereas noise is an agent to which Council Directive 80/1107/EEC of 27 November 1980 on the protection of workers from the risks related to exposure to chemical, physical and biological agents at work applies; whereas Articles 3 and 4 of the said Directive provide for the possibility of laying down limit values and other special measures in respect of the agents being considered;

Whereas certain technical aspects must be defined and may be reviewed in the light of experience and progress made in the technical and scientific field;

Whereas the current situation in the Member States does not make it possible to fix a noise-exposure value below which there is no longer any risk to workers' hearing;

Whereas current scientific knowledge about the effects that exposure to noise may have on health, other than on hearing, does not enable precise safety levels to be set; whereas, however, reduction of noise will lower the risk of illnesses unrelated to auditory complaints; whereas this Directive contains provisions which will be reviewed in the light of experience and developments in scientific and technical knowledge in this field,

HAS ADOPTED THIS DIRECTIVE:

Article 1

1.   This Directive, which is the third individual Directive within the meaning of Directive 80/1107/EEC, has as its aim the protection of workers against risks to their hearing and, in so far as this Directive expressly so provides, to their health and safety, including the prevention of such risks arising or likely to arise from exposure to noise at work.

2.   This Directive shall apply to all workers, including those exposed to radiation covered by the scope of the EAEC Treaty, with the exception of workers engaged in sea transport and in air transport.

For the purpose of this Directive, the expression 'workers engaged in sea transport and in air transport shall refer to personnel on board.

On a proposal from the Commission the Council shall examine, before 1 January 1990, the possibility of applying this Directive to workers engaged in sea transport and in air transport.

3.   This Directive shall not prejudice the right of Member States to apply or introduce, subject to compliance with the Treaty, laws, regulations or administrative provisions ensuring, where possible, greater protection for workers and/or intended to reduce the level of noise experienced at work by taking action at source, particularly in order to achieve exposure values which prevent unnecessary nuisance.

Article 2

For the purposes of this Directive, the following terms shall have the meaning hereby assigned to them:

1.   *Daily personal noise exposure of a worker* $L_{EP,d}$

The daily personal noise exposure of a worker is expressed in dB (A) using the formula:

$$L_{EP,d} = L_{Aeq,Te} + 10 \log_{10} \frac{T_e}{T_o}$$

where:

$$L_{Aeq,Te} = 10 \log_{10} \left\{ \frac{1}{T_e} \int_o^{T_e} \left[\frac{p_A(t)}{p_o} \right]^2 dt \right\}$$

Te = daily duration of a worker's personal exposure to noise,

To = 8 h = 28 800 s,

Po = 20 UPa,

PA = 'A'-weighted instantaneous sound pressure in pascals to which is exposed, in air at atmospheric pressure, a person who might or might not move from one place to another while at work; it is determined from measurements made at the position occupied by the person's ears during work, preferably in the person's absence, using a technique which minimizes the effect on the sound field.

If the microphone has to be located *very close* to the person's body, appropriate adjustments should be made to determine an equivalent undisturbed field pressure.

The daily personal noise exposure does not take account of the effect of any personal ear protector used.

2. *Weekly average of the daily values* $L_{EP, w}$

The weekly average of the daily values is found using the following formula:

$$L_{EP,w} = 10 \log_{10} \left[\tfrac{1}{5} \sum_{k=1}^{m} 10^{0.1(L_{EP, d})k} \right]$$

where $(L_{EP, d})k$ are the values of $L_{EP, d}$ for each of the m working days in the week being considered.

Article 3

1. Noise experienced at work shall be assessed and, when necessary, measured in order to identify the workers and workplaces referred to in this Directive and to determine the conditions under which the specific provisions of this Directive shall apply.

2. The assessment and measurement mentioned in paragraph 1 shall be competently planned and carried out at suitable intervals under the responsibility of the employers.

Any sampling must be representative of the daily personal exposure of a worker to noise.

The methods and apparatus used must be adapted to the prevailing conditions in the light, particularly, of the characteristics of the noise to be measured, the length of exposure, ambient factors and the characteristics of the measuring apparatus.

These methods and this apparatus shall make it possible to determine the parameters defined in Article 2 and to decide whether, in a given case, the values fixed in this Directive have been exceeded.

3. Member States may lay down that personal exposure to noise shall be replaced by noise recorded at the workplace. In that event the criterion of personal exposure to noise shall be replaced, for the purposes of Articles 4 to 10, by that of noise exposure during the daily work period, such period being at least eight hours, at the places where the workers are situated.

Member States may also lay down that, when the noise is measured, special consideration shall be given to impulse noise.

4. The workers and/or their representatives in the undertaking or establishment shall be associated, according to national law and practice, with the assessment and measurement provided for in paragraph 1. These shall be revised where there is reason to believe that they are incorrect or that a material change has taken place in the work.

5. The recording and preservation of the data obtained pursuant to this Article shall be carried out in a suitable form, in accordance with national law and practice.

    The doctor and/or the authority responsible and the workers and/or their representatives in the undertaking shall have access to these data, in accordance with national law and practice.

Article 4

1. Where the daily personal exposure of a worker to noise is likely to exceed 85 dB (A) or the maximum value of the unweighted instantaneous sound pressure is likely to be greater than 200 Pa, appropriate measures shall be taken to ensure that:

    (a) workers and/or their representatives in the undertaking or establishment receive adequate information and, when relevant, training concerning:

    —   potential risks to their hearing arising from noise exposure,

    —   the measures taken in pursuance of this Directive,

    —   the obligation to comply with protective and preventive measures, in accordance with national legislation,

    —   the wearing of personal ear protectors and the role of checks on hearing in accordance with Article 7;

    (b) workers and/or their representatives in the undertaking or establishment have access to the results of noise assessments and measurements made pursuant to Article 3 and can be given explanations of the significance of those results.

2. At workplaces where the daily personal noise exposure of a worker is likely to exceed 85 dB (A), appropriate information must be provided to workers as to where and when Article 6 applies.

    At workplaces where the daily personal noise exposure of a worker is likely to exceed 90 dB (A) or where the maximum value of the unweighted instantaneous sound pressure is likely to exceed 200 Pa, the information provided for in the first sub-paragraph must, where reasonably practicable, take the form of appropriate signs. The areas in question must also be delimited and access to them must be restricted, where the risk of exposure so justifies and where these measures are reasonably practicable.

Article 5

1. The risks resulting from exposure to noise must be reduced to the lowest level reasonably practicable, taking account of technical progress and the availability of measures to control the noise, in particular at source.

2. Where the daily personal noise exposure of a worker exceeds 90 dB (A), or the maximum value of the unweighted instantaneous sound pressure is greater than 200 Pa:

(a) the reasons for the excess level shall be identified and the employer shall draw up and apply a programme of measures of a technical nature and/or of organization of work with a view to reducing as far as reasonably practicable the exposure of workers to noise;

(b) workers and their representatives in the undertaking or establishment shall receive adequate information on the excess level and on the measures taken pursuant to sub-paragraph (a).

Article 6

1. Without prejudice to Article 5, where the daily personal noise exposure of a worker exceeds 90 dB (A) or the maximum value of the unweighted instantaneous sound pressure is greater than 200 Pa, personal ear protectors must be used.

2. Where the exposure referred to in paragraph 1 is likely to exceed 85 dB (A), personal ear protectors must be made available to workers.

3. Personal ear protectors must be supplied in sufficient numbers by the employer, the models being chosen in association, according to national law and practice, with the workers concerned.

The ear protectors must be adapted to the individual worker and to his working conditions, taking account of his safety and health. They are deemed, for the purposes of this Directive, suitable and adequate if, when properly worn, the risk to hearing can reasonably be expected to be kept below the risk arising from the exposure referred to in paragraph 1.

4. Where application of this Article involves a risk of accident, such risk must be reduced as far as is reasonably practicable by means of appropriate measures.

Article 7

1. Where it is not reasonably practicable to reduce the daily personal noise exposure of a worker to below 85 dB (A), the worker exposed shall be able to have his hearing checked by a doctor or on the responsibility of the doctor and, if judged necessary by the doctor, by a specialist.

The manner in which this check is carried out shall be established by the Member States in accordance with national law and practice.

2. The purpose of the check shall be the diagnosis of any hearing impairment by noise and the preservation of hearing.

3. The results of checks on workers' hearing shall be kept in accordance with national law and practice.

Workers shall have access to the results which apply to them in so far as national law and practice allow.

4. Member States shall take the necessary measures with a view to the doctor and/or the authority responsible giving, as part of the check, appropriate indications on any individual protective or preventive measures to be taken.

Article 8

1. Member States shall take appropriate measures to ensure that:

(a) the design, building and/or construction of new plant (new factories, plant or machinery, substantial extensions or modifications to existing factories or plant and replacement of plant or machinery) comply with Article 5 (1);

(b) where a new article (tool, machine, apparatus, *etc.*) which is intended for use at work is likely to cause, for a worker who uses it properly for a conventional eight-hour period, a daily personal noise exposure equal to or greater than 85 dB (A) or an unweighted instantaneous sound pressure the maximum value of which is equal to or greater than 200 Pa, adequate information is made available about the noise produced in conditions of use to be specified.

2. The Council shall establish, on a proposal from the Commission, requirements according to which, so far as is reasonably practicable, the articles referred to in paragraph 1(b), when properly used, do not produce noise likely to constitute a risk to hearing.

Article 9

1. In the case of workplaces where the noise exposure of a worker varies markedly from one working day to the next, Member States may, for workers performing special operations, exceptionally grant derogations from Article 5(2), Article 6(1) and Article 7(1), but only on condition that the average weekly noise exposure of a worker, as shown by adequate monitoring, complies with the value laid down in these provisions.

2. (a) In exceptional situations where it is not reasonably practicable, by technical measures or organization of work, to reduce daily personal noise exposure to below 90 dB (A) or to ensure that the personal ear protectors provided for in Article 6 of this Directive are suitable and adequate within the meaning of the second sub-paragraph of Article 6(3), the Member States may grant derogations from this provision for limited periods, such derogations being renewable.

In such a case, however, personal ear protectors affording the highest degree of protection which is reasonably practicable must be used.

(b) In addition, for workers performing special operations, Member States may exceptionally grant derogations from Article 6(1) if its application involves an increase in the overall risk to the health and/or safety of the workers concerned and if it is not reasonably practicable to reduce this risk by any other means.

(c) The derogations referred to in (a) and (b) shall be subject to conditions which, in view of the individual circumstances, ensure that the risks resulting from such derogations are reduced to a minimum. The derogations shall be reviewed periodically and be revoked as soon as is reasonably practicable.

(d) Member States shall forward to the Commission every two years an adequate overall account of the derogations referred to in (a) and (b). The Commission shall inform the Member States thereof in an appropriate manner.

Article 10

The Council, acting on a proposal from the Commission, shall re-examine this Directive before 1 January 1994, taking into account in particular progress made in scientific knowledge and technology as well as experience gained in the application of this Directive, with a view to reducing the risks arising from exposure to noise.

In the context of this re-examination, the Council, acting on a proposal from the Commission, shall endeavour to lay down indications for measuring noise which are more precise than those given in Annex I.

Article 11

Member States shall see to it that workers' and employers' organizations are consulted before the provisions for the implementation of the measures referred to in this Directive are adopted, and that where workers' representatives exist in the undertaking or establishments they can check that such provisions are applied or can be involved in their application.

Article 12

1. For the measurement of noise and checking workers' hearing, any methods may be used which at least satisfy the provisions contained in Articles 3 and 7.

2. Indications for measuring noise and for checking workers' hearing are given in Annexes I and II.

    Annexes I and II shall be adapted to technical progress in accordance with Directive 80/1107/EEC and under the procedure set out in Article 10 thereof.

Article 13

1. Member States shall bring into force the laws, regulations and administrative provisions necessary to comply with this Directive by 1 January 1990. They shall forthwith inform the Commission thereof.

    However, in the case of the Hellenic Republic and the Portuguese Republic the relevant date shall be 1 January 1991.

2. Member States shall communicate to the Commission the provisions of national law which they adopt in the field covered by this Directive. The Commission shall inform the other Member States thereof.

Article 14

This Directive is addressed to the Member States.

Done at Brussels, 12 May 1986.

ANNEX I

INDICATIONS FOR MEASURING NOISE

A.1. General

The quantities defined in Article 2 can be either:

    (i)   measured directly by integrating sonometers, or

    (ii)   calculated from measurements of sound pressure and exposure duration.

Measurements may be made at the work place(s) occupied by workers, or by using instruments attached to the person.

The location and duration of the measurements must be sufficient to ensure that exposure to noise during the working day can be recorded.

2. Instrumentation

2.1.   If integrating averaging sonometers are used, they shall comply with IEC standard 804.

If sonometers are used, they shall comply with IEC standard 651. Instruments incorporating an overload indication are preferred.

If data are stored on tape as an intermediate step of the measurement procedure, potential errors caused by the process of sorting and replay shall be taken into account when analyzing the data.

2.2.   An instrument used to measure directly the maximum (peak) value of the unweighted instantaneous sound pressure shall have an onset time constant not exceeding 100 μs.

2.3.   All equipment shall be calibrated in a laboratory at suitable intervals.

3. Measurement

3.1.   An on-site check shall be made at the beginning and end of each day of measurement.

3.2.   Measurement of workplace sound pressure should preferably be made in the undisturbed sound field in the workplace (*i.e.* with the person concerned being absent) and with the microphone located at the position (s) normally occupied by the ear exposed to the highest value of exposure.

If it is necessary for the person to be present, either:

    (i)   the microphone should be located at a distance from the person's head which will reduce, as far as possible, the effects of diffraction and distance on the measured value (a suitable distance is 0·10 m), or

    (ii)   if the microphone must be located very close to the person's body, appropriate adjustments should be made to determine an equivalent undisturbed pressure field.

3.3. Generally, time weightings 'S' and 'F' are valid as long as the measurement time interval is long compared with the time constant of the weighting chosen, but they are not suitable for determining L_{Aeq}, T_e when the noise level fluctuates very rapidly.

3.4. *Indirect measurement of exposure*

The result of the direct measurement of L_{Aeq}, T_e can be approximated with a knowledge of the exposure time and the measurement of clearly distinguishable sound-pressure-level ranges; a sampling method and a statistical distribution may be useful.

4. Accuracy of measuring noise and determining the exposure

The type of the instrument and the standard deviation of the results influence the accuracy of measurement. For comparison with a noise limit, the measuring accuracy determines the range of readings where no decision can be made as to whether the value is exceeded; if no decision can be taken, the measurement must be repeated with a higher accuracy.

Measurements of the highest accuracy enable a decision to be taken in all cases.

B. Short-term measurements with ordinary sonometers are quite satisfactory for workers performing, at a fixed location, repetitive activities which generate roughly the same levels of broad-band noise throughout the day. But when the sound pressure to which a worker is exposed shows fluctuations spread over a wide range of levels and/or of irregular time characteristics, determining the daily personal noise exposure of a worker becomes increasingly complex; the most accurate method of measurement is therefore to monitor exposure throughout the entire shift, using an integrating averaging sonometer.

When an integrating averaging sonometer conforming to IEC standard 804 (which is well suited for measurement of the equivalent continuous sound pressure level of impulse noise) complies at least with the specifications of type I and has recently been fully calibrated in a laboratory, and the microphone is properly located (see 3.2 above), the results make it possible, with certain exceptions, to determine whether a given exposure has been exceeded (see 4) even in complex situations; that method is thus generally applicable, and is well suited for reference purposes.

ANNEX II

INDICATIONS FOR CHECKING WORKERS' HEARING

In the framework of checking workers' hearing the following points are taken into consideration:

1. The check should be carried out in accordance with occupational medical practice and should comprise:

    —   where appropriate, an initial examination, to be carried out before or at the beginning of exposure to noise,

    —   regular examinations at intervals which are commensurate with the seriousness of the risk and are determined by the doctor.

2. Each examination should consist of at least an otoscopy combined with an audiometric test including pure-tone air conduction threshold audiometry in accordance with 6 below.

3. The initial examination should include a medical history; the initial otoscopy and the audiometric test should be repeated within a period of 12 months.

4. The regular examination should be carried out at least every five years where the worker's daily personal noise exposure remains less than 90 dB (A).

5. The examinations should be carried out by suitably qualified persons in accordance with national law and practice and may be organized in successive stages (screening, specialist examination).

6. The audiometric test should comply with the specifications of ISO standard 6189-1983, supplemented as follows:

    Audiometry also covers the frequency of 8 000 Hz; the ambient sound level enables a hearing threshold level equal to 0 dB in relation to ISO standard 389-1975 to be measured.

    However, other methods may be used if they give comparable results.

COUNCIL DIRECTIVE

of 9 June 1988

on the protection of workers by the banning of certain specified agents and/or
certain work activities

(fourth individual Directive within the meaning of Article 8 of Directive 80/1107/EEC)

[88/364/EEC]

THE COUNCIL OF THE EUROPEAN COMMUNITIES,

Having regard to the Treaty establishing the European Economic Community,
and in particular Article 118a thereof,

Having regard to the proposal from the Commission,

In co-operation with the European Parliament,

Having regard to the opinion of the Economic and Social Committee,

Whereas the Council adopts, by means of Directives, minimum progressively
applicable provisions, with a view towards promoting the improvement, in particular, of the working environment, so as to protect the safety and health of workers;

Whereas the Council resolution of 27 February 1984 on a second programme of
action of the European Communities on safety and health at work provides for the
development of protective measures for substances recognized as being carcinogenic and other dangerous substances and processes which may have serious
harmful effects on health;

Whereas certain differences are revealed by an examination of the measures taken
by Member States to protect workers against the risks related to exposure to
specified work agents and work activities; whereas, therefore, in the interest of
balanced development, these measures should be harmonized and improved as
progress is made; whereas this harmonization and improvement should be based
on common principles;

Whereas, to this end, Council Directive 80/1107/EEC of 27 November 1980 on the
protection of workers from the risks related to exposure to chemical, physical and
biological agents at work contains such principles;

Whereas, under the terms of the said Directive, such protection must as far as
possible be ensured by measures to prevent exposure or to keep it at as low a level
as is reasonably practicable; whereas, also under these terms, for the provision of
adequate protection of workers it is necessary to ban in the workplace certain
specified agents and/or work activities which can give rise to serious effects on
health in cases where use of other means does not make it possible to ensure adequate protection;

Whereas provision should be made in these circumstances to ban certain specified
agents and/or certain work activities in the workplace, subject to certain exceptions and derogations;

Whereas representatives of employers and workers have a role to play in the protection of workers;

Whereas these principles need to be applied uniformly and speedily to encourage wherever possible the early development of alternative non-dangerous agents and/or work activities,

HAS ADOPTED THIS DIRECTIVE:

Article 1

1. The purpose of this Directive is to protect workers against risks to their health by means of a ban on certain specific agents and or certain work activities.

   The ban which is the subject of this Directive including the Annex is based on the following factors:

   — there are serious health and safety risks for workers,

   — precautions are not sufficient to ensure a satisfactory level of health and safety protection for workers,

   — the ban does not lead to the use of substitute products which may involve equal or greater health and safety risks for workers.

2. This Directive shall not apply to:

   — sea transport,

   — air transport.

3. This Directive shall not prejudice the right of Member States to apply or introduce, subject to compliance with the Treaty, laws, regulations or administrative provisions ensuring greater protection for workers.

Article 2

For the purposes of this Directive:

(a) 'substances' means chemical elements and their compounds as they occur in the natural state or as produced by industry, including any additives required for the purpose of placing them on the market;

(b) 'agents' means any chemical, physical or biological agents present at work and likely to be harmful to health;

(c) 'preparations' means mixtures or solutions composed of two or more substances;

(d) 'impurities' means substances which are *a priori* present in insignificant amounts in other substances;

(e) 'intermediates' means substances which are formed during a chemical reaction, are converted and therefore disappear by the end of the reaction or process;

(f) 'by-products' means substances which are formed during a chemical reaction and which remain at the end of the reaction or process;

(g) 'waste products' means the remains of a chemical reaction which need to be disposed of at the end of the reaction or process.

Article 3

1. To prevent the exposure of workers to health risks from certain specific agents and/or certain work activities in the cases referred to in Article 1, Member States shall impose a ban in accordance with the procedures laid down in the Annex.

2. The Council, acting by a qualified majority on a proposal from the Commission, in co-operation with the European Parliament and after consulting the Economic and Social Committee, may amend the Annex, in particular to include further agents or activities.

Article 4

In the case of the derogations provided for in the Annex, the Member States shall be obliged to ensure that employers comply with the following procedures and measures:

(a) an employer must take adequate precautions to protect the health and safety of the workers concerned; and

(b) an employer must submit at least the following information to the competent authority:

— the quantities used annually,

— the activities and/or reactions or processes involved,

— the number of workers exposed,

— the technical and organizational measures taken to prevent the exposure of workers.

In addition, the Member States may provide for systems of individual authorizations.

Article 5

1. Workers and/or their representatives in undertakings or establishments shall have access, in accordance with national law, to the documents submitted pursuant to Article 4 in regard to their undertaking or establishment.

2. The documents referred to in paragraph 1 shall contain the information necessary to ensure that workers and/or their representatives in undertakings or establishments are made fully aware of the health and safety risks connected with the agent or work activity to which they are or are likely to be exposed, together with the measures to be taken against such risks.

Article 6

1. Before 1 January 1995 the Commission shall submit to the European Parliament, the Council and the Economic and Social Committee a report concerning in particular experience gained in the application of this Directive and progress in scientific knowledge and technology.

2. The Council shall re-examine this Directive before 1 January 1996 on the basis of the report referred to in paragraph 1.

Article 7

1. Member States shall adopt the laws, regulations and administrative provisions necessary to comply with this Directive by 1 January 1990 at the latest. They shall forthwith inform the Commission thereof.

2. Member States shall communicate to the Commission the provisions of national law which they adopt in the field governed by this Directive.

Article 8

This Directive is addressed to the Member States.

Done at Luxembourg, 9 June 1988.

ANNEX

1. Subject to the conditions listed below, the following may not be produced or used:

    — 2-naphthylamine and its salts   (CAS No 91-59-8),

    — 4-aminobiphenyl and its salts   (CAS No 92-67-1),

    — benzidine and its salts   (CAS No 92-87-5),

    — 4-nitrodiphenyl   (CAS No 92-93-3).

2. This ban does not apply if the agents are present in a substance or a preparation in the form of impurities or by-products, or as a constituent of waste products, provided that their individual concentration therein is less than 0·1 % w/w.

3. Derogations from point 1 laid down by the Member States shall only be permitted:

    — for the sole purpose of scientific research and testing, including analysis,

    — for work activities intended to eliminate the agents that are present in the form of by-products or waste products,

    — for the production of the substances referred to in paragraph 1 for use as intermediates, and for such use.

4. The exposure of workers to the substances referred to in paragraph 1 must be prevented, in particular by providing that the production and earliest possible use of these substances as intermediates must take place in a single closed system, from which the aforesaid substances may be removed only to the extent necessary to monitor the process or service the system.

COUNCIL DIRECTIVE

of 12 June 1989

on the introduction of measures to encourage improvements in the safety
and health of workers at work
[89/391/EEC]

THE COUNCIL OF THE EUROPEAN COMMUNITIES,

Having regard to the Treaty establishing the European Economic Community,
and in particular Article 118a thereof,

Having regard to the proposal from the Commission, drawn up after consultation
with the Advisory Committee on Safety, Hygiene and Health Protection at Work,
in co-operation with the European Parliament,

Having regard to the opinion of the Economic and Social Committee,

Whereas Article 118a of the Treaty provides that the Council shall adopt, by means
of directives, minimum requirements for encouraging improvements, especially
in the working environment, to guarantee a better level of protection of the safety
and health of workers;

Whereas this Directive does not justify any reduction in levels of protection already
achieved in individual Member States, the Member States being committed, under
the Treaty, to encouraging improvements in conditions in this area and to
harmonizing conditions while maintaining the improvements made;

Whereas it is known that workers can be exposed to the effects of dangerous environmental factors at the workplace during the course of their working life;

Whereas, pursuant to Article 118a of the Treaty, such directives must avoid
imposing administrative, financial and legal constraints which would hold back the
creations and development of small and medium-sized undertakings;

Whereas the communication from the Commission on its programme concerning
safety, hygiene and health at work provides for the adoption of directives designed
to guarantee the safety and health of workers;

Whereas the Council, in its resolution of 21 December 1987, on safety, hygiene and
health at work, took note of the Commission's intention to submit to the Council
in the near future a directive on the organization of the safety and health of workers
at the workplace;

Whereas in February 1988, the European Parliament adopted four resolutions
following the debate on the internal market and worker protection; whereas these
resolutions specifically invited the Commission to draw up a framework directive
to serve as a basis for more specific directives covering all the risks connected with
safety and health at the workplace;

Whereas Member States have a responsibility to encourage improvements in the safety and health of workers on their territory; whereas taking measures to protect the health and safety of workers at work also helps, in certain cases, to preserve the health and possibly the safety of persons residing with them;

Whereas Member States' legislative systems covering safety and health at the workplace differ widely and need to be improved; whereas national provisions on the subject, which often include technical specifications and/or self-regulatory standards, may result in different levels of safety and health protection and allow competition at the expense of safety and health;

Whereas the incidence of accidents at work and occupational diseases is still too high; whereas preventive measures must be introduced or improved without delay in order to safeguard the safety and health of workers and ensure a higher degree of protection;

Whereas, in order to ensure an improved degree of protection, workers and/or their representatives must be informed of the risks to their safety and health and of the measures required to reduce or eliminate these risks; whereas they must also be in a position to contribute, by means of balanced participation in accordance with national laws and/or practices, to seeing that the necessary protective measures are taken;

Whereas information, dialogue and balanced participation on safety and health at work must be developed between employers and workers and/or their representatives by means of appropriate procedures and instruments, in accordance with national laws and/or practices;

Whereas the improvement of workers' safety, hygiene and health at work is an objective which should not be subordinated to purely economic considerations;

Whereas employers shall be obliged to keep themselves informed of the latest advances in technology and scientific findings concerning workplace design, account being taken of the inherent dangers in their undertaking, and to inform accordingly the workers' representatives exercising participation rights under this Directive, so as to be able to guarantee a better level of protection of workers' health and safety;

Whereas the provisions of this Directive apply, without prejudice to more stringent present or future Community provisions, to all risks, and in particular to those arising from the use at work of chemical, physical and biological agents covered by Directive 80/1107/EEC, as last amended by Directive 88/642/EEC;

Whereas, pursuant to Decision 74/325/EEC, the Advisory Committee on Safety, Hygiene and Health Protection at Work is consulted by the Commission on the drafting of proposals in this field;

Whereas a committee composed of members nominated by the Member States needs to be set up to assist the Commission in making the technical adaptations to the individual directives provided for in this Directive,

HAS ADOPTED THIS DIRECTIVE:

SECTION I
General provisions

Article 1
Object

1. The object of this Directive is to introduce measures to encourage improvements in the safety and health of workers at work.

2. To that end it contains general principles concerning the prevention of occupational risks, the protection of safety and health, the elimination of risk and accident factors, the informing, consultation, balanced participation in accordance with national laws and/or practices and training of workers and their representatives, as well as general guidelines for the implementation of the said principles.

3. This Directive shall be without prejudice to existing or future national and Community provisions which are more favourable to protection of the safety and health of workers at work.

Article 2
Scope

1. This Directive shall apply to all sectors of activity, both public and private (industrial, agricultural, commercial, administrative, service, educational, cultural, leisure, *etc.*).

2. This Directive shall not be applicable where characteristics peculiar to certain specific public service activities, such as the armed forces or the police, or to certain specific activities in the civil protection services inevitably conflict with it.

   In that event, the safety and health of workers must be ensured as far as possible in the light of the objectives of this Directives.

Article 3
Definitions

For the purposes of this Directive, the following terms shall have the following meanings:

   (a) worker: any person employed by an employer, including trainees and apprentices but excluding domestic servants;

   (b) employer: any natural or legal person who has an employment relationship with the worker and has responsibility for the undertaking and/or establishment;

   (c) workers' representative with specific responsibility for the safety and health of workers: any person elected, chosen or designated in accordance with national laws and/or practices to represent workers where problems arise relating to the safety and health protection of workers at work.

(d) prevention: all the provisions or measures taken or provided for at each stage of the activities performed within the undertaking with a view to avoiding or reducing the occupational risks.

(e) occupational risk: any work-related situation liable to damage the physical or psychological safety and/or health of the worker, excluding accidents on the way to and from work.

Article 4

1. Member States shall take the necessary steps to ensure that employers, workers and workers' representatives are subject to the legal provisions necessary for the implementation of this Directive.

2. In particular, Member States shall ensure adequate controls and supervision.

SECTION II
EMPLOYERS' OBLIGATIONS

Article 5
General provision

1. The employer shall have a duty to ensure the safety and health of workers in every aspect related to the work.

2. Where, pursuant to Article 7 (3), an employer enlists competent external services or persons, this shall not discharge him from his responsibilities in this area.

3. The workers' obligations in the field of safety and health at work shall not affect the principle of the responsibility of the employer.

4. This Directive shall not restrict the possibility of Member States to provide for the exclusion or the limitation of employers' responsibility where facts are due to circumstances unknown to them, exceptional and unforeseeable, or to exceptional events, the consequences of which could not have been avoided despite the exercise of all due care.

    Member States need not exercise the right referred to in the first sub-paragraph.

Article 6
General obligations on employers

1. Within the context of his responsibilities, the employer shall take the measures necessary for the safety and health protection of workers, including prevention of occupational risks and provision of information and training, as well as provision of the necessary organization and means.

    The employer shall be alert to the need to adjust these measures to take account of changing circumstances and the aim to improve existing situations.

2. The employer shall implement the measures referred to in the first sub-paragraph of paragraph 1 on the basis of the following general principles of prevention:

    (a) avoiding risks;

    (b) evaluating the risks which cannot be avoided;

    (c) combating the risks at source;

    (d) adapting the work to the individual, especially as regards the design of workplaces, the choice of work equipment and the choice of working and production methods, with a view, in particular, to alleviating monotonous work and work at a pre-determined work-rate, thus reducing the effects on health;

    (e) adapting to technical progress;

    (f) replacing the dangerous by the non-dangerous or the less dangerous;

    (g) developing a coherent overall prevention policy which covers technology, organization of work, working conditions, social relationships and the influence of factors related to the working environment;

    (h) giving collective protective measures priority over individual protective measures;

    (i) giving appropriate instructions to the workers.

3. Without prejudice to the other provisions of this Directive, the employer shall, taking into account the nature of the activities of the enterprise and/or establishment:

    (a) evaluate the risks to the safety and health of workers, *inter alia* in the choice of work equipment, the chemical substances or preparations used, and the fitting out of workplaces.

    Subsequent to this evaluation and as necessary, the preventive measures and the working and production methods implemented by the employer must:

    — assure an improvement in the level of protection afforded to workers with regard to safety and health,

    — be integrated into all the activities of the undertaking and/or establishment and at all hierarchical levels;

    (b) where he entrusts tasks to a worker, take into account the capabilities and, where appropriate, the handicaps of the worker concerned as regards health and safety;

    (c) ensure that the planning and introduction of new technologies are the subject of consultation with the workers and/or their representatives, as regards the consequences of the choice of equipment, the working conditions and the working environment for the safety and health of workers;

4. Without prejudice to the other provisions of this Directive, where several undertakings share a workplace, the employers shall co-operate in implementing the safety, health and occupational hygiene provisions and, taking into account the nature of

the activities, shall co-ordinate their actions in matters of the protection and prevention of occupational risks, and shall inform one another and their respective workers and/or workers' representatives of these risks.

5. Measures related to safety, hygiene and health at work may in no circumstances involve the workers in financial cost.

Article 7
Protective and preventive services

1. Without prejudice to the obligations referred to in Articles 5 and 6, the employer shall designate with the balanced participation of the workers and/or their representatives, in accordance with national legislation and/or practices, one or more workers to carry out activities related to the protection and prevention of occupational risks for the undertaking and/or establishment.

2. Designated workers may not be placed at any disadvantage whatsoever, also as regards their career in the enterprise, because of their activities related to protection and/or the prevention of occupational risks.

   Designated workers shall be allowed adequate time and the necessary means to enable them to fulfil their obligations arising from this Directive.

3. If such protective and preventive measures cannot be organized for lack of competent personnel in the undertaking and/or establishment, the employer shall enlist competent external senices or persons after consulting the workers, as referred to in paragraph 1.

4. Where the employer enlists such services or persons, he shall inform them of the factors known to affect, or suspected of affecting, the safety and health of the workers and they must have access to the information referred to in Article 10(2).

5. In all cases the workers designated from within the undertaking must have the necessary capabilities and the external services or persons consulted must have the necessary skills, staff and occupational resources and be sufficient in number to deal with the organization of protective and preventive measures, taking into account the size of the undertaking and/or establishment, the hazards to which the workers are exposed and their distribution throughout the entire undertaking and/or establishment.

6. The protection from, and prevention of, the health and safety risks which form the subject of this Article shall be the responsibility of one or more workers, of one service of of separate services whether from inside or outside the undertaking and/or establishment.

   The worker(s) and/or agency(ies) must work together whenever necessary.

7. Member States may define, in the light of the nature of the activities and size of the undertakings, the categories of undertakings in which the employer, provided he is competent, may himself take responsibility for the measures referred to in paragraph 1.

8.   Member States shall define the necessary capabilities referred to in paragraph 5.

They may determine the sufficient number referred to in paragraph 5.

Article 8
First-aid, fire-fighting and evacuation of workers, serious and imminent danger

1.   The employer shall:

—   take the necessary measures for first-aid, fire-fighting and evacuation of workers, adapted to the nature of the activities and the size of the undertaking and/or establishment and taking into account other persons present,

—   arrange any necessary contacts with external services, particularly as regards first-aid, emergency medical care, rescue work and fire-fighting.

2.   Pursuant to paragraph 1, the employer shall, *inter alia* for first-aid, fire-fighting and the evacuation of workers, designate the workers required to implement such measures.

The number of such workers, their training and the equipment available to them shall be adequate, taking account of the size and/or specific hazards of the undertaking and/or establishment.

3.   The employer shall:

(a)   as soon as possible, inform all workers who are, or may be, exposed to serious and imminent danger of the risk involved and of the steps taken or to be taken as regards protection;

(b)   take action and give instructions to enable workers in the event of serious, imminent and unavoidable danger to stop work and/or immediately to leave the workplace and proceed to a place of safety;

(c)   save in exceptional cases for reasons duly substantiated, refrain from asking workers to resume work in a working situation where there is still a serious and imminent danger.

4.   Workers who, in the event of serious, imminent and unavoidable danger, leave their workstation and/or a dangerous area may not be placed at any disadvantage because of their action and must be protected against any harmful and unjustified consequences, in accordance with national laws and/or practices.

5.   The employer shall ensure that all workers are able, in the event of serious and imminent danger to their own safety and/or that of other persons, and where the immediate superior responsible cannot be contacted, to take the appropriate steps, in the light of their knowledge and the technical means at their disposal, to avoid the consequences of such danger.

Their actions shall not place them at any disadvantage, unless they acted carelessly or there was negligence on their part.

Article 9
Various obligations on employers

1. The employer shall:

    (a) be in possession of an assessment of the risks to safety and health at work and of the situation of groups of workers who are exposed to particular risks;

    (b) decide on the protective measures to be taken and, if necessary, the protective equipment to be used;

    (c) draw up a list of accidents which resulted in a worker being unfit for work for more than three working days, and a list of occupational illnesses contracted;

    (d) draw up, for the responsible authorities and in accordance with national laws and/or practices, reports on occupational accidents and occupational illnesses suffered by his workers.

2. Member States shall define, in the light of the nature of the activities and size of the undertakings, the obligations to be met by the different categories of undertakings in respect of the drawing-up of the documents provided for in paragraph 1(a) and (b) and when preparing the documents provided for in paragraph 1(c) and (d).

Article 10
Information of workers

1. The employer shall take appropriate measures so that workers and/or their representatives in the undertaking and/or establishment receive, in accordance with national laws and/or practices which may take account *inter alia* of the size of the undertaking and/or establishment, all necessary information concerning:

    (a) the safety and health risks and protective and preventive measures and activities in respect of both the undertaking and/or establishment in general and each type of workstation and/or job;

    (b) the measures taken pursuant to Article 8(2).

    (c) such information shall also be provided in a suitable form to temporary workers and hired workers present in the establishment or enterprise.

2. The employer shall take appropriate measures so that employers of workers from any outside undertakings and/or establishments engaged in work in his undertaking and/or establishment receive, in accordance with national laws and/or practices, adequate information concerning the points referred to in paragraph 1(a) and (b) which is to be provided to the workers in question.

3. The employer shall take appropriate measures so that workers with specific functions in protecting the safety and health of workers, or workers' representatives with specific responsibility for the safety and health of workers shall have access, in order to perform their duties and in accordance with national laws and/or practices, to:

(a) the risk assessment and protective measures referred to in Article 9(1)(a) and (b);

(b) the list and reports referred to in Article 9(1)(c) and (d);

(c) the information yielded by protective and preventive measures, inspection agencies and bodies responsible for safety and health.

Article 11
Consultation and participation of workers

1. Employers shall consult workers and/or their representatives and allow them to take part in discussions on all questions relating to safety and health at work.

   This pre-supposes:

   — the consultation of workers,

   — the right of workers and/or their representatives to make proposals,

   — balanced participation in accordance with national laws and/or practices.

2. Workers with specific functions in protecting the safety and health of workers or workers' representatives with specific responsibility for the safety and health of workers shall participate in a balanced way, in accordance with national laws and/or practices, or be consulted in advance and in due time by the employer with regard to:

   (a) any measure which may substantially affect safety and health;

   (b) the designation of workers referred to in Article 8(2) and the activities referred to in Article 7(1);

   (c) the information referred to in Article 9(1) and Article 10;

   (d) the planning and organization of the training referred to in Article 12.

3. Workers with specific functions for the protection of the safety and health of workers at work, and workers' representatives, may call on the employer to take appropriate measures and submit to him relevant proposals by means of which all risks to workers may be reduced and/or sources of danger eliminated.

4. The workers and workers' representatives referred to in the introductory part of paragraph 2 may not be placed at any disadvantage because of their specific responsibility for the safety and health of workers.

5. Employers must allow workers' representatives with specific responsibility for the safety and health of workers adequate time off work, without loss of pay, and provide them with the necessary means to enable such representatives to exercise their rights and functions deriving from this Directive.

6. Workers and/or their representatives are entitled to appeal, in accordance with national law and/or practice, to the authority responsible for safety and health protection at work if they consider that the measures taken and the means employed by the employer are inadequate for the purposes of ensuring safety and health at work.

Workers' representatives and the workers designated in accordance with Article 7(1) must be invited to — and be able to submit their observations at — inspection visits by the competent authorities.

Article 12
Training of workers

1. The employer shall ensure that each worker receives adequate safety and health training, in particular in the form of information and instructions specific to his workstation or job:

   — on recruitment,

   — in the event of a transfer or a change of job,

   — in the event of the introduction of new working equipment or a change in equipment,

   — in the event of the introduction of any new technology.

   The training shall be:

   — adopted to take account of new or changed risks, and

   — repeated periodically if necessary.

2. The employer shall ensure that workers from outside undertakings and/or establishments engaged in work in his undertaking and/or establishment have in fact received appropriate instructions regarding health and safety risks during their activities in his undertaking and/or establishment.

3. Workers' representatives with a specific role in protecting the safety and health of workers shall be entitled to appropriate training.

4. The training referred to in paragraphs 1 and 3 may not be at the workers' expense or at that of the workers' representatives.

   The training referred to in paragraph 1 must take place during working hours.

   The training referred to in paragraph 3 must take place during working hours or in accordance with national practices, either inside or outside the enterprise and/or establishment.

SECTION III
WORKERS' OBLIGATIONS

Article 13

1. It shall be the responsibility of each worker to take care as far as possible of his own safety and health and that of other persons affected by his acts or omissions at work in accordance with his training and the instructions given by his employer.

2. To this end, workers must, in particular, in accordance with their training and the instructions given by their employer:

    (a) make correct use of machinery, apparatus, tools, dangerous substances, transport equipment and other means of production;

    (b) make correct use of the personal protective equipment supplied to them and, after use, return it to its proper place;

    (c) refrain from disconnecting, changing or removing arbitrarily safety devices fitted *eg.* to machinery, apparatus, tools, plant and buildings, and use such safety devices correctly;

    (d) immediately inform the employer and/or the workers with specific responsibility for the safety and health of workers of any work situation they have reasonable grounds for considering represents a serious and immediate danger to safety and health and of any shortcomings in the protection arrangements;

    (e) co-operate, in accordance with national practice, with the employer and/or workers with specific responsibility for the safety and health of workers, for as long as may be necessary to enable any tasks or requirements imposed by the competent authority to protect the safety and health of workers at work to be carried out;

    (f) co-operate, in accordance with national practice, with the employer and/or workers with specific responsibility for the safety and health of workers, for as long as may be necessary to enable the employer to ensure that the working environment and working conditions are safe and pose no risk to safety and health within their field of activity.

Article 14
Health surveillance

1. To ensure that workers receive health surveillance appropriate to the health and safety risks they incur at work, measures shall be introduced in accordance with national law and/or practices.

2. The measures referred to in paragraph 1 shall be such that each worker, if he so wishes, may receive health checks at regular intervals.

3. Health surveillance may be provided as part of a national health system.

Article 15
Risk groups

Particularly sensitive risk groups must be protected against the dangers which specifically affect them.

Article 16
Individual Directives
Amendments
General scope of this Directive

1. The Council, acting on a proposal from the Commission based on Article 118a of the Treaty, shall adopt individual directives *inter alia* in the areas listed in the Annex.

2. This Directive and, without prejudice to the procedure referred to in Article 17 concerning technical adjustments, the individual directives, may be amended in accordance with the procedure provided for in Article 118a of the Treaty.

3. The provisions of this Directive shall apply in full to all the areas covered by the individual directives, without prejudice to more stringent and/or specific provisions contained in these individual directives.

Article 17
Committee

1. Fcr the purely technical adjustments to the individual directives provided for in Article 16(1) to take account of:

    — the adoption of directives in the field of technical harmonization and standardization, and/or

    — technical progress, changes in international regulations, or

    — specifications, and new findings,

    the Commission shall be assisted by a committee composed of the representatives of the Member States and chaired by the representative of the Commission.

2. The representative of the Commission shall submit to the committee a draft of the measures to be taken.

    The committee shall deliver its opinion on the draft within a time limit which the Chairman may lay down according to the urgency of the matter.

    The opinion shall be delivered by the majority laid down in Article 148(2) of the Treaty in the case of decisions which the Council is required to adopt on a proposal from the Commission.

    The votes of the representatives of the Member States within the committee shall be weighted in the manner set out in that Article. The Chairman shall not vote.

3. The Commission shall adopt the measures envisaged if they are in accordance with the opinion of the committee.

    If the measures envisaged are not in accordance with the opinion of the committee, or if no opinion is delivered, the Commission shall, without delay, submit to the Council a proposal relating to the measures to be taken. The Council shall act by a qualified majority.

If, on the expiry of three months from the date of the referral to the Council, the Council has not acted, the proposed measures shall be adopted by the Commission.

Article 18
Final provisions

1. Member States shall bring into force the laws, regulations and administrative provisions necessary to comply with this Directive by 31 December 1992.

    They shall forthwith inform the Commission thereof.

2. Member States shall communicate to the Commission the texts of the provisions of national law which they have already adopted or adopt in the field covered by this Directive.

3. Member States shall report to the Commission every three years on the application of the provisions of this Directive, having regard to the point of view of employers and workers.

    The Commission shall inform the European Parliament, the Council, the Economic and Social Committee and the Advisory Committee on Safety, Hygiene and Health Protection at Work.

4. The Commission shall submit periodically to the European Parliament, the Council and the Economic and Social Committee a report on the implementation of this Directive, taking into account paragraphs 1 to 3.

Article 19

This Directive is addressed to the Member States.

Done at Luxembourg, 12 June 1989

ANNEX

List of areas referred to in Article 16(1)

— Work places

— Work equipment

— Personal protective equipment

— Work with visual display units

— Handling of heavy loads involving risk of back injury

— Temporary or mobile work sites

— Fisheries and agriculture

COUNCIL DIRECTIVE

of 14 June 1989

on the approximation of the laws of the Member States relating to machinery
[89/392/EEC]

THE COUNCIL OF THE EUROPEAN COMMUNITIES,

Having regard to the Treaty establishing the European Economic Community, and in particular Article 100a thereof,

Having regard to the proposal from the Commission,

In co-operation with the European Parliament,

Having regard to the opinion of the Economic and Social Committee,

Whereas Member States are responsible for ensuring the health and safety on their territory of their people and, where appropriate, of domestic animals and goods and, in particular, of workers notably in relation to the risks arising out of the use of machinery;

Whereas, in the Member States, the legislative systems regarding accident prevention are very different; whereas the relevant compulsory provisions, frequently supplemented by de facto mandatory technical specifications and/or voluntary standards, do not necessarily lead to different levels of health and safety, but nevertheless, owing to their disparities, constitute barriers to trade within the Community; whereas, furthermore, conformity certification and national certification systems for machinery differ considerably;

Whereas the maintenance or improvement of the level of safety attained by the Member States constitutes one of the essential aims of this Directive and of the principle of safety as defined by the essential requirements;

Whereas existing national health and safety provisions providing protection against the risks caused by machinery must be approximated to ensure free movement of machinery without lowering existing justified levels of protection in the Member States; whereas the provisions of this Directive concerning the design and construction of machinery, essential for a safer working environment shall be accompanied by specific provisions concerning the prevention of certain risks to which workers can be exposed at work, as well as by provisions based on the organization of safety of workers in the working environment;

Whereas the machinery sector is an important part of the engineering industry and is one of the industrial mainstays of the Community economy;

Whereas paragraphs 65 and 68 of the White Paper on the completion of the internal market, approved by the European Council in June 1985, provide for a new approach to legislative harmonization;

Whereas the social cost of the large number of accidents caused directly by the use of machinery can be reduced by inherently safe design and construction of machinery and by proper installations and maintenance;

Whereas the field of application of this Directive must be based on a general definition of the term 'machinery' so as to allow the technical development of products; whereas the development of 'complex installations' and the risks they involve are of an equivalent nature and their express inclusion in the Directive is therefore justified;

Whereas specific Directives containing design and construction provisions for certain categories of machinery are now envisaged; whereas the very broad scope of this Directive must be limited in relation to these Directives and also existing Directives where they contain design and construction provisions;

Whereas Community law, in its present form, provides — by way of derogation from one of the fundamental rules of the Community, namely the free movement of goods — that obstacles to movement within the Community resulting from disparities in national legislation relating to the marketing of products must be accepted in so far as the provisions concerned can be recognized as being necessary to satisfy imperative requirements; whereas, therefore, the harmonization of laws in this case must be limited only to those requirements necessary to satisfy the imperative and essential health and safety requirements relating to machinery; whereas these requirements must replace the relevant national provisions because they are essential;

Whereas the essential health and safety requirements must be observed in order to ensure that machinery is safe; whereas these requirements must be applied with discernment to take account of the state of the art at the time of construction and of technical and economic requirements;

Whereas the putting into service of machinery within the meaning of this Directive can relate only to the use of the machinery itself as intended by the manufacturer; whereas this does not preclude the laying-down of conditions of use external to the machinery, provided that it is not thereby modified in a way not specified in this Directive;

Whereas, for trade fairs, exhibitions, etc, it must be possible to exhibit machinery which does not conform to this Directive; whereas, however, interested parties should be properly informed that the machinery does not conform and cannot be purchased in that condition;

Whereas, therefore, this Directive defines only the essential health and safety requirements of general application, supplemented by a number of more specific requirements for certain categories of machinery; whereas, in order to help manufacturers to prove conformity to these essential requirements and in order to allow inspection for conformity to the essential requirements, it is desirable to have standards harmonized at European level for the prevention of risks arising out of the design and construction of machinery; whereas these standards harmonized at European level are drawn up by private-law bodies and must retain their non-binding status; whereas for this purpose the European Committee for Standardization (CEN) and the European Committee for Electrotechnical Standardization (Cenelec) are the bodies recognized as competent to adopt harmonized standards in accordance with the general guidelines for co-operation between the

Commission and these two bodies signed on 13 November 1984; whereas, within the meaning of this Directive, a harmonized standard is a technical specification (European standard or harmonization document) adopted by either or both of these bodies, on the basis of a remit from the Commission in accordance with the provisions of Council Directive 83/189/EEC of 28 March 1983 laying down a procedure for the provision of information in the field of technical standards and regulations, as last amended by Directive 88/182/EEC, and on the basis of general guidelines referred to above;

Whereas the legislative framework needs to be improved in order to ensure an effective and appropriate contribution by employers and employees to the standardization process; whereas such improvement should be completed at the latest by the time this Directive is implemented;

Whereas, as is currently the practice in Member States, manufacturers should retain the responsibility for certifying the conformity of their machinery to the relevant essential requirements; whereas conformity to harmonized standards creates a presumption of conformity to the relevant essential requirements; whereas it is left to the sole discretion of the manufacturer, where he feels the need, to have his products examined and certified by a third party;

Whereas, for certain types of machinery having a higher risk factor, a stricter certification procedure is desirable; where the EC type-examination procedure adopted may result in an EC declaration being given by the manufacturer without any stricter requirement such as a guarantee of quality, EC verification or EC supervision;

Whereas it is essential that, before issuing an EC declaration of conformity, the manufacturer or his authorized representative established in the Community should provide a technical construction file; whereas it is not, however, essential that all documentation be permanently available in a material manner but it must be made available on demand; whereas it need not include detailed plans of the sub-assemblies used in manufacturing the machines, unless knowledge of these is indispensable in order to ascertain conformity with essential safety requirements;

Whereas it is necessary not only to ensure the free movement and putting into service of machinery bearing the EC mark and having an EC conformity certificate but also to ensure free movement of machinery not bearing the EC mark where it is to be incorporated into other machinery or assembled with other machinery to form a complex installation;

Whereas the Member States' responsibility for safety, health and the other aspects covered by the essential requirements on their territory must be recognized in a safeguard clause providing for adequate Community protection procedures;

Whereas the addresses of any decision taken under this Directive must be informed on the reasons for such a decision and the legal remedies open to them;

Whereas the measures aimed at the gradual establishment of the internal market must be adopted by 31 December 1992; whereas the internal market consists of an area without internal frontiers within which the free movement of goods persons, services and capital is guaranteed,

HAS ADOPTED THIS DIRECTIVE:

CHAPTER I

SCOPE, PLACING ON THE MARKET AND FREEDOM OF MOVEMENT

Article 1

1. This Directive applies to machinery and lays down the essential health and safety requirements therefor, as defined in Annex I.

2. For the purposes of this Directive, 'machinery' means an assembly of linked parts or components, at least one of which moves, with the appropriate actuators, control and power circuits, *etc*, joined together for a specific application, in particular for the processing, treatment, moving or packaging of a material.

    The term 'machinery' also covers an assembly of machines which, in order to achieve the same end, are arranged and controlled so that they function as an integral whole.

3. The following are excluded from the scope of this Directive:

    — mobile equipment,

    — lifting equipment,

    — machinery whose only power source is directly applied manual effort,

    — machinery for medical use used in direct contact with patients,

    — special equipment for use in fairgrounds and/or amusement parks,

    — steam boilers, tanks and pressure vessels,

    — machinery specially designed or put into service for nuclear purposes which, in the event of failure, may result in an emission of radioactivity,

    — radioactive sources forming part of a machine,

    — firearms,

    — storage tanks and pipelines for petrol, diesel fuel, inflammable liquids and dangerous substances.

4. Where, for machinery, the risks referred to in this Directive are wholly or partly covered by specific Community Directives, this Directive shall not apply, or shall cease to apply, in the case of such machinery and of such risks on the entry into force of these specific Directives.

5. Where, for machinery, the risks are mainly of electrical origin, such machinery shall be covered exclusively by Council Directive 73/23/EEC of 19 February 1973 on the harmonization of the laws of the Member States relating to electrical equipment designed for use within certain voltage limits.

Article 2

1. Member States shall take all appropriate measures to ensure that machinery covered by this Directive may be placed on the market and put into service only if it does not endanger the health or safety of persons and, where appropriate, domestic animals or property, when properly installed and maintained and used for its intended purpose.

2. The provisions of this Directive shall not affect Member States' entitlement to lay down, in due observance of the Treaty, such requirements as they may deem necessary to ensure that persons and in particular workers are protected when using the machines in question, provided that this does not mean that the machinery is modified in a way not specified in the Directive.

3. At trade fairs, exhibitions, demonstrations, *etc*, Member States shall not prevent the showing of machinery which does not conform to the provisions of this Directive, provided that a visible sign clearly indicates that such machinery does not conform and that it is not for sale until it has been brought into conformity by the manufacturer or his authorized representative established in the Community During demonstrations, adequate safety measures shall be taken to ensure the protection of persons.

Article 3

Machinery covered by this Directive shall satisfy the essential health and safety requirements set out in Annex I.

Article 4

1. Member States shall not prohibit, restrict or impede the placing on the market and putting into service in their territory of machinery which complies with the provisions of this Directive.

2. Member States shall not prohibit, restrict or impede the placing on the market of machinery where the manufacturer or his authorized representative established in the Community declares in accordance with Annex II.B that it is intended to be incorporated into machinery or assembled with other machinery to constitute machinery covered by this Directive except where it can function independently.

Article 5

1. Member States shall regard machinery bearing the EC mark and accompanied by the EC declaration of conformity referred to in Annex II as conforming to the essential health and safety requirements referred to in Article 3.

    In the absence of harmonized standards, Member States shall take any steps they deem necessary to bring to the attention of the parties concerned the existing national technical standards and specifications which are regarded as important or relevant to the proper implementation of the essential safety and health requirements in Annex I.

2. Where a national standard transposing a harmonized standard, the reference for which has been published in the *Official Journal of the European Communities*, covers one or more of the essential safety requirements, machinery constructed in accordance with this standard shall be presumed to comply with the relevant essential requirements.

    Member States shall publish the references of national standards transposing harmonized standards.

3. Member States shall ensure that appropriate measures are taken to enable the social partners to have an influence at national level on the process of preparing and monitoring the harmonized standards.

Article 6

1. Where a Member State or the Commission considers that the harmonized standards referred to in Article 5(2) do not entirely satisfy the essential requirements referred to in Article 3, the Commission or the Member State concerned shall bring the matter before the Committee set up under Directive 83/189/EEC, giving the reasons therefor. The Committee shall deliver an opinion without delay.

   Upon receipt of the Committee's opinion, the Commission shall inform the Member States whether or not it is necessary to withdraw those standards from the published information referred to in Article 5(2).

2. A standing committee shall be set up, consisting of representatives appointed by the Member States and chaired by a representative of the Commission.

   The standing committee shall draw up its own rules of procedure.

   Any matter relating to the implementation and practical application of this Directive may be brought before the standing committee, in accordance with the following procedure:

   The representative of the Commission shall submit to the committee a draft of the measures to be taken. The committee shall deliver its opinion on the draft, within a time limit which the chairman may lay down according to the urgency of the matter, if necessary by taking a vote.

   The opinion shall be recorded in the minutes; in addition, each Member State shall have the right to ask to have its position recorded in the minutes.

   The Commission shall take the utmost account of the opinion delivered by the committee. It shall inform the committee of the manner in which its opinion has been taken into account.

Article 7

1. Where a Member State ascertains that machinery bearing the EC mark and used in accordance with its intended purpose is liable to endanger the safety of persons and, where appropriate, domestic animals or property, it shall take all appropriate measures to withdraw such machinery from the market, to prohibit the placing on the market, putting into service or use thereof, or to restrict free movement thereof.

   The Member State shall immediately inform the Commission of any such measure, indicating the reasons for its decision and, in particular, whether non-conformity is due to:

   (a)   failure to satisfy the essential requirements referred to in Article 3;

   (b)   incorrect application of the standards referred to in Article 5(2);

   (c)   shortcomings in the standards referred to in Article 5(2) themselves.

2. The Commission shall enter into consultation with the parties concerned without delay. Where the Commission considers, after this consultation, that the measure is justified, it shall immediately so inform the Member State which took the initiative and the other Member States. Where the Commission considers, after this consultation, that the action is unjustified, it shall immediately so inform the Member State which took the initiative and the manufacturer or his authorized representative established within the Community. Where the decision referred to in paragraph 1 is based on a shortcoming in the standards, and where the Member State at the origin of the decision maintains its position, the Commission shall immediately inform the Committee in order to initiate the procedure referred to in Article 6(1).

3. Where machinery which does not comply bears the EC mark, the competent Member State shall take appropriate action against whomsoever has affixed the mark and shall also inform the Commission and the other Member States.

4. The Commission shall ensure that the Member States are kept informed of the progress and outcome of the procedure.

CHAPTER II

CERTIFICATION PROCEDURE

Article 8

1. The manufacturer, or his authorized representative established in the Community, shall, in order to certify the conformity of machinery with the provisions of this Directive, draw up an EC declaration of conformity based on the model given in Annex II for each machine manufactured, and shall affix to the machinery the EC mark referred to at Article 10.

2. Before placing on the market, the manufacturer, or his authorized representative established in the Community, shall:

   a   if the machinery is not referred to in Annex IV, draw up the file provided for in Annex V;

   b   if the machinery is referred to in Annex IV and its manufacturer does not comply, or only partly complies, with the standards referred to in Article 5(2) or if there are no such standards, submit an example of the machinery for the EC type-examination referred to in Annex VI;

   c   if the machinery is referred to in Annex IV and is manufactured in accordance with the standards referred to in Article 5(2):

   — either draw up the file referred to in Annex VI and forward it to a notified body, which will acknowledge receipt of the file as soon as possible and keep it,

   — submit the file referred to in Annex VI to the notified body, which will simply verify that the standards referred to in Article 5(2) have been correctly applied and will draw up a certificate of adequacy for the file,

   — or submit the example of the machinery for the EC type-examination referred to in Annex VI.

3. Where the first indent of paragraph 2(c) applies, the provisions of the first sentence of paragraph 5 and paragraph 6 of Annex VI shall also apply.

Where the second indent of 2(c) applies, the provisions of paragraphs 5, 6 and 7 of Annex VI shall also apply.

4. Where paragraph 2(a) and the first and second indents of paragraph 2(c) apply, the EC declaration of conformity shall solely state conformity with the essential requirements of the Directive.

Where paragraph 2(b) and (c) apply, the EC declaration of conformity shall state conformity with the example that underwent EC type-examination.

5. Where the machinery is subject to other Community Directives concerning other aspects, the EC mark referred to in Article 10 shall indicate in these cases that the machinery also fulfils the requirements of the other Directives.

6. Where neither the manufacturer nor his authorized representative established in the Community fulfils the obligations of the preceding paragraphs, these obligations shall fall to any person placing the machinery on the market in the Community. The same obligations shall apply to any person assembling machinery or parts thereof of various origins or constructing machinery for his own use.

Article 9

1. Each Member State shall notify the Commission and the other Member States of the approved bodies responsible for carrying out the certification procedures referred to in Article 8(2)(b) and (c) The Commission shall publish a list of these bodies in the *Official Journal of the European Communities* for information and shall ensure that the list is kept up to date.

2. Member States shall apply the criteria laid down in Annex VII in assessing the bodies to be indicated in such notification. Bodies meeting the assessment criteria laid down in the relevant harmonized standards shall be presumed to fulfil those criteria.

3. A Member State which has approved a body must withdraw its notification if it finds that the body no longer meets the criteria referred to in Annex VII. It shall immediately inform the Commission and the other Member States accordingly.

CHAPTER III

EC MARK

Article 10

1. The 'EC' mark shall consist of the EC symbol followed by the last two digits of the year in which the mark was affixed.

Annex III shows the model to be used.

2. The EC mark shall be affixed to machinery distinctly and visibly in accordance with point 1.7.3 of Annex I.

3. Marks or inscriptions liable to be confused with the EC mark shall not be put on machinery.

CHAPTER IV
FINAL PROVISIONS

Article 11

Any decision taken pursuant to this Directive which restricts the marketing and putting into service of machinery shall state the exact grounds on which it is based. Such a decision shall be notified as soon as possible to the party concerned, who shall at the same time be informed of the legal remedies available to him under the laws in force in the Member State concerned and of the time limits to which such remedies are subject.

Article 12

The Commission will take the necessary steps to have information on all the relevant decisions relating to the management of this Directive made available.

Article 13

1. Member States shall adopt and publish the laws, regulations and administrative provisions necessary in order to comply with this Directive by 1 January 1992 at the latest. They shall forthwith inform the Commission thereof.

   They shall apply these provisions with effect from 31 December 1992.

2. Member States shall ensure that the texts of the provisions of national law which they adopt in the field covered by this Directive are communicated to the Commission.

Article 14

This Directive is addressed to the Member States.

Done at Luxembourg, 14 June 1989.

ANNEX I

ESSENTIAL HEALTH AND SAFETY REQUIREMENTS RELATING TO THE DESIGN AND CONSTRUCTION OF MACHINERY

PRELIMINARY OBSERVATIONS

1. The obligations laid down by the essential health and safety requirements apply only when the corresponding hazard exists for the machinery in question when it is used under the conditions foreseen by the manufacturer. In any event, requirements 1.1.2, 1.7.3 and 1.7.4 apply to all machinery covered by this Directive.

2. The essential health and safety requirements laid down in this Directive are mandatory. However, taking into account the state of the art, it may not be possible to meet the objectives set by them. In this case, the machinery must as far as possible be designed and constructed with the purpose of approaching those objectives.

1. ESSENTIAL HEALTH AND SAFETY REQUIREMENTS

1.1. General remarks

1.1.1. *Definitions*

For the purpose of this Directive

1. 'danger zone' means any zone within and/or around machinery in which an exposed person is subject to a risk to his health or safety;

2. 'exposed person' means any person wholly or partially in a danger zone;

3. 'operator' means the person or persons given the task of installing, operating, adjusting, maintaining, cleaning, repairing or transporting machinery.

1.1.2. *Principles of safety integration*

(a) Machinery must be so constructed that it is fitted for its function, and can be adjusted and maintained without putting persons at risk when these operations are carried out under the conditions foreseen by the manufacturer.

The aim of measures taken must be to eliminate any risk of accident throughout the foreseeable lifetime of the machinery, including the phases of assembly and dismantling, even where risks of accident arise from foreseeable abnormal situations.

(b) In selecting the most appropriate methods, the manufacturer must apply the following principles, in the order given:

— eliminate or reduce risks as far as possible (inherently safe machinery design and construction),

— take the necessary protection measures in relation to risks that cannot be eliminated,

— inform users of the residual risks due to any shortcomings of the protection measures adopted, indicate whether any particular training is required and specify any need to provide personal protection equipment.

(c)   When designing and constructing machinery, and when drafting the instructions, the manufacturer must envisage not only the normal use of the machinery but also uses which could reasonably be expected.

The machinery must be designed to prevent abnormal use if such use would engender a risk. In other cases the instructions must draw the user's attention to ways — which experience has shown might occur — in which the machinery should not be used.

(d)   Under the intended conditions of use, the discomfort, fatigue and psychological stress faced by the operator must be reduced to the minimum possible taking ergonomic principles into account.

(e)   When designing and constructing machinery, the manufacturer must take account of the constraints to which the operator is subject as a result of the necessary or foreseeable use of personal protection equipment (such as footwear, gloves, *etc.*).

(f)   Machinery must be supplied with all the essential special equipment and accessories to enable it to be adjusted, maintained and used without risk.

1.1.3. *Materials and products*

The materials used to construct machinery or products used and created during its use must not endanger exposed persons' safety or health.

In particular, where fluids are used, machinery must be designed and constructed for use without risks due to filling, use, recovery or draining.

1.1.4. *Lighting*

The manufacturer must supply integral lighting suitable for the operations concerned where its lack is likely to cause a risk despite ambient lighting of normal intensity.

The manufacturer must ensure that there is no area of shadow likely to cause nuisance, that there is no irritating dazzle and that there are no dangerous stroboscopic effects due to the lighting provided by the manufacturer.

Internal parts requiring frequent inspection, and adjustment and maintenance areas, must be provided with appropriate lighting.

1.1.5. *Design of machinery to facilitate its handling*

Machinery or each component part thereof must:

— be capable of being handled safely,

— be packaged or designed so that it can be stored safely and without damage (e.g. adequate stability, special supports, *etc.*).

Where the weight, size or shape of machinery or its various component parts prevents them from being moved by hand, the machinery or each component part must:

— either be fitted with attachments for lifting gear, or

— be designed so that it can be fitted with such attachments (e.g. threaded holes, or

— be shaped in such a way that standard lifting gear can easily be attached.

Where machinery or one of its component parts is to be moved by hand, it must:

— either be easily movable, or

— be equipped for picking up (e.g. hand-grips, *etc.*) and moving in complete safety.

Special arrangements must be made for the handling of tools and/or machinery parts, even if lightweight, which could be dangerous (shape, material, *etc.*).

1.2. Controls

1.2.1. *Safety and reliability of control systems*

Control systems must be designed and constructed so that they are safe and reliable, in a way that will prevent a dangerous situation arising. Above all they must be designed and constructed in such a way that:

— they can withstand the rigours of normal use and external factors,

— errors in logic do not lead to dangerous situations.

1.2.2. *Control devices*

Control devices must be:

— clearly visible and identifiable and appropriately marked where necessary,

— positioned for safe operation without hesitation or loss of time, and without ambiguity,

— designed so that the movement of the control is consistent with its effect,

— located outside the danger zones, except for certain controls where necessary, such as emergency stop, console for training of robots,

— positioned so that their operation cannot cause additional risk,

— designed or protected so that the desired effect, where a risk is involved, cannot occur without an intentional operation,

— made so as to withstand foreseeable strain; particular attention must be paid to emergency stop devices liable to be subjected to considerable strain.

Where a control is designed and constructed to perform several different actions, namely where there is no one-to-one correspondence (e.g. keyboards, *etc.*), the action to be performed must be clearly displayed and subject to confirmation where necessary.

Controls must be so arranged that their layout, travel and resistance to operation are compatible with the action to be performed, taking account of ergonomic principles. Constraints due to the necessary or foreseeable use of personal protection equipment (such as footwear, gloves, *etc.*) must be taken into account.

Machinery must be fitted with indicators (dials, signals, *etc.*) as required for safe operation. The operator must be able to read them from the control position.

From the main control position the operator must be able to ensure that there are no exposed persons in the danger zones.

If this is impossible, the control system must be designed and constructed so that an acoustic and/or visual warning signal is given whenever the machinery is about to start. The exposed person must have the time and the means to take rapid action to prevent the machinery starting up.

1.2.3. *Starting*

It must be possible to start machinery only be voluntary actuation of a control provided for the purpose.

The same requirement applies:

— when restarting the machinery after a stoppage, whatever the cause,

— when effecting a significant change in the operating conditions (e.g. speed, pressure, *etc.*),

unless such restarting or change in operating conditions is without risk to exposed persons.

This essential requirement does not apply to the restarting of the machinery or to the change in operating conditions resulting from the normal sequence of an automatic cycle.

Where machinery has several starting controls and the operators can therefore put each other in danger, additional devices (e.g. enabling devices or selectors allowing only one part of the starting mechanism to be actuated at any one time) must be fitted to rule out such risks.

It must be possible for automated plant functioning in automatic mode to be restarted easily after a stoppage once the safety conditions have been fulfilled.

1.2.4. *Stopping device*

Normal stopping

Each machine must be fitted with a control whereby the machine can be brought safely to a complete stop.

Each workstation must be fitted with a control to stop some or all of the moving parts of the machinery, depending on the type of hazard, so that the machinery is rendered safe. The machinery's stop control must have priority over the start controls.

Once the machinery or its dangerous parts have stopped, the energy supply to the actuators concerned must be cut off.

Emergency stop

Each machine must be fitted with one or more emergency stop devices to enable actual or impending danger to be averted. The following exceptions apply:

— machines in which an emergency stop device would not lessen the risk, either because it would not reduce the stopping time or because it would not enable the special measures required to deal with the risk to be taken,

— hand-held portable machines and hand-guided machines.

This device must:

— have clearly identifiable, clearly visible and quickly accessible controls,

— stop the dangerous process as quickly as possible, without creating additional hazards,

— where necessary, trigger or permit the triggering of certain safeguard movements.

The emergency stop control must remain engaged; it must be possible to disengage it only by an appropriate operation; disengaging the control must not restart the machinery, but only permit restarting; the stop control must not trigger the stopping function before being in the engaged position.

Complex installations

In the case of machinery or parts of machinery designed to work together, the manufacturer must so design and construct the machinery that the stop controls, including the emergency stop, can stop not only the machinery itself but also all equipment upstream and/or downstream if its continued operation can be dangerous.

1.2.5. *Mode selection*

The control mode selected must override all other control systems with the exception of the emergency stop.

If machinery has been designed and built to allow for its use in several control or operating modes presenting different safety levels (e.g. to allow for adjustment, maintenance, inspection, *etc.*), it must be fitted with a mode selector which can be locked in each position. Each position of the selector must correspond to a single operating or control mode.

The selector may be replaced by another selection method which restricts the use of certain functions of the machinery to certain categories of operator (e.g. access codes for certain numerically controlled functions, *etc.*).

If, for certain operations, the machinery must be able to operate with its protection devices neutralized, the mode selector must simultaneously:

— disable the automatic control mode,

— permit movements only by controls requiring sustained action,

— permit the operation of dangerous moving parts only in enhanced safety conditions (e.g. reduced speed, reduced power, step-by-step, or other adequate provision) while preventing hazards from linked sequences,

— prevent any movement liable to pose a danger by acting voluntarily or involuntarily on the machine's internal sensors.

In addition, the operator must be able to control operation of the parts he is working on at the adjustment point.

1.2.6. *Failure of the power supply*

The interruption, re-establishment after an interruption or fluctuation in whatever manner of the power supply to the machinery must not lead to a dangerous situation.

In particular:

— the machinery must not start unexpectedly,

— the machinery must not be prevented from stopping if the command has already been given,

— no moving part of the machinery or piece held by the machinery must fall or be ejected,

— automatic or manual stopping of the moving parts whatever they may be must be unimpeded,

— the protection devices must remain fully effective.

1.2.7. *Failure of the control circuit*

A fault in the control circuit logic, or failure of or damage to the control circuit must not lead to dangerous situations.

In particular:

— the machinery must not start unexpectedly,

— the machinery must not be prevented from stopping if the command has already been given,

— no moving part of the machinery or piece held by the machinery must fall or be ejected,

— automatic or manual stopping of the moving parts whatever they may be must be unimpeded,

— the protection devices must remain fully effective.

1.2.8. *Software*

Interactive software between the operator and the command or control system of a machine must be user-friendly.

1.3. Protection against mechanical hazards

1.3.1. *Stability*

Machinery, components and fittings thereof must be so designed and constructed that they are stable enough, under the foreseen operating conditions (if necessary taking climatic conditions into account) for use without risk of overturning, falling or unexpected movement.

If the shape of the machinery itself or its intended installation does not offer sufficient stability, appropriate means of anchorage must be incorporated and indicated in the instructions.

1.3.2. *Risk of break-up during operation*

The various parts of machinery and their linkages must be able to withstand the stresses to which they are subject when used as foreseen by the manufacturer.

The durability of the materials used must be adequate for the nature of the work place foreseen by the manufacturer, in particular as regards the phenomena of fatigue, ageing, corrosion and abrasion.

The manufacturer must indicate in the instructions the type and frequency of inspection and maintenance required for safety reasons. He must, where appropriate, indicate the parts subject to wear and the criteria for replacement.

Where a risk of rupture or disintegration remains despite the measures taken (e.g. as with grinding wheels) the moving parts must be mounted and positioned in such a way that in case of rupture their fragments will be contained.

Both rigid and flexible pipes carrying fluids, particularly those under high pressure, must be able to withstand the foreseen internal and external stresses and must be firmly attached and/or protected against all manner of external stresses and strains; precautions must be taken to ensure that no risk is posed by a rupture (sudden movement, high-pressure jets, *etc.*).

Where the material to be processed is fed to the tool automatically, the following conditions must be fulfilled to avoid risks to the persons exposed (e.g. tool breakage):

— when the workpiece comes into contact with the tool the latter must have attained its normal working conditions,

— when the tool starts and/or stops (intentionally or accidentally) the feed movement and the tool movement must be co-ordinated.

1.3.3. *Risks due to falling or ejected objects*

Precautions must be taken to prevent risks from falling or ejected objects (e.g. work-pieces, tools, cuttings, fragments, waste, *etc.*).

1.3.4. *Risks due to surfaces, edges or angles*

In so far as their purpose allows, accessible parts of the machinery must have no sharp edges, no sharp angles, and no rough surfaces likely to cause injury.

1.3.5. *Risks related to combined machinery*

Where the machinery is intended to carry out several different operations with the manual removal of the piece between each operation (combined machinery), it must be designed and constructed in such a way as to enable each element to be used separately without the other elements constituting a danger or risk for the exposed person.

For this purpose, it must be possible to start and stop separately any elements that are not protected.

1.3.6 *Risks relating to variations in the rotational speed of tools*

When the machine is designed to perform operations under different conditions of use (e.g. different speeds or energy supply), it must be designed and constructed in such a way that selection and adjustment of these conditions can be carried out safely and reliably.

1.3.7. *Prevention of risks related to moving parts*

The moving parts of machinery must be designed, built and laid out to avoid hazards or, where hazards persist, fixed with guards or protective devices in such a way as to prevent all risk of contact which could lead to accidents.

1.3.8. *Choice of protection against risks related to moving parts*

Guards or protection devices Used to protect against the risks related to moving parts must be selected on the basis of the type of risk. The following guidelines must be used to help make the choice.

A.   Moving transmission parts

Guards designed to protect exposed persons against the risks associated with moving transmission parts (such as pulleys, belts, gears, rack and pinions, shafts, *etc.*) must be:

— either fixed, complying with requirements 1.4.1 and 1.4.2.1, or

— movable, complying with requirements 1.4.1 and 1.4.2.2.A.

Movable guards should be used where frequent access is foreseen.

B.   Moving parts directly involved in the process

Guards or protection devices designed to protect exposed persons against the risks associated with moving parts contributing to the work (such as cutting tools, moving parts of presses, cylinders, parts in the process of being machined, *etc.*) must be:

— wherever possible fixed guards complying with requirements 1.4.1 and 1.4.2.1,

— otherwise, movable guards complying with requirements 1.4.1 and 1.4.2.2.B or protection devices such as sensing devices (e.g. non-material barriers, sensor mats), remote-hold protection devices (e.g. two-hand controls), or protection devices intended automatically to prevent all or part of the operator's body from encroaching on the danger zone in accordance with requirements 1.4.1 and 1.4.3.

However, when certain moving parts directly involved in the process cannot be made completely or partially inaccessible during operation owing to operations requiring nearby operator intervention, where technically possible such parts must be fitted with:

— fixed guards, complying with requirements 1.4.1 and 1.4.2.1 preventing access to those sections of the parts that are not used in the work,

— adjustable guards, complying with requirements 1.4.1 and 1.4.2.3 restricting access to those sections of the moving parts that are strictly for the work.

1.4.   Required characteristics of guards and protection devices

1.4.1. *General requirement*

Guards and protection devices must:

— be of robust construction,

— not give rise to any additional risk,

— not be easy to by-pass or render non-operational,

— be located at an adequate distance from the danger zone,

— cause minimum obstruction to the view of the production process,

— enable essential work to be carried out on installation and/or replacement of tools and also for maintenance by restricting access only to the area where the work has to be done, if possible without the guard or protection device having to be dismantled.

1.4.2. *Special requirements for guards*

1.4.2.1. Fixed guards

Fixed guards must be securely held in place.

They must be fixed by systems that can be opened only with tools.

Where possible, guards must be unable to remain in place without their fixings.

1.4.2.2. Movable guards

A.   Type A movable guards must:

— as far as possible remain fixed to the machinery when open,

— be associated with a locking device to prevent moving parts starting up as long as these parts can be accessed and to give a stop command whenever they are no longer closed.

B.   Type B movable guards must be designed and incorporated into the control system so that:

— moving parts cannot start up while they are within the operator's reach,

— the exposed person cannot reach moving parts once they have started up,

— they can be adjusted only by means of an intentional action, such as the use of a tool, key, *etc,*

— the absence or failure of one of their components prevents starting or stops the moving parts,

— protection against any risk of ejection is proved by means of an appropriate barrier.

1.4.2.3. Adjustable guards restricting access

Adjustable guards restricting access to those areas of the moving parts strictly necessary for the work must:

— be adjustable manually or automatically according to the type of work involved,

— be readily adjustable without the use of tools,

— reduce as far as possible the risk of ejection.

1.4.3. Special requirements for protection devices

Protection devices must be designed and incorporated into the control system so that:

— moving parts cannot start up while they are within the operator's reach,

— the exposed person cannot reach moving parts once they have started up,

— they can be adjusted only be means of an intentional action, such as the use of a tool, key, *etc.*,

— the absence or failure of one of their components prevents starting or stops the moving parts.

1.5. Protection against other hazards

1.5.1. *Electricity supply*

Where machinery has an electricity supply it must be designed, constructed and equipped so that all hazards of an electrical nature are or can be prevented.

The specific rules in force relating to electrical equipment designed for use within certain voltage limits must apply to machinery which is subject to those limits.

1.5.2. *Static electricity*

Machinery must be so designed and constructed as to prevent or limit the build-up of potentially dangerous electrostatic charges and/or be fitted with a discharging system.

1.5.3. *Energy supply other than electricity*

Where machinery is powered by an energy other than electricity (e.g. hydraulic, pneumatic or thermal energy, *etc.*), it must be so designed, constructed and equipped as to avoid all potential hazards associated with these types of energy.

1.5.4. *Errors of fitting*

Errors likely to be made when fitting or refitting certain parts which could be a source of risk must be made impossible by the design of such parts or, failing this, by information given on the parts themselves and/or the housings. The same information must be given on moving parts and/or their housings where the direction of movement must be known to avoid a risk. Any further information that may be necessary must be given in the instructions.

Where a faulty connection can be the source of risk, incorrect fluid connections, including electrical conductors, must be made impossible by the design or, failing this, by information given on the pipes, cables, *etc.* and/or connector blocks.

1.5.5. *Extreme temperatures*

Steps must be taken to eliminate any risk of injury caused by contact with or proximity to machinery parts or materials at high or very low temperatures.

The risk of hot or very cold material being ejected should be assessed. Where this risk exists, the necessary steps must be taken to prevent it or, if this is not technically possible, to render it non-dangerous.

1.5.6. *Fire*

Machinery must, be designed and constructed to avoid all risk of fire or overheating posed by the machinery itself or by gases, liquids, dusts, vapours or other substances produced or used by the machinery.

1.5.7. *Explosion*

Machinery must be designed and constructed to avoid any risk of explosion posed by the machinery itself or by gases, liquids, dusts, vapours or other substances produced or used by the machinery.

To that end the manufacturer must take steps to:

— avoid a dangerous concentration of products,

— prevent combustion of the potentially explosive atmosphere,

— minimize any explosion which may occur so that it does not endanger the surroundings.

The same precautions must be taken if the manufacturer foresees the use of the machinery in a potentially explosive atmosphere.

Electrical equipment forming part of the machinery must conform, as far as the risk from explosion is concerned, to the provision of the specific Directives in force.

1.5.8. *Noise*

Machinery must be so designed and constructed that risks resulting from the emission of airborne noise are reduced to the lowest level taking account of technical progress and the availability of means of reducing noise, in particular at source.

1.5.9. *Vibration*

Machinery must be so designed and constructed that risks resulting from vibrations produced by the machinery are reduced to the lowest level, taking account of technical progress and the availability of means of reducing vibration, in particular at source.

1.5.10. *Radiation*

Machinery must be so designed and constructed that any emission of radiation is limited to the extent necessary for its operation and that the effects on exposed persons are non-existent or reduced to non-dangerous proportions.

1.5.11. *External radiation*

Machinery must be so designed and constructed that external radiation does not interfere with its operation.

1.5.12. *Laser equipment*

Where laser equipment is used, the following provisions should be taken into account:

— laser equipment on machinery must be designed and constructed so as to prevent any accidental radiation,

— laser equipment on machinery must be protected so that effective radiation, radiation produced by reflection or diffusion and secondary radiation do not damage health,

— optical equipment for the observation or adjustment of laser equipment on machinery must be such that no health risk is created by the laser rays.

1.5.13. *Emissions of dust, gases, etc.*

Machinery must be so designed, constructed and/or equipped that risks due to gases, liquids, dust, vapours and other waste materials which it produces can be avoided.

Where a hazard exists, the machinery must be so equipped that the said substances can be contained and/or evacuated.

Where machinery is not enclosed during normal operation, the devices for containment and/or evacuation must be situated as close as possible to the source emission.

1.6. Maintenance

1.6.1. *Machinery maintenance*

Adjustment, lubrication and maintenance points must be located outside danger zones. It must be possible to carry out adjustment, maintenance, repair, cleaning and servicing operations while machinery is at a standstill.

If one or more of the above conditions cannot be satisfied for technical reasons, these operations must be possible without risk (see 1.2.5).

In the case of automated machinery and, where necessary, other machinery, the manufacturer must make provision for a connecting device for mounting diagnostic fault-finding equipment.

Automated machine components which have to be changed frequently, in particular for a change in manufacture or where they are liable to wear or likely to deteriorate following an accident, must be capable of being removed and replaced easily and in safety. Access to the components must enable these tasks to be carried out with the necessary technical means (tools, measuring instruments, *etc.*) in accordance with an operating method specified by the manufacturer.

1.6.2. *Access to operating position and servicing points*

The manufacturer must provide means of access (stairs, ladders, catwalks, *etc.*) to allow access in safety to all areas used for production, adjustment and maintenance operations.

Parts of the machinery where persons are liable to move about or stand must be designed and constructed to avoid falls.

1.6.3. *Isolation of energy sources*

All machinery must be fitted with means to isolate it from all energy sources. Such isolators must be clearly identified. They must be capable of being locked if reconnection could endanger exposed persons. In the case of machinery supplied with electricity through a plug capable of being plugged into a circuit, separation of the plug is sufficient.

The isolator must be capable of being locked also where an operator is unable, from any of the points to which he has access, to check that the energy is still cut off.

After the energy is cut off, it must be possible to dissipate normally any energy remaining or stored in the circuits of the machinery without risk to exposed persons.

As an exception to the above requirements, certain circuits may remain connected to their energy sources in order, for example, to hold parts, protect information, light interiors, *etc.* In this case, special steps must be taken to ensure operator safety.

1.6.4. *Operator intervention*

Machinery must be so designed, constructed and equipped that the need for operator intervention is limited.

If operator intervention cannot be avoided, it must be possible to carry it out easily and in safety.

1.7. Indicators

1.7.0 *Information devices*

The information needed to control machinery must be unambiguous and easily understood.

It must not be excessive to the extent of overloading the operator.

1.7.1. *Warning devices*

Where machinery is equipped with warning devices (such as signals, *etc.*), these must be unambiguous and easily perceived.

The operator must have facilities to check the operation of such warning devices at all times.

The requirements of the specific Directives concerning colours and safety signals must be complied with.

1.7.2. *Warning of residual risks*

Where risks remain despite all the measures adopted or in the case of potential risks which are not evident (e.g. electrical cabinets, radioactive sources, bleeding of a hydraulic circuit, hazard in an unseen area, *etc.*), the manufacturer must provide warnings.

Such warnings should preferably use readily understandable diagrams and/or be drawn up in one of the languages of the country in which the machinery is to be used, accompanied, on request, by the languages understood by the operators.

1.7.3. *Marking*

All machinery must be marked legibly and indelibly with the following minimum particulars:

— name and address of the manufacturer,

— EC mark, which includes the year of construction (see Annex III),

— designation of series or type,

— serial number, if any.

Furthermore, where the manufacturer constructs machinery intended for use in a potentially explosive atmosphere, this must be indicated on the machinery.

Machinery must also bear full information relevant to its type and essential to its safe use (e.g. maximum speed of certain rotating parts, maximum diameter of tools to be fitted, mass, *etc.*).

1.7.4. *Instructions*

(a) All machinery must be accompanied by instructions including at least the following:

— a repeat of the information with which the machinery is marked (see 1.7.3), together with any appropriate additional information to facilitate maintenance (e.g. addresses of the importer, repairers, *etc.*),

— foreseen use of the machinery within the meaning of 1.1.2 (c),

— workstation(s) likely to be occupied by operators,

— instructions for safe:

— putting into service,

— use,

— handling, giving the mass of the machinery and its various parts where they are regularly to be transported separately,

— assembly, dismantling,

— adjustment,

— maintenance (servicing and repair),

— where necessary, training instructions.

Where necessary, the instructions should draw attention to ways in which the machinery should not be used.

(b) The instructions must be drawn up by the manufacturer or his authorized representative established in the Community in one of the languages of the country in which the machinery is to be used and should preferably be accompanied by the same instructions drawn up in another Community language, such as that of the country in which the manufacturer or his authorized representative is established. By way of derogation from this requirement, the maintenance instructions for use by the specialized personnel frequently employed by the manufacturer or his authorized representative may be drawn up in only one of the official Community languages.

(c) The instructions must contain the drawings and diagrams necessary for putting into service, maintenance, inspection, checking of correct operation and, where appropriate, repair of the machinery, and all useful instructions in particular with regard to safety.

(d) Any sales literature describing the machinery must not contradict the instructions as regards safety aspects; it must give information regarding the airborne noise emissions referred to in (f) and, in the case of hand-held and/or hand-guided machinery, information regarding vibration as referred to in 2.2.

(e) Where necessary, the instructions must give the requirements relating to installation and assembly for reducing noise or vibration (e.g. use of dampers, type and mass of foundation block, *etc.*).

(f) The instructions must give the following information concerning airborne noise emissions by the machinery, either the actual value or a value established on the basis of measurements made on identical machinery:

— equivalent continuous A-weighted sound pressure level at workstations, where this exceeds 70 dB(A); where this level does not exceed 70 dB(A), this fact must be indicated,

— peak C-weighted instantaneous sound pressure value at workstations, where this exceeds 63 Pa (130 dB in relation to 20 μPa),

— sound power level emitted by the machinery where the equivalent continuous A-weighted sound pressure level at workstations exceeds 85 dB(A).

In the case of very large machinery, instead of the sound power level, the equivalent continuous sound pressure levels at specified positions around the machinery may be indicated.

Sound levels must be measured using the most appropriate method for the machinery.

The manufacturer must indicate the operating conditions of the machinery during measurement and what methods have been used for the measurement.

Where the workstation(s) are undefined or cannot be defined, sound pressure levels must be measured at a distance of 1 metre from the surface of the machinery and at height of 1,60 metres from the floor or access platform. The position and value of the maximum sound pressure must be indicated.

(g) If the manufacturer foresees that the machinery will be used in a potentially explosive atmosphere, the instructions must give all the necessary information.

(h) In the case of machinery which may also be intended for use by non-professional operators, the wording and layout of the instructions for use, whilst respecting the other essential requirements mentioned above, must take into account the level of general education and acumen that can reasonably be expected from such operators.

2. ADDITIONAL ESSENTIAL HEALTH AND SAFETY REQUIREMENTS FOR CERTAIN CATEGORIES OF MACHINERY

2.1 Agri-foodstuffs machinery

In addition to the essential health and safety requirements set out in 1 above, where machinery is intended to prepare and process foodstuffs (e.g. cooking, refrigeration, thawing, washing, handling, packaging, storage, transport or distribution), it must be so designed and constructed as to avoid any risk of infection, sickness or contagion and the following hygiene rules must be observed:

(a) materials in contact, or intended to come into contact, with the foodstuffs must satisfy the conditions set down in the relevant Directives. The machinery must be so designed and constructed that these materials can be clean before each use;

(b) all surfaces including their joinings must be smooth, and must have neither ridges nor crevices which could harbour organic materials;

(c) assemblies must be designed in such a way as to reduce projections, edges and recesses to a minimum. They should preferably be made by welding or continuous bonding. Screws, screwheads and rivets may not be used except where technically unavoidable;

(d) all surfaces in contact with the foodstuffs must be easily cleaned and disinfected, where possible after removing easily dismantled parts. The inside surfaces must have curves of a radius sufficient to allow thorough cleaning;

(e) liquid deriving from foodstuffs as well as cleaning, disinfecting and rinsing fluids should be able to be discharged from the machine without impediment (possibly in a 'clean' position);

(f) machinery must be so designed and constructed as to prevent any liquids or living creatures, in particular insects, entering, or any organic matter accumulating in areas that cannot be cleaned (e.g. for machinery not mounted on feet or casters, by placing a seal between the machinery and its base, by the use of sealed units, *etc.*);

(g) machinery must be so designed and constructed that no ancillary substances (e.g. lubricants, *etc.*) can come into contact with foodstuffs. Where necessary, machinery must be designed and constructed so that continuing compliance with this requirement can be checked.

Instructions

In addition to the information required in section 1, the instructions must indicate recommended products and methods for cleaning, disinfecting and rinsing (not only for easily accessible areas but also where areas to which access is impossible or inadvisable, such as piping, have to be cleaned *in situ*).

2.2. Portable hand-held and/or hand-guided machinery

In addition to the essential health and safety requirements set out in 1 above, portable hand-held and/or hand-guided machinery must conform to the following essential health and safety requirements:

— according to the type of machinery, it must have a supporting surface of sufficient size and have a sufficient number of handles and supports of an appropriate size and arranged to ensure the stability of the machinery under the operating conditions foreseen by the manufacturer,

— except where technically impossible or where there is an independent control, in the case of handles which cannot be released on complete safety, it must be fitted with start and stop controls arranged in such a way that the operator can operate them without releasing the handles,

— it must be designed, constructed or equipped to eliminate the risks of accidental starting and/or continued operation after the operator has released the handles. Equivalent steps must be taken if this requirement is not technically feasible,

— portable hand-held machinery must be designed and constructed to allow, where necessary, a visual check of the contact of the tool with the material being processed.

Instructions

The instructions must give the following information concerning vibrations transmitted by hand-held and hand-guided machinery:

— the weighted root mean square acceleration value to which the arms are subjected, if it exceeds 2,5 m/s^2 as determined by the appropriate test code. Where the acceleration does not exceed 2,5 m/s^2, this must be mentioned.

If there is no applicable test code, the manufacturer must indicate the measurement methods and conditions under which measurements were made.

2.3. Machinery for working wood and analogous materials

In addition to the essential health and safety requirements set out in 1 above, machinery for working wood and machinery for working materials with physical and technological characteristics similar to those of wood, such as cork, bone, hardened rubber, hardened plastic material and other similar stiff material must conform to the following essential health and safety requirements:

(a) the machinery must be designed, constructed or equipped so that the piece being machined can be placed and guided in safety; where the piece is hand-held on a work-bench the latter must be sufficiently stable during the work and must not impede the movement of the piece;

(b) where the machinery is likely to be used in conditions involving the risk of ejection of pieces of wood, it must be designed, constructed or equipped to eliminate this ejection, or, if this is not the case, so that the ejection does not engender risks for the operator and/or exposed persons;

(c) the machinery must be equipped with an automatic brake that stops the tool in a sufficiently short time if there is a risk of contact with the tool whilst it runs down;

(d) where the tool is incorporated into a non-fully automated machine, the latter must be so designed and constructed as to eliminate or reduce the risk of serious accidental injury, for example by using cylindrical cutter blocks, restricting depth of cut, *etc.*

ANNEX II

A.   Contents of the EC declaration of conformity [1]

The EC declaration of conformity must contain the following particulars:

— name and address of the manufacturer or his authorized representative established in the Community [2],

— description of the machinery [3],

— all relevant provisions complied with by the machinery,

— where appropriate, name and address of the notified body and number of the EC type-examination certificate,

— where appropriate, the name and address of the notified body to which the file has been forwarded in accordance with the first indent of Article 8(2)(c),

— where appropriate, the name and address of the notified body which has carried out the verification referred to in the second indent of Article 8(2)(c),

— where appropriate, a reference to the harmonized standards,

— where appropriate, the national technical standards and specifications used,

— identification of the person empowered to sign on behalf of the manufacturer or his authorized representatives.

B.   Contents of the declaration by the manufacturer or his authorized representatives established in the Community (Article 4(2))

The manufacturer's declaration referred to in Article 4(2) must contain the following particulars:

— name and address of the manufacturer or the authorized representative,

— description of the machinery or machinery parts,

— a statement that the machinery must not be put into service until the machinery into which it is to be incorporated has been declared in conformity with the provisions of the Directive,

— identification of the person signing.

———————————

(1)   This declaration must be drawn up in the same language as the instructions (see Annex I, point 1.7.4) and must be either typewritten or handwritten in block capitals.

(2)   Business name and full address; authorized representatives must also give the business name and address of the manufacturer.

(3)   Description of the machinery (make, type, serial number, etc).

ANNEX III

EC MARK

The EC mark consists of the symbol shown below and the last two figures of the year in which the mark was affixed.

The different elements of the EC mark should have materially the same vertical dimensions, which should not be less than 5 mm.

ANNEX IV

TYPES OF MACHINES FOR WHICH THE PROCEDURE REFERRED TO IN ARTICLE 8 (2) (b) AND (c) MUST BE APPLIED

1. Circular saws (single- or multi-blade) for working with wood and meat.

1.1 Sawing machines with fixed tool during operation, having a fixed bed with manual feed of the workpiece or with a demountable power feed.

1.2 Sawing machines with fixed tool during operation, having a manually operated reciprocating saw-bench or carriage.

1.3 Sawing machines with fixed tool during operation, having a built-in mechanical feed device for the workpieces, with manual loading and/or unloading.

1.4 Sawing machines with movable tool during operation, with a mechanical feed device and manual loading and/or unloading.

2. Hand-fed surface planing machines for woodworking.

3. Thicknessers for one-side dressing with manual loading and/or unloading for woodworking.

4. Band-saws with a mobile bed or carriage and manual loading and/or unloading for working with wood and meat.

5. Combined machines of the types referred to in 1 to 4 and 7 for woodworking.

6. Hand-fed tenoning machines with several tool holders for woodworking.

7. Hand-fed vertical spindle moulding machines.

8. Portable chain saws for woodworking.

9. Presses, including press-brakes, for the cold working of metals, with manual loading and/or unloading, whose movable working parts may have a travel exceeding 6 mm and a speed exceeding 30 mm/s.

10. Injection or compression plastics-moulding machines with manual loading or unloading.

11. Injection or compression rubber-moulding machines with manual loading or unloading.

12. Cartridge-operated fixing guns.

ANNEX V

EC DECLARATION OF CONFORMITY

1. The EC declaration of conformity is the procedure by which the manufacturer, or his authorized representative established in the Community declares that the machinery being placed on the market complies with all the essential health and safety requirements applying to it.

2. Signature of the EC declaration of conformity authorizes the manufacturer, or his authorized representative in the Community, to affix the EC mark to the machinery.

3. Before drawing up the EC declaration of conformity, the manufacturer, or his authorized representative in the Community, shall have ensured and be able to guarantee that the documentation listed below is and will remain available on his premises for any inspection purposes:

    (a) a technical construction file comprising:

    — an overall drawing of the machinery together with drawings of the control circuits,

    — full detailed drawings, accompanied by any calculation notes, test results, etc., required to check the conformity of the machinery with the essential health and safety requirements,

    — a list of:

    — the essential requirements of this Directive, .

    — standards, and

    — other technical specifications, which were used when the machinery was designed,

    — a description of methods adopted to eliminate hazards presented by the machinery,

    — if he so desires, any technical report or certificate obtained from a competent body or laboratory [1],

    — if he declares conformity with a harmonized standard which provides therefor, any technical report giving the results of tests carried out at his choice either by himself or by a competent body or laboratory [1],

    — a copy of the instructions for the machinery;

    (b) for series manufacture, the internal measures that will be implemented to ensure that the machinery remains in conformity with the provisions of the Directive.

    The manufacturer must carry out necessary research or tests on components, fittings or the completed machine to determine whether by its design or construction, the machine is capable of being erected and put into service safely.

    Failure to present the documentation in response to a duly substantiated request by the competent national authorities may constitute sufficient grounds for doubting the presumption of conformity with the requirements of the Directive.

4.   (a)   The documentation referred to in 3 above need not permanently exist in a material manner but it must be possible to assemble it and make it available within a period of time commensurate with its importance. It does not have to include detailed plans or any other specific information as regards the sub-assemblies used for the manufacture of the machinery unless a knowledge of them is essential for verification of conformity with the basic safety requirements.

   (b)   The documentation referred to in 3 above shall be retained and kept available for the competent national authorities for at least 10 years following the date of manufacture of the machinery or of the last unit produced, in the case of series manufacture.

   (c)   The documentation referred to in 3 above shall be drawn up in one of the official languages of the Communities, with the exception of the instructions for the machinery.

----

(1)   A body or laboratory is presumed competent if it meets the assessment criteria laid down in the relevant harmonized standards.

ANNEX VI

EC TYPE-EXAMINATION

1. EC type-examination is the procedure by which a notified body ascertains and certifies that an example of machinery satisfies the provisions of this Directive which apply to it.

2. The application for EC type-examination shall be lodged by the manufacturer or by his authorized representative established in the Community, with a single notified body in respect of an example of the machinery.

    The application shall include:

    — the name and address of the manufacturer or his authorized representative established in the Community and the place of manufacture of the machinery,

    — a technical file comprising at least:

    — an overall drawing of the machinery together with drawings of the control circuits,

    — full detailed drawings, accompanied by any calculation notes, test results, etc., required to check the conformity of the machinery with the essential health and safety requirements,

    — a description of methods adopted to eliminate hazards presented by the machinery and a list of standards used,

    — a copy of the instructions for the machinery,

    — for series manufacture, the internal measures that will be implemented to ensure that the machinery remains in conformity with the provisions of the Directive.

    It shall be accompanied by a machine representative of the production planned or, where appropriate, a statement of where the machine may be examined.

    The documentation referred to above does not have to include detailed plans or any other specific information as regards the sub-assemblies used for the manufacture of the machinery unless a knowledge of them is essential for verification of conformity with the basic safety requirements.

3. The notified body shall carry out the EC type-examination in the manner described below:

    — it shall examine the technical construction file to verify its appropriateness and the machine supplied or made available to it.

    — during the examination of the machine, the body shall

    (a) ensure that it has been manufactured in conformity which the technical construction file and may safely be used under its intended working conditions;

    (b) check that standards, if used, have been properly applied;

(c) perform appropriate examinations and tests to check that the machine complies with the essential health and safety requirements applicable to it.

4. If the example complies with the provisions applicable to it the body shall draw up an EC type-examination certificate which shall be forwarded to the applicant. That certificate shall state the conclusions of the examination, indicate any conditions to which its issue may be subject and be accompanied by the descriptions and drawings necessary for identification of the approved example.

   The Commission, the Member States and the other approved bodies may obtain a copy of the certificate and on a reasoned request, a copy of the technical construction file and of the reports on the examinations and tests carried out.

5. The manufacturer or his authorized representative established in the Community shall inform the notified body of any modifications, even of a minor nature, which he has made or plans to make to the machine to which the example relates. The notified body shall examine those modifications and inform the manufacturer or his authorized representative established in the Community whether the EC type-examination certificate remains valid.

6. A body which refuses to issue an EC type-examination certificate shall so inform the other notified bodies. A body which withdraws an EC type-examination certificate shall so inform the Member State which notified it. The latter shall inform the other Member States and the Commission thereof, giving the reasons for the decision.

7. The files and correspondence referring to the EC type-examination procedures shall be drawn up in an official language of the Member State where the notified body is established or in a language acceptable to it.

ANNEX VII

MINIMUM CRITERIA TO BE TAKEN INTO ACCOUNT BY MEMBER STATES FOR THE NOTIFICATION OF BODIES

1. The body, its director and the staff responsible for carrying out the verification tests shall not be the designer, manufacturer, supplier or installer of machinery which they inspect, nor the authorized representative of any of these parties. They shall not become either involved directly or as authorized representatives in the design, construction, marketing or maintenance of the machinery. This does not preclude the possibility of exchanges of technical information between the manufacturer and the body.

2. The body and its staff shall carry out the verification tests with the highest degree of professional integrity and technical competence and shall be free from all pressures and inducements, particularly financial, which might influence their judgement or the results of the inspection, especially from persons or groups of persons with an interest in the result of verifications.

3. The body shall have at its disposal the necessary staff and possess the necessary facilities to enable it to perform properly the administrative and technical tasks connected with verification; it shall also have access to the equipment required for special verification.

4. The staff responsible for inspection shall have:

   — sound technical and professional training,

   — satisfactory knowledge of the requirements of the tests they carry out and adequate experience of such tests,

   — the ability to draw up the certificates, records and reports required to authenticate the performance of the tests .

5. The impartiality of inspection staff shall be guaranteed. Their remuneration shall not depend on the number of tests carried out or on the results of such tests.

6. The body shall take out liability insurance unless its liability is assumed by the State in accordance with national law, or the Member State itself is directly responsible for the tests.

7. The staff of the body shall be bound to observe professional secrecy with regard to all information gained in carrying out its tasks (except vis-à-vis the competent administrative authorities of the State in which its activities are carried out) under this Directive or any provision of national law giving effect to it.

COUNCIL DIRECTIVE

of 30 November 1989

concerning the minimum safety and health requirements for the workplace
(first individual Directive within the meaning of Article 16(1) of Directive 89/391/EEC)
[89/654/EEC]

THE COUNCIL OF THE EUROPEAN COMMUNITIES,

Having regard to the Treaty establishing the European Economic Community, and in particular Article 118a thereof,

Having regard to the proposal from the Commission, submitted after consulting the Advisory Committee on Safety, Hygiene and Health Protection at Work,

In co-operation with the European Parliament,

Having regard to the opinion of the Economic and Social Committee,

Whereas Article 118a of the Treaty provides that the Council shall adopt, by means of directives, minimum requirements for encouraging improvements, especially in the working environment, to ensure a better level of protection of the safety and health of workers;

Whereas, under the terms of that Article, those directives are to avoid imposing administrative, financial and legal constraints in a way which would hold back the creation and development of small and medium-sized undertakings;

Whereas the communication from the Commission on its programme concerning safety, hygiene and health at work provides for the adoption of a directive designed to guarantee the safety and health of workers at the workplace;

Whereas, in its resolution of 21 December 1987 on safety, hygiene and health at work, the Council took note of the Commission's intention of submitting to the Council in the near future minimum requirements concerning the arrangement of the place of work;

Whereas compliance with the minimum requirements designed to guarantee a better standard of safety and health at work is essential to ensure the safety and health of workers;

Whereas this Directive is an individual directive within the meaning of Article 16(1) of Council Directive 89/391/EEC of 12 June 1989 on the introduction of measures to encourage improvements in the safety and health of workers at work; whereas the provisions of the latter are therefore fully applicable to the workplace without prejudice to more stringent and/or specific provisions contained in the present Directive;

Whereas this Directive is a practical contribution towards creating the social dimension of the internal market;

Whereas, pursuant to Decision 74/325/EEC, as last amended by the 1985 Act of Accession, the Advisory Committee on Safety, Hygiene and Health Protection at Work is consulted by the Commission on the drafting of proposals in this field,

HAS ADOPTED THIS DIRECTIVE:

SECTION I
GENERAL PROVISIONS

Article 1
Subject

1. This Directive, which is the first individual directive within the meaning of Article 16(1) of Directive 89/391/EEC, lays down minimum requirements for safety and health at the workplace, as defined in Article 2.

2. This Directive shall not apply to:

    (a) means of transport used outside the undertaking and/or the establishment, or workplaces' inside means of transport;

    (b) temporary or mobile work sites;

    (c) extractive industries;

    (d) fishing boats;

    (e) fields, woods and other land forming part of an agricultural or forestry undertaking but situated away from the undertaking's buildings.

3. The provisions of Directive 89/391/EEC are fully applicable to the whole scope referred to in paragraph 1, without prejudice to more restrictive and/or specific provisions contained in this Directive.

Article 2
Definition

For the purposes of this Directive, 'workplace' means the place intended to house workstations on the premises of the undertaking and/or establishment and any other place within the area of the undertaking and/or establishment to which the worker has access in the course of his employment.

SECTION II
EMPLOYERS' OBLIGATIONS

Article 3
Workplaces used for the first time

Workplaces used for the first time after 31 December 1992 must satisfy the minimum safety and health requirements laid down in Annex I.

Article 4
Workplaces already in use

Workplaces already in use before 1 January 1993 must satisfy the minimum safety and health requirements laid down in Annex II at the latest three years after that date.

However, as regards the Portuguese Republic, workplaces used before 1 January 1993 must satisfy, at the latest four years after that date, the minimum safety and health requirements appearing in Annex II.

Article 5
Modifications to workplaces

When workplaces undergo modifications, extensions and/or conversions after 31 December 1992, the employer shall take the measures necessary to ensure that those modifications, extensions and/or conversions are in compliance with the corresponding minimum requirements laid down in Annex I.

Article 6
General requirements

To safeguard the safety and health of workers, the employer shall see to it that:

— traffic routes to emergency exits and the exits themselves are kept clear at all times,

— technical maintenance of the workplace and of the equipment and devices, and in particular those referred to in Annexes I and II, is carried out and any faults found which are liable to affect the safety and health of workers are rectified as quickly as possible,

— the workplace and the equipment and devices, and in particular those referred to in Annex I, point 6, and Annex II, point 6, are regularly cleaned to an adequate level of hygiene,

— safety equipment and devices intended to prevent or eliminate hazards, and in particular those referred to in Annexes I and II, are regularly maintained and checked.

Article 7
Information of workers

Without prejudice to Article 10 of Directive 89/391/EEC, workers and/or their representatives shall be informed of all measures to be taken concerning safety and health at the workplace.

Article 8
Consultation of workers and workers' participation

Consultation and participation of workers and/or of their representatives shall take place in accordance with Article 11 of Directive 89/391/EEC on the matters covered by this Directive, including the Annexes thereto.

SECTION III
MISCELLANEOUS PROVISIONS

Article 9
Amendments to the Annexes

Strictly technical amendments to the Annexes as a result of:

— the adoption of Directives on technical harmonization and standardization of the design, manufacture or construction of parts of workplaces, and/or

— technical progress, changes in international regulations or specifications and knowledge with regard to workplaces,

shall be adopted in accordance with the procedure laid down in Article 17 of Directive 89/391/EEC.

Article 10
Final provisions

1.   Member States shall bring into force the laws, regulations and administrative provisions necessary to comply with this Directive by 31 December 1992. They shall forthwith inform the Commission thereof.

   However, the date applicable for the Hellenic Republic shall be 31 December 1994.

2.   Member States shall communicate to the Commission the texts of the provisions of national law which they have already adopted or adopt in the field governed by this Directive.

3.   Member States shall report to the Commission every five years on the practical implementation of the provisions of this Directive, indicating the points of view of employers and workers.

   The Commission shall inform the European Parliament, the Council, the Economic and Social Committee and the Advisory Council on Safety, Hygiene and Health Protection at Work.

4.   The Commission shall submit periodically to the European Parliament, the Council and the Economic and Social Committee a report on the implementation of this Directive, taking into account paragraphs 1 to 3.

Article 11

This Directive is addressed to the Member States.

Done at Brussels, 30 November 1989.

ANNEX I

MINIMUM SAFETY AND HEALTH REQUIREMENTS FOR WORKPLACES USED FOR THE FIRST TIME, AS REFERRED TO IN ARTICLE 3 OF THE DIRECTIVE

1. *Preliminary note*

   The obligations laid down in this Annex apply whenever required by the features of the workplace, the activity, the circumstances or a hazard.

2. *Stability and solidity*

   Buildings which house workplaces must have a structure and solidity appropriate to the nature of their use.

3. *Electrical installations*

   Electrical installations must be designed and constructed so as not to present a fire or explosion hazard; persons must be adequately protected against the risk of accidents caused by direct or indirect contact.

   The design, construction and choice of material and protection devices must be appropriate to the voltage, external conditions and the competence of persons with access to parts of the installation.

4. *Emergency routes and exits*

4.1. Emergency routes and exits must remain clear and lead as directly as possible to the open air or to a safe area.

4.2. In the event of danger, it must be possible for workers to evacuate all workstations quickly and as safely as possible.

4.3. The number, distribution and dimensions of the emergency routes and exits depend on the use, equipment and dimensions of the workplaces and the maximum number of persons that may be present.

4.4. Emergency doors must open outwards.

   Sliding or revolving doors are not permitted if they are specifically intended as emergency exits.

   Emergency doors should not be so locked or fastened that they cannot be easily and immediately opened by any person who may require to use them in an emergency.

4.5. Specific emergency routes and exits must be indicated by signs in accordance with the national regulations transposing Directive 77/576/EEC into law.

   Such signs must be placed at appropriate points and be made to last.

4.6. Emergency doors must not be locked.

   The emergency routes and exits, and the traffic routes and doors giving access to them, must be free from obstruction so that they can be used at any time without hindrance.

4.7. Emergency routes and exits requiring illumination must be provided with emergency lighting of adequate intensity in case the lighting fails.

5.   *Fire detection and fire fighting*

5.1.   Depending on the dimensions and use of the buildings, the equipment they contain, the physical and chemical properties of the substances present and the maximum potential number of people present, workplaces must be equipped with appropriate fire-fighting equipment and, as necessary, with fire detectors and alarm systems.

Non-automatic fire-fighting equipment must be easily accessible and simple to use.

The equipment must be indicated by signs in accordance with the national regulations transposing Directive 77/576/EEC into law.

Such signs must be placed at appropriate points and be made to last.

6.   *Ventilation of enclosed workplaces*

Steps shall be taken to see to it that there is sufficient fresh air in enclosed workplaces, having regard to the working methods used and the physical demands placed on the workers.

If a forced ventilation system is used, it shall be maintained in working order.

Any breakdown must be indicated by a control system where this is necessary for workers' health.

If air-conditioning or mechanical ventilation installations are used, they must operate in such a way that workers are not exposed to draughts which cause discomfort.

Any deposit or dirt likely to create an immediate danger to the health of workers by polluting the atmosphere must be removed without delay.

7.   *Room temperature*

During working hours, the temperature in rooms containing workplaces must be adequate for human beings, having regard to the working methods being used and the physical demands placed on the workers.

The temperature in rest areas, rooms for duty staff, sanitary facilities, canteens and first aid rooms must be appropriate to the particular purpose of such areas.

Windows, skylights and glass partitions should allow excessive effects of sunlight in workplaces to be avoided, having regard to the nature of the work and of the workplace.

8.   *Natural and artificial room lighting*

Workplaces must as far as possible receive sufficient natural light and be equipped with artificial lighting adequate for the protection of workers' safety and health.

Lighting installations in rooms containing workplaces and in passageways must be placed in such a way that there is no risk of accident to workers as a result of the type of lighnng fitted.

Workplaces in which workers are especially exposed to risks in the event of failure of artificial lighting must be provided with emergency lighting of adequate intensity.

9.   *Floors, walls, ceilings and roofs of rooms*

The floors of workplaces must have no dangerous bumps, holes or slopes and must be fitted, stable and not slippery.

Workplaces containing workstations must be adequately insulated, bearing in mind the type of undertaking involved and the physical activity of the workers.

The surfaces of floors, walls and ceilings in rooms must be such that they can be cleaned or refurbished to an appropriate standard of hygiene.

Transparent or translucent walls, in particular all-glass partitions, in rooms or in the vicinity of workplaces and traffic routes must be clearly indicated and made of safety material or be shielded from such places or traffic routes to prevent workers from coming into contact with walls or being injured should the walls shatter.

Access to roofs made of materials of insufficient strength must not be permitted unless equipment is provided to ensure that the work can be carried out in a safe manner.

10.1.   It must be possible for workers to open, close, adjust or secure windows, skylights and ventilators in a safe manner. When open, they must not be positioned so as to constitute a hazard to workers.

10.2.   Windows and skylights must be designed in conjunction with equipment or otherwise fitted with devices allowing them to be cleaned without risk to the workers carrying out this work or to workers present in and around the building.

11.   *Doors and gates*

11.1.   The position, number and dimensions of doors and gates, and the materials used in their construction, are determined by the nature and use of the rooms or areas.

11.2.   Transparent doors must be appropriately marked at a conspicuous level.

11.3.   Swing doors and gates must be transparent or have see-through panels.

11.4.   If transparent or translucent surfaces in doors and gates are not made of safety material and if there is a danger that workers may be injured if a door or gate should shatter, the surfaces must be protected against breakage.

11.5.   Sliding doors must be fitted with a safety device to prevent them from being derailed and falling over.

11.6.   Doors and gates opening upwards must be fitted with a mechanism to secure them against falling back.

11.7.   Doors along escape routes must be appropriately marked.

It must be possible to open them from the inside at any time without special assistance.

It must be possible to open the doors when the workplaces are occupied.

11.8.   Doors for pedestrians must be provided in the immediate vicinity of any gates intended essentially for vehicle traffic, unless it is safe for pedestrians to pass through; such doors must be clearly marked and left permanently unobstructed.

11.9.   Mechanical doors and gates must function in such a way that there is no risk of accident to workers.

They must be fitted with easily identifiable and accessible emergency shut-down devices and, unless they open automatically in the event of a power failure, it must also be possible to open them manually.

12.   *Traffic routes — danger areas*

12.1.   Traffic routes, including stairs, fixed ladders and loading bays and ramps, must be located and dimensioned to ensure easy, safe and appropriate access for pedestrians or vehicles in such a way as not to endanger workers employed in the vicinity of these traffic routes.

12.2.   Routes used for pedestrian traffic and/or goods traffic must be dimensioned in accordance with the number of potential users and the type of undertaking.

If means of transport are used on traffic routes, a sufficient safety clearance must be provided for pedestrians.

12.3.   Sufficient clearance must be allowed between vehicle traffic routes and doors, gates, passages for pedestrians, corridors and staircases.

12.4.   Where the use and equipment of rooms so requires for the protection of workers, traffic routes must be clearly identified.

12.5.   If the workplaces contain danger areas in which, owing to the nature of the work, there is a risk of the worker or objects falling, the places must be equipped, as far as possible, with devices preventing unauthorized workers from entering those areas.

Appropriate measures must be taken to protect workers authorized to enter danger areas.

Danger areas must be clearly indicated.

13.   *Specific measures for escalators and travelators*

Escalators and travelators must function safely. They must be equipped with any necessary safety devices. They must be fitted with easily identifiable and accessible emergency shut-down devices.

14.   *Loading bays and ramps*

14.1.   Loading bays and ramps must be suitable for the dimensions of the loads to be transported.

14.2.   Loading bays must have at least one exit point. Where technically feasible, bays over a certain length must have an exit point at each end.

14.3.   Loading ramps must as far as possible be safe enough to prevent workers from falling off.

15. *Room dimensions and air space in rooms — freedom of movement at the workstation*

15.1. Workrooms must have sufficient surface area, height and air space to allow workers to perform their work without risk to their safety, health or well-being.

15.2. The dimensions of the free unoccupied area at the workstation must be calculated to allow workers sufficient freedom of movement to perform their work.

If this is not possible for reasons specific to the workplace, the worker must be provided with sufficient freedom of movement near his workstation.

16. *Rest rooms*

16.1. Where the safety or health of workers, in particular because of the type of activity carried out or the presence of more than a certain number of employees, so require, workers must be provided with an easily accessible rest room.

This provision does not apply if the workers are employed in offices or similar workrooms providing equivalent relaxation during breaks.

16.2. Rest rooms must be large enough and equipped with an adequate number of tables and seats with backs for the number of workers.

16.3. In rest rooms appropriate measures must be introduced for the protection of non-smokers against discomfort caused by tobacco smoke.

16.4. If working hours are regularly and frequently interrupted and there is no rest room, other rooms must be provided in which workers can stay during such interruptions, wherever this is required for the safety or health of workers.

Appropriate measures should be taken for the protection of non-smokers against discomfort caused by tobacco smoke.

17. *Pregnant women and nursing mothers*

Pregnant women and nursing mothers must be able to lie down to rest in appropriate conditions.

18. **Sanitary equipment**

18.1. *Changing rooms and lockers*

18.1.1. Appropriate changing rooms must be provided for workers if they have to wear special work clothes and where, for reasons of health or propriety, they cannot be expected to change in another room.

Changing rooms must be easily accessible, be of sufficient capacity and be provided with seating.

18.1.2. Changing rooms must be sufficiently large and have facilities to enable each worker to lock away his clothes during working hours.

If circumstances so require (*eg.* dangerous substances, humidity, dirt), lockers for work clothes must be separate from those for ordinary clothes.

18.1.3. Provision must be made for separate changing rooms or separate use of changing rooms for men and women.

18.1.4.   If changing rooms are not required under 18.1.1, each worker must be provided with a place to store his clothes.

18.2.   *Showers and washbasins*

18.2.1.   Adequate and suitable showers must be provided for workers if required by the nature of the work or for health reasons.

Provision must be made for separate shower rooms or separate use of shower rooms for men and women.

18.2.2.   The shower rooms must be sufficiently large to permit each worker to wash without hindrance in conditions of an appropriate standard of hygiene.

The showers must be equipped with hot and cold running water.

18.2.3.   Where showers are not required under the first sub-paragraph of 18.2.1, adequate and suitable washbasins with running water (hot water if necessary) must be provided in the vicinity of the workstations and the changing rooms.

Such washbasins must be separate for, or used separately by, men and women when so required for reasons of propriety.

18.2.4.   Where the rooms housing the showers or washbasins are separate from the changing rooms, there must be easy communication between the two.

18.3.   *Lavatories and washbasins*

Separate facilities must be provided in the vicinity of workstations, rest rooms, changing rooms and rooms housing showers or washbasins, with an adequate number of lavatories and washbasins.

Provision must be made for separate lavatories or separate use of lavatories for men and women.

19.   *First aid rooms*

19.1.   One or more first aid rooms must be provided where the size of the premises, type of activity being carried out and frequency of accidents so dictate.

19.2.   First aid rooms must be fitted with essential first aid installations and equipment and be easily accessible to stretchers.

They must be signposted in accordance with the national regulations transposing Directive 77/576/EEC into law.

19.3.   In addition, first aid equipment must be available in all places where working conditions require it.

This equipment must be suitably marked and easily accessible.

20.   *Handicapped workers*

Workplaces must be organized to take account of handicapped workers, if necessary.

This provision applies in particular to the doors, passageways, staircases, showers, washbasins, lavatories and workstations used or occupied directly by handicapped persons.

21.   *Outdoor workplaces (special provisions)*

21.1.   Workstations, traffic routes and other areas or installations outdoors which are used or occupied by the workers in the course of their activity must be organized in such a way that pedestrians and vehicles can circulate safely.

Sections 12, 13 and 14 also apply to main traffic routes on the site of the undertaking (traffic routes leading to fixed workstations), to traffic routes used for the regular maintenance and supervision of the undertaking's installations and to loading bays.

Section 12 is also applicable to outdoor workplaces.

21.2.   Workplaces outdoors must be adequately lit by artificial lighting if daylight is not adequate.

21.3.   When workers are employed at workstations outdoors, such workstations must as far as possible be arranged so that workers:

(a)   are protected against inclement weather conditions and if necessary against falling objects;

(b)   are not exposed to harmful noise levels nor to harmful external influences such as gases, vapours or dust;

(c)   are able to leave their workstations swiftly in the event of danger or are able to be rapidly assisted;

(d)   cannot slip or fall.

ANNEX II

MINIMUM HEALTH AND SAFETY REQUIREMENTS FOR WORKPLACES ALREADY IN USE, AS REFERRED TO IN ARTICLE 4 OF THE DIRECTIVE

1. *Preliminary note*

   The obligations laid down in this Annex apply wherever required by the features of the workplace, the activity, the circumstances or a hazard.

2. *Stability and solidity*

   Buildings which have workplaces must have a structure and solidity appropriate to the nature of their use.

3. *Electrical installations*

   Electrical installations must be designed and constructed so as not to present a fire or explosion hazard; persons must be adequately protected against the risk of accidents caused by direct or indirect contact.

   Electrical installations and protection devices must be appropriate to the voltage, external conditions and the competence of persons with access to parts of the installation.

4. *Emergency routes and exits*

4.1. Emergency routes and exits must remain clear and lead as directly as possible to the open air or to a safe area.

4.2. In the event of danger, it must be possible for workers to evacuate all workstations quickly and as safely as possible.

4.3. There must be an adequate number of escape routes and emergency exits.

4.4. Emergency exit doors must open outwards.

   Sliding or revolving doors are not permitted if they are specifically intended as emergency exits.

   Emergency doors should not be so locked or fastened that they cannot be easily and immediately opened by any person who may require to use them in an emergency.

4.5. Specific emergency routes and exits must be indicated by signs in accordance with the national regulations transposing Directive 77/576/EEC into law.

   Such signs must be placed at appropriate points and be made to last.

4.6. Emergency doors must not be locked.

   The emergency routes and exits, and the traffic routes and doors giving access to them, must be free from obstruction so that they can be used at any time without hindrance.

4.7. Emergency routes and exits requiring illumination must be provided with emergency lighting of adequate intensity in case the lighting fails.

5.   *Fire detection and fire fighting*

5.1.   Depending on the dimensions and use of the buildings, the equipment they contain, the physical and chemical characteristics of the substances present and the maximum potential number of people present, workplaces must be equipped with appropriate fire-fighting equipment, and, as necessary, fire detectors and an alarm system.

5.2.   Non-automatic fire-fighting equipment must be easily accessible and simple to use.

It must be indicated by signs in accordance with the national regulations transposing Directive 77/576/EEC into law.

Such signs must be placed at appropriate points and be made to last.

6.   *Ventilation of enclosed workplaces*

Steps shall be taken to see to it that there is sufficient fresh air in enclosed workplaces, having regard to the working methods used and the physical demands placed on the workers.

If a forced ventilation system is used, it shall be maintained in working order.

Any breakdown must be indicated by a control system where this is necessary for the workers' health.

7.   *Room temperature*

7.1.   During working hours, the temperature in rooms containing workplaces must be adequate for human beings, having regard to the working methods being used and the physical demands placed on the workers.

7.2.   The temperature in rest areas, rooms for duty staff, sanitary facilities, canteens and first aid rooms must be appropriate to the particular purpose of such areas.

8.   *Natural and artificial room lighting*

8.1.   Workplaces must as far as possible receive sufficient natural light and be equipped with artificial lighting adequate for workers' safety and health.

8.2.   Workplaces in which workers are especially exposed to risks in the event of failure of artificial lighting must be provided with emergency lighting of adequate intensity.

9.   *Doors and gates*

9.1.   Transparent doors must be appropriately marked at a conspicuous level.

9.2.   Swing doors and gates must be transparent or have see-through panels.

10.   *Danger areas*

If the workplaces contain danger areas in which, owing to the nature of the work, there is a risk of the worker or objects falling, the places must be equipped, as far as possible, with devices preventing unauthorized workers from entering those areas.

Appropriate measures must be taken to protect workers authorized to enter danger areas.

Danger areas must be clearly indicated.

11. *Rest rooms and rest areas*

11.1. Where the safety or health of workers, in particular because of the type of activity carried out or the presence of more than a certain number of employees, so require, workers must be provided with an easily accessible rest room or appropriate rest area.

This provision does not apply if the workers are employed in offices or similar workrooms providing equivalent relaxation during breaks.

11.2. Rest rooms and rest areas must be equipped with tables and seats with backs.

11.3. In rest rooms and rest areas appropriate measures must be introduced for the protection of non-smokers against discomfort caused by tobacco smoke.

12. *Pregnant women and nursing mothers*

Pregnant women and nursing mothers must be able to lie down to rest in appropriate conditions.

13. **Sanitary equipment**

13.1. *Changing rooms and lockers*

13.1.1. Appropriate changing rooms must be provided for workers if they have to wear special work clothes and where, for reasons of health or propriety, they cannot be expected to change in another room.

Changing rooms must be easily accessible and of sufficient capacity.

13.1.2. Changing rooms must have facilities to enable each worker to lock away his clothes during working hours.

If circumstances so require (*eg*. dangerous substances, humidity, dirt), lockers for work clothes must be separate from those for ordinary clothes.

13.1.3. Provision must be made for separate changing rooms or separate use of changing rooms for men and women.

13.2. *Showers, lavatories and washbasins*

13.2.1. Workplaces must be fitted out in such a way that workers have in the vicinity:

— showers, if required by the nature of their work,

— special facilities equipped with an adequate number of lavatories and washbasins.

13.2.2. The showers and washbasins must be equipped with running water (hot water if necessary).

13.2.3. Provision must be made for separate showers or separate use of showers for men and women. Provision must be made for separate lavatories or separate use of lavatories for men and women.

14. *First aid equipment*

Workplaces must be fitted with first aid equipment. The equipment must be suitably marked and easily accessible.

15.   *Handicapped workers*

Workplaces must be organized to take account of handicapped workers, if necessary.

This provision applies in particular to the doors, passageways, staircases, showers, washbasins, lavatories and workstations used or occupied directly by handicapped persons.

16.   *Movement of pedestrians and vehicles*

Outdoor and indoor workplaces must be organized in such a way that pedestrians and vehicles can circulate in a safe manner.

17.   *Outdoor workplaces (special provisions)*

When workers are employed at workstations outdoors, such workstations must as far as possible be organized so that workers:

(a)   are protected against inclement weather conditions and if necessary against falling objects;

(b)   are not exposed to harmful noise levels nor to harmful external influences such as gases, vapours or dust;

(c)   are able to leave their workstations swiftly in the event of danger or are able to be rapidly assisted;

(d)   cannot slip or fall.

COUNCIL DIRECTIVE

of 30 November 1989

concerning the minimum safety and health requirements for the use of work equipment by workers at work

(second individual Directive within the meaning of Article 16(1) of Directive 89/391/EEC)

[89/655/EEC]

THE COUNCIL OF THE EUROPEAN COMMUNITIES,

Having regard to the Treaty establishing the European Economic Community, and in particular Article 118a thereof,

Having regard to the proposal from the Commission, submitted after consulting the Advisory Committee on Safety, Hygiene and Health Protection at Work,

In co-operation with the European Parliament,

Having regard to the opinion of the Economic and Social Committee,

Whereas Article 118a of the Treaty provides that the Council shall adopt, by means of directives, minimum requirements for encouraging improvements, especially in the working environment, to guarantee a better level of protection of the safety and health of workers;

Whereas, pursuant to the said Article, such directives must avoid imposing administrative, financial and legal constraints in a way which would hold back the creation and development of small and medium-sized undertakings;

Whereas the communication from the Commission on its programme concerning safety, hygiene and health at work provides for the adoption of a directive on the use of work equipment at work;

Whereas, in its resolution of 21 December 1987 on safety, hygiene and health at work, the Council took note of the Commission's intention of submitting to the Council in the near future minimum requirements concerning the organization of safety and health at work;

Whereas compliance with the minimum requirements designed to guarantee a better standard of safety and health in the use of work equipment is essential to ensure the safety and health of workers;

Whereas this Directive is an individual directive within the meaning of Article 16(1) of Council Directive 89/391/EEC of 12 June 1989 on the introduction of measures to encourage improvements in the safety and health of workers at work; whereas, therefore, the provisions of the said Directive are fully applicable to the scope of the use of work equipment by workers at work without prejudice to more restrictive and/or specific provisions contained in this Directive;

Whereas this Directive constitutes a practical aspect of the realization of the social dimension of the internal market;

Whereas, pursuant to Directive 83/189/EEC, Member States are required to notify the Commission of any draft technical regulations relating to machines, equipment and installations;

Whereas, pursuant to Decision 74/325/EEC, as last amended by the 1985 Act of Accession, the Advisory Committee on Safety, Hygiene and Health Protection at Work is consulted by the Commission on the drafting of proposals in this field,

HAS ADOPTED THIS DIRECTIVE:

SECTION I
GENERAL PROVISIONS

Article 1
Subject

1. This Directive, which is the second individual directive within the meaning of Article 16 (1) of Directive 89/391/EEC, lays down minimum safety and health requirements for the use of work equipment by workers at work, as defined in Article 2.

2. The provisions of Directive 89/391/EEC are fully applicable to the whole scope referred to in paragraph 1, without prejudice to more restrictive and/or specific provisions contained in this Directive.

Article 2
Definitions

For the purposes of this Directive, the following terms shall have the following meanings:

(a)	'work equipment':	any machine, apparatus, tool or installation used at work;
(b)	'use of work equipment':	any activity involving work equipment such as starting or stopping the equipment, its use, transport, repair, modification, maintenance and servicing, including, in particular, cleaning;
(c)	'danger zone':	any zone within and/or around work equipment in which an exposed worker is subject to a risk to his health or safety;
(d)	'exposed worker':	any worker wholly or partially in a danger zone;
(e)	'operator':	the worker or workers given the task of using work equipment.

SECTION II
EMPLOYERS' OBLIGATIONS

Article 3
General obligations

1. The employer shall take the measures necessary to ensure that the work equipment made available to workers in the undertaking and/or establishment is suitable for the work to be carried out or properly adapted for that purpose and may be used by workers without impairment to their safety or health.

   In selecting the work equipment which he proposes to use, the employer shall pay attention to the specific working conditions and characteristics and to the hazards which exist in the undertaking and/or establishment, in particular at the workplace, for the safety and health of the workers, and/or any additional hazards posed by the use of work equipment in question.

2. Where it is not possible fully so to ensure that work equipment can be used by workers without risk to their safety or health, the employer shall take appropriate measures to minimize the risks.

Article 4
Rules concerning work equipment

1. Without prejudice to Article 3, the employer must obtain and/or use:

   (a) work equipment which, if provided to workers in the undertaking and/or establishment for the first time after 31 December 1992, complies with:

       (i) the provisions of any relevant Community directive which is applicable;

       (ii) the minimum requirements laid down in the Annex, to the extent that no other Community directive is applicable or is so only partially;

   (b) work equipment which, if already provided to workers in the undertaking and/or establishment by 31 December 1992, complies with the minimum requirements laid down in the Annex no later than four years after that date.

2. The employer shall take the measures necessary to ensure that, throughout its working life, work equipment is kept, by means of adequate maintenance, at a level such that it complies with the provisions of paragraph 1(a) or (b) as applicable.

Article 5
Work equipment involving specific risks

When the use of work equipment is likely to involve a specific risk to the safety or health of workers, the employer shall take the measures necessary to ensure that:

— the use of work equipment is restricted to those persons given the task of using it;

— in the case of repairs, modifications, maintenance or servicing, the workers concerned are specifically designated to carry out such work.

Article 6
Informing workers

1. Without prejudice to Article 10 of Directive 89/391/EEC, the employer shall take the measures necessary to ensure that workers have at their disposal adequate information and, where appropriate, written instructions on the work equipment used at work.

2. The information and the written instructions must contain at least adequate safety and health information concerning:

    —   the conditions of use of work equipment,

    —   foreseeable abnormal situations,

    —   the conclusions to be drawn from experience, where appropriate, in using work equipment.

3. The information and the written instructions must be comprehensible to the workers concerned.

Article 7
Training of workers

Without prejudice to Article 12 of Directive 89/391/EEC the employer shall take the measures necessary to ensure that:

—   workers given the task of using work equipment receive adequate training, including training on any risks which such use may entail,

—   workers referred to in the second indent of Article 5 receive adequate specific training.

Article 8
Consultation of workers and workers' participation

Consultation and participation of workers and/or of their representatives shall take place in accordance with Article 11 of Directive 89/391/EEC on the matters covered by this Directive, including the Annexes thereto.

SECTION III
MISCELLANEOUS PROVISIONS

Article 9
Amendment to the Annex

1. Addition to the Annex of the supplementary minimum requirements applicable to specific work equipment referred to in point 3 thereof shall be adopted by the Council in accordance with the procedure laid down in Article 118a of the Treaty.

2. Strictly technical adaptations of the Annex as a result of:

    —   the adoption of directives on technical harmonization and standardization of work equipment, and/or

— technical progress, changes in international regulations or specifications or knowledge in the field of work equipment

shall be adopted, in accordance with the procedure laid down in Article 17 of Directive 89/391/EEC.

Article 10
Final provisions

1. Member States shall bring into force the laws, regulations and administrative provisions necessary to comply with this Directive by 31 December 1992. They shall forthwith inform the Commission thereof.

2. Member States shall communicate to the Commission the texts of the provisions of national law which they have already adopted or adopt in the field governed by this Directive.

3. Member States shall report to the Commission every five years on the practical implementation of the provisions of this Directive, indicating the points of view of employers and workers.

   The Commission shall accordingly inform the European Parliament, the Council, the Economic and Social Committee, and the Advisory Committee on Safety, Hygiene and Health Protection at Work.

4. The Commission shall submit periodically to the European Parliament, the Council and the Economic and Social Committee a report on the implementation of this Directive, taking into account paragraphs 1 to 3.

Article 11

This Directive is addressed to the Member States.

Done at Brussels, 30 November 1989.

ANNEX

MINIMUM REQUIREMENTS REFERRED TO IN ARTICLE 4(1)(a)(ii) and (b)

1. **General comment**

    The obligations laid down in this Annex apply having regard to the provisions of the Directive and where the corresponding risk exists for the work equipment in question.

2. **General minimum requirements applicable to work equipment**

2.1. Work equipment control devices which affect safety must be clearly visible and identifiable and appropriately marked where necessary.

    Except where necessary for certain control devices, control devices must be located outside danger zones and in such a way that their operation cannot cause additional hazard. They must not give rise to any hazard as a result of any unintentional operation.

    If necessary, from the main control position, the operator must be able to ensure that no person is present in the danger zones. If this is impossible, a safe system such as an audible and/or visible warning signal must be given automatically whenever the machinery is about to start. An exposed worker must have the time and/or the means quickly to avoid hazards caused by the starting and/or stopping of the work equipment.

    Control systems must be safe. A breakdown in, or damage to, control systems must not result in a dangerous situation.

2.2. It must be possible to start work equipment only by deliberate action on a control provided for the purpose.

    The same shall apply:

    — to restart it after a stoppage for whatever reason,

    — for the control of a significant change in the operating conditions (e.g. speed, pressure, *etc.*), unless such a restart or change does not subject exposed workers to any hazard.

    This requirement does not apply to restarting or a change in operating conditions as a result of the normal operating cycle of an automatic device.

2.3. All work equipment must be fitted with a control to stop it completely and safely.

    Each work station must be fitted with a control to stop some or all of the work equipment, depending on the type of hazard, so that the equipment is in a safe state. The equipment's stop control must have priority over the start controls. When the work equipment or the dangerous parts of it have stopped, the energy supply of the actuators concerned must be switched off.

2.4. Where appropriate, and depending on the hazards the equipment presents and its normal stopping time, work equipment must be fitted with an emergency stop device.

2.5. Work equipment presenting risk due to falling objects or projections must be fitted with appropriate safety devices corresponding to the risk.

Work equipment presenting hazards due to emissions of gas, vapour, liquid or dust must be fitted with appropriate containment and/or extraction devices near the sources of the hazard.

2.6. Work equipment and parts of such equipment must, where necessary for the safeq and health of workers be stabilized by clamping or some other means.

2.7. Where there is a risk of rupture or disintegration of parts of the work equipment, likely to pose significant danger to the safety and health of workers, appropriate protection measures must be taken.

2.8. Where there is a risk of mechanical contact with moving parts of work equipment which could lead to accidents, those parts must be provided with guards or devices to prevent access to danger zones or to halt movements of dangerous parts before the danger zones are reached.

The guards and protection devices must

— be of robust construction,

— not give rise to any additional hazard,

— not be easily removed or rendered inoperative,

— be situated at sufficient distance from the danger zone,

— not restrict more than necessary the view of the operating cycle of the equipment,

— allow operations necessary to fit or replace parts and for maintenance work, restricting access only to the area where the work is to be carried out and, if possible, without removal of the guard or protection device.

2.9. Areas and points for working on, or maintenance of, work equipment must be suitably lit in line with the operation to be carried out.

2.10. Work equipment parts at high or very low temperature must, where appropriate, be protected to avoid the risk of workers coming into contact or coming too close.

2.11. Warning devices on work equipment must be unambiguous and easily perceived and understood.

2.12. Work equipment may be used only for operations and under conditions for which it is appropriate.

2.13. It must be possible to carry out maintenance operations when the equipment is shut down. If this is not possible, it must be possible to take appropriate protection measures for the carrying out of such operations or for such operations to be carried out outside the danger zones.

If any machine has a maintenance log, it must be kept up to date.

2.14. All work equipment must be fitted with clearly identifiable means to isolate it from all its energy sources.

Reconnection must be presumed to pose no risk to the workers concerned.

2.15. Work equipment must bear the warnings and markings essential to ensure the safety of workers.

2.16. Workers must have safe means of access to, and be able to remain safely in, all the areas necessary for production, adjustment and maintenance operations.

2.17. All work equipment must be appropriate for protecting workers against the risk of the work equipment catching fire or overheating, or of discharges of gas, dust, liquid, vapour or other substances produced, used or stored in the work equipment.

2.18. All work equipment must be appropriate for preventing the risk of explosion of the work equipment or of substances produced, used or stored in the work equipment.

2.19. All work equipment must be appropriate for protecting exposed workers against the risk of direct or indirect contact with electricity.

3. **Minimum additional requirements applicable to specific work equipment,**

as referred to in Article 9(1) of the Directive.

COUNCIL DIRECTIVE

of 30 November 1989

on the minimum health and safety requirements for the use by workers of
personal protective equipment at the workplace
(third individual Directive within the meaning of Article 16(1) of Directive 89/391/EEC)
[89/656/EEC]

THE COUNCIL OF THE EUROPEAN COMMUNITIES,

Having regard to the Treaty establishing the European Economic Community and
in particular Article 118a thereof,

Having regard to the Commission proposal, submitted after consultation with the
Advisory Committee on Safety, Hygiene and Health Protection at Work,

In co-operation with the European Parliament,

Having regard to the opinion of the Economic and Social Committee,

Whereas Article 118a of the Treaty provides that the Council shall adopt, by means
of directives, minimum requirements designed to encourage improvements,
especially in the working environment, to guarantee greater protection of the
health and safety of workers;

Whereas, under the said Article, such directives shall avoid imposing administrative, financial and legal constraints in a way which would hold back the creation and
development of small and medium-sized undertakings;

Whereas the Commission communication on its programme concerning safety,
hygiene and health at work provides for the adoption of a directive on the use of
personal protective equipment at work;

Whereas the Council, in its resolution of 21 December 1987 concerning safety,
hygiene and health at work, noted the Commission's intention of submitting to it
in the near future minimum requirements concerning the organization of the
safety and health of workers at work;

Whereas compliance with the minimum requirements designed to guarantee
greater health and safety for the user of personal protective equipment is essential
to ensure the safety and health of workers;

Whereas this Directive is an individual directive within the meaning of Article
16(1) of Council Directive 89/391/EEC of 12 June 1989 on the introduction of
measures to encourage improvements in the safety and health of workers at work;
whereas, consequently, the provisions of the said Directive apply fully to the use by
workers of personal protective equipment at the workplace, without prejudice to
more stringent and/or specific provisions contained in this Directive;

Whereas this Directive constitutes a practical step towards the achievement of the
social dimension of the internal market;

Whereas collective means of protection shall be accorded priority over individual protective equipment; whereas the employer shall be required to provide safety equipment and take safety measures;

Whereas the requirements laid down in this Directive should not entail alterations to personal protective equipment whose design and manufacture complied with Community directives relating to safety and health at work;

Whereas provision should be made for descriptions which Member States may use when laying down general rules for the use of individual protective equipment;

Whereas, pursuant to Decision 74/325/EEC, as last amended by the 1985 Act of Accession, the Advisory Committee on Safety, Hygiene and Health Protection at Work is consulted by the Commission with a view to drawing up proposals in this field,

HAS ADOPTED THIS DIRECTIVE:

SECTION I

GENERAL PROVISIONS

Article 1

Subject

1. This Directive, which is the third individual directive within the meaning of Article 16(1) of Directive 89/391/EEC, lays down minimum requirements for personal protective equipment used by workers at work.

2. The provisions of Directive 89/391/EEC are fully applicable to the whole scope referred to in paragraph 1, without prejudice to more restrictive and/or specific provisions contained in this Directive.

Article 2

Definition

1. For the purposes of this Directive, personal protective equipment shall mean all equipment designed to be worn or held by the worker to protect him against one or more hazards likely to endanger his safety and health at work, and any addition or accessory designed to meet this objective.

2. The definition in paragraph 1 excludes:

    (a) ordinary working clothes and uniforms not specifically designed to protect the safety and health of the worker;

    (b) equipment used by emergency and rescue services;

    (c) personal protective equipment worn or used by the military, the police and other public order agencies;

(d)   personal protective equipment for means of road transport;

(e)   sports equipment;

(f)   self-defence or deterrent equipment;

(g)   portable devices for detecting and signalling risks and nuisances.

Article 3

General rule

Personal protective equipment shall be used when the risks cannot be avoided or sufficiently limited by technical means of collective protection or by measures, methods or procedures of work organization.

SECTION II

EMPLOYERS' OBLIGATIONS

Article 4

General provisions

1.   Personal protective equipment must comply with the relevant Community provisions on design and manufacture with respect to safety and health.

   All personal protective equipment must:

   (a)   be appropriate for the risks involved, without itself leading to any increased risk;

   (b)   correspond to existing conditions at the workplace;

   (c)   take account of ergonomic requirements and the worker's state of health;

   (d)   fit the wearer correctly after any necessary adjustment.

2.   Where the presence of more than one risk makes it necessary for a worker to wear simultaneously more than one item of personal protective equipment, such equipment must be compatible and continue to be effective against the risk or risks in question.

3.   The conditions of use of personal protective equipment, in particular the period for which it is worn, shall be determined on the basis of the seriousness of the risk, the frequency of exposure to the risk, the characteristics of the workstation of each worker and the performance of the personal protective equipment.

4.   Personal protective equipment is, in principle, intended for personal use.

   If the circumstances require personal protective equipment to be worn by more than one person, appropriate measures shall be taken to ensure that such use does not create any health or hygiene problem for the different users.

5. Adequate information on each item of personal protective equipment, required under paragraphs 1 and 2, shall be provided and made available within the undertaking and/or establishment.

6. Personal protective equipment shall be provided free of charge by the employer, who shall ensure its good working order and satisfactory hygienic condition by means of the necessary maintenance, repair and replacements.

    However, Member States may provide, in accordance with their national practice, that the worker be asked to contribute towards the cost of certain personal protective equipment in circumstances where use of the equipment is not exclusive to the workplace.

7. The employer shall first inform the worker of the risks against which the wearing of the personal protective equipment protects him.

8. The employer shall arrange for training and shall, if appropriate, organize demonstrations in the wearing of personal protective equipment.

9. Personal protective equipment may be used only for the purposes specified, except in specific and exceptional circumstances.

    It must be used in accordance with instructions.

    Such instructions must be understandable to the workers.

Article 5

Assessment of personal protective equipment

1. Before choosing personal protective equipment, the employer is required to assess whether the personal protective equipment he intends to use satisfies the requirements of Article 4(1) and (2).

    This assessment shall involve:

    (a) an analysis and assessment of risks which cannot be avoided by other means;

    (b) the definition of the characteristics which personal protective equipment must have in order to be effective against the risks referred to in (a), taking into account any risks which this equipment itself may create;

    (c) comparison of the characteristics of the personal protective equipment available with the characteristics referred to in (b).

2. The assessment provided for in paragraph 1 shall be reviewed if any changes are made to any of its elements.

Article 6

Rules for use

1. Without prejudice to Articles 3, 4 and 5, Member States shall ensure that general rules are established for the use of personal protective equipment and/or rules covering cases and situations where the employer must provide the personal protective equipment, taking account of Community legislation on the free movement of such equipment.

These rules shall indicate in particular the circumstances or the risk situations in which, without prejudice to the priority to be given to collective means of protection, the use of personal protective equipment is necessary.

Annexes I, II and III, which constitute a guide, contain useful information for establishing such rules.

2.   When Member States adapt the rules referred to in paragraph 1, they shall take account of any significant changes to the risk, collective means of protection and personal protective equipment brought about by technological developments.

3.   Member States shall consult the employers' and workers' organization on the rules referred to in paragraphs 1 and 2.

Article 7

Information for workers

Without prejudice to Article 10 of Directive 89/391/EEC, workers and/or their representatives shall be informed of all measures to be taken with regard to the health and safety of workers when personal protective equipment is used by workers at work.

Article 8

Consultation of workers and workers' participation

Consultation and participation of workers and/or of their representatives shall take place in accordance with Article 11 of Directive 89/391/EEC on the matters covered by this Directive, including the Annexes thereto.

SECTION III

MISCELLANEOUS PROVISIONS

Article 9

Adjustment of the Annexes

Alterations of a strictly technical nature to Annexes I, II and III resulting from:

— the adoption of technical harmonization and standardization directives relating to personal protective equipment, and/or

— technical progress and changes in international regulations and specifications or knowledge in the field of personal protective equipment,

shall be adopted in accordance with the procedure provided for in Article 17 of Directive 89/391/EEC.

Article 10

Final provisions

1. Member States shall bring into force the laws, regulations and administrative provisions necessary to comply with this Directive not later than 31 December 1992. They shall immediately inform the Commission thereof.

2. Member States shall communicate to the Commission the text of the provisions of national law which they adopt, as well as those already adopted, in the field covered by this Directive.

3. Member States shall report to the Commission every five years on the practical implementation of the provisions of this Directive, indicating the points of view of employers and workers.

    The Commission shall inform the European Parliament, the Council, the Economic and Social Committee, and the Advisory Committee on Safety, Hygiene and Health Protection at Work.

4. The Commission shall report periodically to the European Parliament, the Council and the Economic and Social Committee on the implementation of the Directive in the light of paragraphs 1, 2 and 3.

Article 11

This Directive is addressed to the Member States.

Done at Brussels, 30 November 1989.

ANNEX I

SPECIMEN RISK SURVEY TABLE FOR THE USE OF PERSONAL PROTECTIVE EQUIPMENT

		RISKS																			
		PHYSICAL										CHEMICAL						BIOLOGICAL			
		MECHANICAL					THERMAL		ELEC-TRI-CAL	RADIATION		NOISE	AEROSOLS		LIQUIDS			GASES, VA-POURS	Harm-ful bacteria	Harm-ful viruses	Mycotic fungi	Non-microbe bio-logical antigens
PARTS OF THE BODY		Falls from a height	Blows, cuts, impact, crushing	Stabs, cuts, grazes	Vibra-tion	Slip-ping, falling over	Heat, fire	Cold		Non-ion-izing	Ion-izing		Dust, fibres	Fumes	Vapours	Im-mer-sion	Splashes, spurts					
HEAD	Cranium																					
	Ears																					
	Eyes																					
	Respiratory tract																					
	Face																					
	Whole head																					
UPPER LIMBS	Hands																					
	Arms (parts)																					
LOWER LIMBS	Foot																					
	Legs (parts)																					
VARIOUS	Skin																					
	Trunk/abdomen																					
	Parenteral passages																					
	Whole body																					

ANNEX II

NON-EXHAUSTIVE GUIDE LIST OF ITEMS OF PERSONAL PROTECTIVE EQUIPMENT

HEAD PROTECTION

— Protective helmets for use in industry (mines, building sites, other industrial uses).

— Scalp protection (caps, bonnets, hairnets — with or without eye shade).

— Protective headgear (bonnets, caps, sou'westers, *etc.* in fabric, fabric with proofing, *etc.*).

HEARING PROTECTION

— Earplugs and similar devices.

— Full acoustic helmets.

— Earmuffs which can be fitted to industrial helmets.

— Ear defenders with receiver for LF induction loop.

— Ear protection with intercom equipment.

EYE AND FACE PROTECTION

— Spectacles.

— Goggles.

— X-ray goggles, laser-beam goggles, ultra-violet, infra-red, visible radiation goggles.

— Face shields.

— Arc-welding masks and helmets (hand masks, headband masks or masks which can be fitted to protective helmets).

RESPIRATORY PROTECTION

— Dust filters, gas filters and radioactive dust filters.

— Insulating appliances with an air supply.

— Respiratory devices including a removable welding mask.

— Diving equipment.

— Diving suits.

HAND AND ARM PROTECTION

— Gloves to provide protection:

    — from machinery (piercing, cuts, vibrations, *etc.*),

    — from chemicals,

    — for electricians and from heat.

— Mittens.

— Finger stalls.

— Oversleeves.

— Wrist protection for heavy work.

— Fingerless gloves.

— Protective gloves.

FOOT AND LEG PROTECTION

— Low shoes, ankle boots, calf-length boots, safety boots.

— Shoes which can be unlaced or unhooked rapidly.

— Shoes with additional protective toe-cap.

— Shoes and overshoes with heat-resistant soles.

— Heat-resistant shoes, boots and overboots.

— Thermal shoes, boots and overboots.

— Vibration-resistant shoes, boots and overboots.

— Anti-static shoes, boots and overboots.

— Insulating shoes, boots and overboots.

— Protective boots for chain saw operators.

— Clogs.

— Kneepads.

— Removable instep protectors.

— Gaiters.

— Removable soles (heat-proof, pierce-proof or sweat-proof).

— Removable spikes for ice, snow or slippery flooring.

SKIN PROTECTION

— Barrier creams/ointments.

TRUNK AND ABDOMEN PROTECTION

— Protective waistcoats, jackets and aprons to provide protection from machinery (piercing, cutting, molten metal splashes, *etc.*).

— Protective waistcoats, jackets and aprons to provide protection from chemicals.

— Heated waistcoats.

— Life jackets.

— Protective X-ray aprons.

— Body belts.

WHOLE BODY PROTECTION

— **Equipment designed to prevent falls**

   — Fall-prevention equipment (full equipment with all necessary accessories).

   — Braking equipment to absorb kinetic energy (full equipment with all necessary accessories).

   — Body-holding devices (safety harness).

— **Protective clothing**

   — 'Safety' working clothing (two-piece and overalls).

   — Clothing to provide protection from machinery (piercing, cutting, *etc.*).

   — Clothing to provide protection from chemicals.

   — Clothing to provide protection from molten metal splashes and infra-red radiation.

   — Heat-resistant clothing.

   — Thermal clothing.

   — Clothing to provide protection from radioactive contamination.

   — Dust-proof clothing.

   — Gas-proof clothing.

   — Fluorescent signalling, retro-reflecting clothing and accessories (armbands, gloves, *etc.*).

   — Protective coverings.

ANNEX III

NON-EXHAUSTIVE GUIDE LIST OF ACTIVITIES AND SECTORS OF ACTIVITY WHICH MAY REQUIRE THE PROVISION OF PERSONAL PROTECTIVE EQUIPMENT

1. HEAD PROTECTION (SKULL PROTECTION)

Protective helmets

— Building work, particularly work on, underneath or in the vicinity of scaffolding and elevated workplaces, erection and stripping of formwork, assembly and installation work, work on scaffolding and demolition work.

— Work on steel bridges, steel building construction, masts, towers, steel hydraulic structures, blast furnaces, steel works and rolling mills, large containers, large pipelines, boiler plants and power stations .

— Work in pits, trenches, shafts and tunnels.

— Earth and rock works.

— Work in underground workings, quarries, open diggings, coal stock removal.

— Work with bolt-driving tools.

— Blasting work.

— Work in the vicinity of lifts, lifting gear, cranes and conveyors.

— Work with blast furnaces, direct reduction plants, steelworks, rolling mills, metalworks, forging, drop forging and casting.

— Work with industrial furnaces, containers, machinery, silos, bunkers and pipelines.

— Shipbuilding.

— Railway shunting work.

— Slaughterhouses.

2. FOOT PROTECTION

— Safety shoes with puncture-proof socks

— Carcase work, foundation work and roadworks.

— Scaffolding work.

— The demolition of carcase work.

— Work with concrete and pre-fabricated parts involving formwork erection and stripping.

— Work in contractors' yards and warehouses.

— Roof work. Safety shoes without pierce-proof soles

— Work on steel bridges, steel building construction, masts, towers, lifts, steel hydraulic structures, blast furnaces, steelworks and rolling mills, large containers, large pipelines, cranes, boiler plants and power stations.

— Furnace construction, heating and ventilation installation and metal assembly work.

— Conversion and maintenance work.

— Work with blast furnaces, direct reduction plants, steelworks, rolling mills, metal-works, forging, drop forging, hot pressing and drawing plants.

— Work in quarries and open diggings, coal stock removal.

— Working and processing of rock.

— Flat glass products and container glassware manufacture, working and processing.

— Work with moulds in the ceramics industry.

— Lining of kilns in the ceramics industry.

— Moulding work in the cemmic ware and building materials industry.

— Transport and storage.

— Work with frozen meat blocks and preserved foods packaging.

— Shipbuilding.

— Railway shunting work.

Safety shoes with heels or wedges and pierce-proof soles

— Roof work.

Protective shoes with insulated soles

— Work with and on very hot or very cold materials.

Safety shoes which can easily be removed

— Where there is a risk of penetration by molten substances.

3.   EYE OR FACE PROTECTION

Protective goggles, face shields or screens

— Welding, grinding and separating work.

— Caulking and chiselling.

— Rock working and processing.

— Work with bolt-driving tools.

— Work on stock removing machines for small chippings.

— Drop forging.

— The removal and breaking up of fragments.

— Spraying of abrasive substances.

— Work with acids and caustic solutions, disinfectants and corrosive cleaning products.

— Work with liquid sprays.

— Work with and in the vicinity of molten substances.

— Work with radiant heat.

— Work with lasers.

4. RESPIRATORY PROTECTION

Respirators / breathing apparatus

— Work in containers, restricted areas and gas-fired industrial furnaces where there may be gas or insufficient oxygen.

— Work in the vicinity of the blast furnace charge.

— Work in the vicinity of gas converters and blast furnace gas pipes.

— Work in the vicinity of blast furnace taps where there may be heavy metal fumes.

— Work on the lining of furnaces and ladles where there may be dust.

— Spray painting where dedusting is inadequate.

— Work in shafts, sewers and other underground areas connected with sewage.

— Work in refrigeration plants where there is a danger that the refrigerant may escape.

5. HEARING PROTECTION

Ear protectors

— Work with metal presses.

— Work with pneumatic drills.

— The work of ground staff at airports.

— Pile-driving work.

— Wood and textile working.

6. BODY, ARM AND HAND PROTECTION

Protective clothing

— Work with acids and caustic solutions, disinfectants and corrosive cleaning substances.

— Work with or in the vicinity of hot materials and where the effects of heat are felt.

— Work on flat glass products.

— Shot blasting.

— Work in deep-freeze rooms.

Fire-resistant protective clothing

— Welding in restricted areas.

Pierce-proof aprons

— Boning and cutting work.

— Work with hand knives involving drawing the knife towards the body.

Leather aprons

— Welding.

— Forging.

— Casting.

Forearm protection

— Boning and cutting.

Gloves

— Welding.

— Handling of sharp-edged objects, other than machines where there is a danger of the glove's being caught.

— Unprotected work with acids and caustic solutions.

Metal mesh gloves

— Boning and cutting.

— Regular cutting using a hand knife for production and slaughtering.

— Changing the knives of cutting machines.

7.   WEATHERPROOF CLOTHING

— Work in the open air in rain and cold weather.

8.   REFLECTIVE CLOTHING

— Work where the workers must be clearly visible.

9.   SAFETY HARNESSES

— Work on scaffolding.

— Assembly of prefabricated parts.

— Work on masts.

10. SAFETY ROPES

— Work in high crane cabs.

— Work in high cabs of warehouse stacking and retrieval equipment.

— Work in high sections of drilling towers.

— Work in shafts and sewers.

11. SKIN PROTECTION

— Processing of coating materials.

— Tanning.

COUNCIL DIRECTIVE

of 29 May 1990

on the minimum health and safety requirements for the manual handling of loads where there is a risk particularly of back injury to workers

(fourth individual Directive within the meaning of Article 16(1) of Directive 89/391/EEC)

[90/269/EEC]

THE COUNCIL OF THE EUROPEAN COMMUNITIES,

Having regard to the Treaty establishing the European Economic Community, and in particular Article 118a thereof

Having regard to the Commission proposal submitted after consultation with the Advisory Committee on Safety, Hygiene and Health Protection at Work,

In co-operation with the European Parliament,

Having regard to the opinion of the Economic and Social Committee,

Whereas Article 118a of the Treaty provides that the Council shall adopt, by means of Directives, minimum requirements for encouraging improvements, especially in the working environment, to guarantee a better level of protection of the health and safety of workers;

Whereas, pursuant to that Article, such Directives must avoid imposing administrative, financial and legal constraints in a way which would hold back the creation and development of small and medium-sized undertakings;

Whereas the Commission communication on its programme concerning safety, hygiene and health at work, provides for the adoption of Directives designed to guarantee the health and safety of workers at the workplace;

Whereas the Council, in its resolution of 21 December 1987 on safety, hygiene and health at work, took note of the Commission's intention of submitting to the Council in the near future a Directive on protection against the risks resulting from the manual handling of heavy loads;

Whereas compliance with the minimum requirements designed to guarantee a better standard of health and safety at the workplace is essential to ensure the health and safety of workers;

Whereas this Directive is an individual Directive within the meaning of Article 16(1) of Council Directive 89/391/EEC of 12 June 1989 on the introduction of measures to encourage improvements in the health and safety of workers at work; whereas therefore the provisions of the said Directive are fully applicable to the field of the manual handling of loads where there is a risk particularly of back injury to workers, without prejudice to more stringent and/or specific provisions set out in this Directive;

Whereas this Directive constitutes a practical step towards the achievement of the social dimension of the internal market;

Whereas, pursuant to Decision 74/325/EEC, the Advisory Committee on Safety, Hygiene and Health Protection at Work shall be consulted by the Commission with a view to drawing up proposals in this field,

HAS ADOPTED THIS DIRECTIVE:

SECTION I
GENERAL PROVISIONS

Article 1
Subject

1. This Directive, which is the fourth individual Directive within the meaning of Article 16(1) of Directive 89/391/EEC, lays down minimum health and safety requirements for the manual handling of loads where there is a risk particularly of back injury to workers.

2. The provisions of Directive 89/391/EEC shall be fully applicable to the whole sphere referred to in paragraph 1, without prejudice to more restrictive and/or specific provisions contained in this Directive.

Article 2
Definition

For the purposes of this Directive, 'manual handling of loads' means any transporting or supporting of a load, by one or more workers, including lifting, putting down, pushing, pulling, carrying or moving of a load, which, by reason of its characteristics or of unfavourable ergonomic conditions, involves a risk particularly of back injury to workers.

SECTION II
EMPLOYERS' OBLIGATIONS

Article 3
General provision

1. The employer shall take appropriate organizational measures, or shall use the appropriate means, in particular mechanical equipment, in order to avoid the need for the manual handling of loads by workers.

2. Where the need for the manual handling of loads by workers cannot be avoided, the employer shall take the appropriate organizational measures, use the appropriate means or provide workers with such means in order to reduce the risk involved in the manual handling of such loads, having regard to Annex I.

Article 4
Organization of workstations

Wherever the need for manual handling of loads by workers cannot be avoided, the employer shall organize workstations in such a way as to make such handling as safe and healthy as possible and:

(a) assess, in advance if possible, the health and safety conditions of the type of work involved, and in particular examine the characteristics of loads, taking account of Annex I;

(b) take care to avoid or reduce the risk particularly of back injury to workers, by taking appropriate measures, considering in particular the characteristics of the working environment and the requirements of the activity, taking account of Annex I.

Article 5
Reference to Annex II

For the implementation of Article 6(3)(b) and Articles 14 and 15 of Directive 89/391/EEC, account should be taken of Annex II.

Article 6
Information for, and training of, workers

1. Without prejudice to Article 10 of Directive 89/391/EEC, workers and/or their representatives shall be informed of all measures to be implemented, pursuant to this Directive, with regard to the protection of safety and of health.

   Employers must ensure that workers and/or their representatives receive general indications and, where possible, precise information on:

   — the weight of a load,

   — the centre of gravity of the heaviest side when a package is eccentrically loaded.

2. Without prejudice to Article 12 of Directive 89/391/EEC, employers must ensure that workers receive in addition proper training and information on how to handle loads correctly and the risks they might be open to particularly if these tasks are not performed correctly, having regard to Annexes I and II.

Article 7
Consultation of workers and workers' participation

Consultation and participation of workers and/or of their representatives shall take place in accordance with Article 11 of Directive 89/391/EEC on matters covered by this Directive, including the Annexes thereto.

SECTION III
MISCELLANEOUS PROVISIONS

Article 8
Adjustment of the Annexes

Alterations of a strictly technical nature to Annexes I and II resulting from technical progress and changes in international regulations and specifications or knowledge in the field of the manual handling of loads shall be adopted in accordance with the procedure provided for in Article 17 of Directive 89/391/EEC.

Article 9
Final provisions

1. Member States shall bring into force the laws, regulations and administrative provisions needed to comply with this Directive not later than 31 December 1992.

    They shall forthwith inform the Commission thereof.

2. Member States shall communicate to the Commission the text of the provisions of national law which they adopt, or have adopted, in the field covered by this Directive.

3. Member States shall report to the Commission every four years on the practical implementation of the provisions of this Directive, indicating the points of view of employers and workers.

    The Commission shall inform the European Parliament, the Council, the Economic and Social Committee and the Advisory Committee on Safety, Hygiene and Health Protection at Work thereof.

4. The Commission shall report periodically to the European Parliament, the Council and the Economic and Social Committee on the implementation of the Directive in the light of paragraphs 1, 2 and 3.

Article 10

This Directive is addressed to the Member States.

Done at Brussels, 29 May 1990.

ANNEX I (*)

REFERENCE FACTORS
(Article 3 (2), Article 4(a) and (b) and Article 6(2))

1. **Characteristics of the load**

   The manual handling of a load may present a risk particularly of back injury if it is:

   — too heavy or too large,

   — unwieldy or difficult to grasp,

   — unstable or has contents likely to shift,

   — positioned in a manner requiring it to be held or manipulated at a distance from the trunk, or with a bending or twisting of the trunk,

   — likely, because of its contours and/or consistency, to result in injury to workers, particularly in the event of a collision.

2. **Physical effort required**

   A physical effort may present a risk particularly of back injury if it is:

   — too strenuous,

   — only achieved by a twisting movement of the trunk,

   — likely to result in a sudden movement of the load,

   — made with the body in an unstable posture.

3. **Characteristics of the working environment**

   The characteristics of the work environment may increase a risk particularly of back injury if:

   — there is not enough room, in particular vertically, to carry out the activity,

   — the floor is uneven, thus presenting tripping hazards, or is slippery in relation to the worker's footwear,

   — the place of work or the working environment prevents the handling of loads at a safe height or with good posture by the worker,

   — the door or foot rest is unstable,

   — the temperature, humidity or ventilation is unsuitable.

4. **Requirements of the activity**

   The activity may present a risk particularly of back injury if it entails one or more of the following requirements:

   — over-frequent or over-prolonged physical effort involving in particular the spine,

— an insufficient bodily rest or recovery period,

— excessive lifting, lowering or carrying distances,

— a rate of work imposed by a process which cannot be altered by the worker.

(*) With a view to making a multi-factor analysis, reference may be made simultaneously to the various factors listed in Annexes I and II.

ANNEX II (*)

INDIVIDUAL RISK FACTORS
(Articles 5 and 6(2))

The worker may be at risk if he/she:

— is physically unsuited to carry out the task in question,

— is wearing unsuitable clothing, footwear or other personal effects,

— does not have adequate or appropriate knowledge or training.

(*) With a view to multi-factor analysis, reference may be made simultaneously to the various factors listed in Annexes I and II.

COUNCIL DIRECTIVE

of 29 May 1990

on the minimum safety and health requirements for work with
display screen equipment

(fifth individual Directive within the meaning of Article 16(1) of Directive 89/391/EEC)

[90/270/EEC]

THE COUNCIL OF THE EUROPEAN COMMUNITIES,

Having regard to the Treaty establishing the European Economic Community, and in particular Article 118a thereof,

Having regard to the Commission proposal drawn up after consultation with the Advisory Committee on Safety, Hygiene and Health Protection at Work,

In co-operation with the European Parliament,

Having regard to the opinion of the Economic and Social Committee,

Whereas Article 118a of the Treaty provides that the Council shall adopt, by means of Directives, minimum requirements designed to encourage improvements, especially in the working environment, to ensure a better level of protection of workers' safety and health;

Whereas, under the terms of that Article, those Directives shall avoid imposing administrative, financial and legal constraints, in a way which would hold back the creation and development of small and medium-sized undertakings;

Whereas the communication from the Commission on its programme concerning safety, hygiene and health at work provides for the adoption of measures in respect of new technologies; whereas the Council has taken note thereof in its resolution of 21 December 1987 on safety, hygiene and health at work;

Whereas compliance with the minimum requirements for ensuring a better level of safety at workstations with display screens is essential for ensuring the safety and health of workers;

Whereas this Directive is an individual Directive within the meaning of Article 16(1) of Council Directive 89/391/EEC of 12 June 1989 on the introduction of measures to encourage improvements in the safety and health of workers at work; whereas the provisions of the latter are therefore fully applicable to the use by workers of display screen equipment, without prejudice to more stringent and/or specific provisions contained in the present Directive;

Whereas employers are obliged to keep themselves informed of the latest advances in technology and scientific findings concerning workstation design so that they can make any changes necessary so as to be able to guarantee a better level of protection of workers' safety and health;

Whereas the ergonomic aspects are of particular importance for a workstation with display screen equipment;

Whereas this Directive is a practical contribution towards creating the social dimension of the internal market;

Whereas, pursuant to Decision 74/325/EEC, the Advisory Committee on Safety, Hygiene and Health Protection at Work shall be consulted by the Commission on the drawing-up of proposals in this field,

HAS ADOPTED THIS DIRECTIVE:

SECTION I
GENERAL PROVISIONS

Article 1
Subject

1. This Directive, which is the fifth individual Directive within the meaning of Article 16(1) of Directive 89/391/EEC, lays down minimum safety and health requirements for work with display screen equipment as defined in Article 2.

2. The provisions of Directive 89/391/EEC are fully applicable to the whole field referred to in paragraph 1, without prejudice to more stringent and/or specific provisions contained in the present Directive.

3. This Directive shall not apply to:

   (a) drivers' cabs or control cabs for vehicles or machinery;

   (b) computer systems on board a means of transport;

   (c) computer systems mainly intended for public use;

   (d) 'portable' systems not in prolonged use at a workstation;

   (e) calculators, cash registers and any equipment having a small data or measurement display required for direct use of the equipment;

   (f) typewriters of traditional design, of the type known as 'typewriter with window'.

Article 2
Definitions

For the purpose of this Directive, the following terms shall have the following meanings:

(a) *display screen equipment:* an alphanumeric or graphic display screen, regardless of the display process employed;

(b) *workstation:* an assembly comprising display screen equipment, which may be provided with a keyboard or input device and/or software determining the operator/machine interface, optional accessories, peripherals including the diskette drive, telephone, modem, printer, document holder, work chair and work desk or work surface, and the immediate work environment;

(c) *worker:* any worker as defined in Article 3 (a) of Directive 89/391/EEC who habitually uses display screen equipment as a significant part of his normal work.

SECTION II
EMPLOYERS' OBLIGATIONS

Article 3
Analysis of workstations

1. Employers shall be obliged to perform an analysis of workstations in order to evaluate the safety and health conditions to which they give rise for their workers, particularly as regards possible risks to eyesight, physical problems and problems of mental stress.

2. Employers shall take appropriate measures to remedy the risks found, on the basis of the evaluation referred to in paragraph 1, taking account of the additional and/ or combined effects of the risks so found.

Article 4
Workstations put into service for the first time

Employers must take the appropriate steps to ensure that workstations first put into service after 31 December 1992 meet the minimum requirements laid down in the Annex.

Article 5
Workstations already put into service

Employers must take the appropriate steps to ensure that workstations already put into service on or before 31 December 1992 are adapted to comply with the minimum requirements laid down in the Annex not later than four years after that date.

Article 6
Information for, and training of, workers

1. Without prejudice to Article 10 of Directive 89/391/EEC, workers shall receive information on all aspects of safety and health relating to their workstation, in particular information on such measures applicable to workstations as are implemented under Articles 3, 7 and 9.

    In all cases, workers or their representatives shall be informed of any health and safety measure taken in compliance with this Directive.

2. Without prejudice to Article 12 of Directive 89/391/EEC, every worker shall also receive training in use of the workstation before commencing this type of work and whenever the organization of the workstation is substantially modified.

Article 7
Daily work routine

The employer must plan the worker's activities in such a way that daily work on a display screen is periodically interrupted by breaks or changes of activity reducing the workload at the display screen.

Article 8
Worker consultation and participation

Consultation and participation of workers and/or their representatives shall take place in accordance with Article 11 of Directive 89/391/EEC on the matters covered by this Directive, including its Annex.

Article 9
Protection of workers' eyes and eyesight

1. Workers shall be entitled to an appropriate eye and eyesight test carried out by a person with the necessary capabilities:

    — before commencing display screen work,

    — at regular intervals thereafter, and

    — if they experience visual difficulties which may be due to display screen work.

2. Workers shall be entitled to an ophthalmological examination if the results of the test referred to in paragraph 1 show that this is necessary.

3. If the results of the test referred to in paragraph 1 or of the examination referred to in paragraph 2 show that it is necessary and if normal corrective appliances cannot be used, workers must be provided with special corrective appliances appropriate for the work concerned.

4. Measures taken pursuant to this Article may in no circumstances involve workers in additional financial cost.

5. Protection of workers' eyes and eyesight may be provided as part of a national health system.

SECTION III
MISCELLANEOUS PROVISIONS

Article 10
Adaptations to the Annex

The strictly technical adaptations to the Annex to take account of technical progress, developments in international regulations and specifications and knowledge in the field of display screen equipment shall be adopted in accordance with the procedure laid down in Article 17 of Directive 89/391/EEC.

Article 11
Final provisions

1. Member States shall bring into force the laws, regulations and administrative provisions necessary to comply with this Directive by 31 December 1992.

    They shall forthwith inform the Commission thereof.

2.   Member States shall communicate to the Commission the texts of the provisions of national law which they adopt, or have already adopted, in the field covered by this Directive.

3.   Member States shall report to the Commission every four years on the practical implementation of the provisions of this Directive, indicating the points of view of employers and workers.

The Commission shall inform the European Parliament, the Council, the Economic and Social Committee and the Advisory Committee on Safety, Hygiene and Health Protection at Work.

4.   The Commission shall submit a report on the implementation of this Directive at regular intervals to the European Parliament, the Council and the Economic and Social Committee, taking into account paragraphs 1, 2 and 3.

Article 12

This Directive is addressed to the Member States.

Done at Brussels, 29 May 1990.

ANNEX

MINIMUM REQUIREMENTS
(Articles 4 and 5)

Preliminary remark

The obligations laid down in this Annex shall apply in order to achieve the objectives of this Directive and to the extent that, firstly, the components concerned are present at the workstation, and secondly, the inherent requirements or characteristics of the task do not preclude it.

1. EQUIPMENT

(a) General comment

The use as such of the equipment must not be a source of risk for workers.

(b) Display screen

The characters on the screen shall be well-defined and clearly formed, of adequate size and with adequate spacing between the characters and lines.

The image on the screen should be stable, with no flickering or other forms of instability.

The brightness and/or the contrast between the characters and the background shall be easily adjustable by the operator, and also be easily adjustable to ambient conditions.

The screen must swivel and tilt easily and freely to suit the needs of the operator. It shall be possible to use a separate base for the screen or an adjustable table.

The screen shall be free of reflective glare and reflections liable to cause discomfort to the user.

(c) Keyboard

The keyboard shall be tiltable and separate from the screen so as to allow the worker to find a comfortable working position avoiding fatigue in the arms or hands.

The space in front of the keyboard shall be sufficient to provide support for the hands and arms of the operator.

The keyboard shall have a matt surface to avoid reflective glare.

The arrangement of the keyboard and the characteristics of the keys shall be such as to facilitate the use of the keyboard.

The symbols on the keys shall be adequately contrasted and legible from the design working position.

(d) Work desk or work surface

The work desk or work surface shall have a sufficiently large, low-reflectance surface and allow a flexible arrangement of the screen, keyboard, documents and related equipment.

The document holder shall be stable and adjustable and shall be positioned so as to minimize the need for uncomfortable head and eye movements.

There shall be adequate space for workers to find a comfortable position.

(e) **Work chair**

The work chair shall be stable and allow the operator easy freedom of movement and a comfortable position.

The seat shall be adjustable in height.

The seat back shall be adjustable in both height and tilt.

A footrest shall be made available to any one who wishes for one.

2. ENVIRONMENT

(a) **Space requirements**

The workstation shall be dimensioned and designed so as to provide sufficient space for the user to change position and vary movements.

(b) **Lighting**

Room lighting and/or spot lighting (work lamps) shall ensure satisfactory lighting conditions and an appropriate contrast between the screen and the background environment, taking into account the type of work and the user's vision requirements.

Possible disturbing glare and reflections on the screen or other equipment shall be prevented by co-ordinating workplace and workstation layout with the positioning and technical characteristics of the artificial light sources.

(c) **Reflections and glare**

Workstations shall be so designed that sources of light, such as windows and other openings, transparent or translucid walls, and brightly coloured fixtures or walls cause no direct glare and no distracting reflections on the screen.

Windows shall be fitted with a suitable system of adjustable covering to attenuate the daylight that falls on the workstation.

(d) **Noise**

Noise emitted by equipment belonging to workstation(s) shall be taken into account when a workstation is being equipped, in particular so as not to distract attention or disturb speech.

(e) **Heat**

Equipment belonging to workstation(s) shall not produce excess heat which could cause discomfort to workers.

(f) **Radiation**

All radiation with the exception of the visible part of the electro-magnetic spectrum shall be reduced to negligible levels from the point of view of the protection of workers' safety and health.

COUNCIL DIRECTIVE

of 28 June 1990

on the protection of workers from the risks related to exposure to
carcinogens at work

(sixth individual Directive within the meaning of Article 16(1) of Directive 89/391/EEC)

[90/394/EEC]

THE COUNCIL OF THE EUROPEAN COMMUNITIES,

Having regard to the Treaty establishing the European Economic Community,
and in particular Article 118a thereof,

Having regard to the proposal from the Commission, drawn up following consultation with the Advisory Committee on Safety, Hygiene and Health Protection at
Work,

In co-operation with the European Parliament,

Having regard to the opinion of the Economic and Social Committee,

Whereas Article 118a of the Treaty provides that the Council is to adopt, by means
of Directives, minimum requirements in order to encourage improvements, especially in the working environment, so as to guarantee better protection of the
health and safety of workers;

Whereas, according to that Article, such Directives must avoid imposing administrative, financial and legal constraint in a way which would hold back the creation
and development of small and medium-sized undertakings;

Whereas the Council resolution of 27 February 1984 on a second action programme of the European Communities on safety and health at work provides for
the development of protective measures for workers exposed to carcinogens;

Whereas the Commission communication on its programme concerning safety,
hygiene and health at work provides for the adoption of Directives to guarantee the
health and safety of workers;

Whereas compliance with the minimum requirements designed to guarantee a
better standard of health and safety as regards the protection of workers from the
risks related to exposure to carcinogens at work is essential to ensure the health
and safety of workers;

Whereas this Directive is an individual Directive within the meaning of Article
16(1) of Council Directive 89/391/EEC of 12 June 1989 on the introduction of
measures to encourage improvements in the health and safety of workers at work;
whereas therefore the provisions of that Directive are fully applicable to the
exposure of workers to carcinogens, without prejudice to more stringent and/or
specific provisions contained in this Directive;

Whereas Council Directive 67/548/EEC of 27 June 1967 on the approximation of
laws, regulations and administrative provisions relating to the classification,
packaging and labelling of dangerous substances, as last amended by Directive 88/
490/EEC, contains a list of dangerous substances, together with particulars on the
classification and labelling procedures in respect of each substance;

Whereas Council Directive 88/379/EEC of 7 June 1988 on the approximation of the laws, regulations and administrative provisions relating to the classification, packaging and labelling of dangerous preparations, as last amended by Directive 89/178/EEC, contains particulars on the classification and labelling procedures in respect of such preparations;

Whereas the plan of action 1987 to 1989 adopted under the 'Europe against cancer' programme provides for support for European studies on the possible cancer risks of certain chemical substances;

Whereas, although current scientific knowledge is not such that a level can be established below which risks to health cease to exist, a reduction in exposure to carcinogens will nonetheless reduce those risks;

Whereas nevertheless, in order to contribute to a reduction in these risks, limit values and other directly related provisions should be established for all those carcinogens for which the available information, including scientific and technical data, make this possible;

Whereas preventive measures must be taken for the protection of the health and safety of workers exposed to carcinogens;

Whereas this Directive lays down particular requirements specific to exposure to carcinogens;

Whereas this Directive constitutes a practical aspect of the realization of the social dimension of the internal market;

Whereas, pursuant to Decision 74/325/EEC, as last amended by the 1985 Act of Accession, the Advisory Committee on Safety, Hygiene and Health Protection at Work is to be consulted by the Commission with a view to drawing up proposals in this field,

HAS ADOPTED THIS DIRECTIVE:

SECTION I
GENERAL PROVISIONS

Article 1
Objective

1.   This Directive, which is the sixth individual Directive within the meaning of Article 16(1) of Directive 89/391/EEC, has as its aim the protection of workers against risks to their health and safety, including the prevention of such risks, arising or likely to arise from exposure to carcinogens at work.

It lays down particular minimum requirements in this area, including limit values.

2.   This Directive shall not apply to workers exposed only to radiation covered by the Treaty establishing the European Atomic Energy Community.

3.   Directive 89/391/EEC shall apply fully to the whole area referred to in paragraph 1, without prejudice to more stringent and/or specific provisions contained in this Directive.

Article 2
Definition

For the purposes of this Directive, 'carcinogen' means:

(a) a substance to which, in Annex I to Directive 67/548/EEC, the risk-phrase R 45 'may cause cancer' is applied;

(b) a preparation which, under Article 3(5)(j) of Directive 88/379/EEC, must be labelled as R45 'may cause cancer';

(c) a substance, a preparation or a process referred to in Annex I as well as a substance or preparation released by a process referred to in Annex I.

Article 3
Scope — determination and assessment of risks

1. This Directive shall apply to activities in which workers are or are likely to be exposed to carcinogens as a result of their work.

2. In the case of any activity likely to involve a risk of exposure to carcinogens, the nature, degree and duration of workers' exposure must be determined in order to make it possible to assess any risk to the workers' health or safety and to lay down the measures to be taken.

   The assessment must be renewed regularly and in any event when any change occurs in the conditions which may affect workers' exposure to carcinogens.

   The employer must supply the authorities responsible at their request with the information used for making the assessment.

3. Furthermore, when assessing the risk, account shall be taken of all other cases of major exposure, such as those with harmful effects on the skin.

4. When the assessment referred to in paragraph 2 is carried out, employers shall give particular attention to any effects concerning the health or safety of workers at particular risk and shall, *inter alia*, take account of the desirability of not employing such workers in areas where they may come into contact with carcinogens.

SECTION II
EMPLOYERS' OBLIGATIONS

Article 4
Reduction and replacement

1. The employer shall reduce the use of a carcinogen at the place of work, in particular by replacing it, in so far as is technically possible, by a substance, preparation or process which, under its conditions of use, is not dangerous or is less dangerous to workers' health or safety, as the case may be.

2. The employer shall, upon request, submit the findings of his investigations to the relevant authorities.

Article 5
Prevention and reduction of exposure

1. Where the results of the assessment referred to in Article 3(2) reveal a risk to workers' health or safety, workers' exposure must be prevented.

2. Where it is not technically possible to replace the carcinogen by a substance, preparation or process which, under its conditions of use, is not dangerous or is less dangerous to health or safety, the employer shall ensure that the carcinogen is, in so far as is technically possible, manufactured and used in a closed system.

3. Where a closed system is not technically possible, the employer shall ensure that the level of exposure of workers is reduced to as low a level as is technically possible.

4. Wherever a carcinogen is used, the employer shall apply all the following measures:

    (a) limitation of the quantities of a carcinogen at the place of work;

    (b) keeping as low as possible the number of workers exposed or likely to be exposed;

    (c) design of work processes and engineering control measures so as to avoid or minimize the release of carcinogens into the place of work;

    (d) evacuation of carcinogens at source, local extraction system or general ventilation, all such methods to be appropriate and compatible with the need to protect public health and the environment;

    (e) use of existing appropriate procedures for the measurement of carcinogens, in particular for the early detection of abnormal exposures resulting from an unforeseeable event or an accident;

    (f) application of suitable working procedures and methods;

    (g) collective protection measures and/or, where exposure cannot be avoided by other means, individual protection measures;

    (h) hygiene measures, in particular regular cleaning of floors, walls and other surfaces;

    (i) information for workers;

    (j) demarcation of risk areas and use of adequate warning and safety signs including 'no smoking' signs in areas where workers are exposed or likely to be exposed to carcinogens;

    (k) drawing up plans to deal with emergencies likely to result in abnormally high exposure;

    (l) means for safe storage, handling and transportation, in particular by using sealed and clearly and visibly labelled containers;

    (m) means for safe collection, storage and disposal of waste by workers, including the use of sealed and clearly and visibly labelled containers.

Article 6
Information for the competent authority

Where the results of the assessment referred to in Article 3(2) reveal a risk to workers' health or safety, employers shall, when requested, make available to the competent authority appropriate information on:

(a) the activities and/or industrial processes carried out, including the reasons for which carcinogens are used;

(b) the quantities of substances or preparations manufactured or used which contain carcinogens;

(c) the number of workers exposed;

(d) the preventive measures taken;

(e) the type of protective equipment used;

(f) the nature and degree of exposure;

(g) the cases of replacement.

Article 7
Unforeseen exposure

1. In the event of an unforeseeable event or an accident which is likely to result in an abnormal exposure of workers, the employer shall inform the workers thereof.

2. Until the situation has been restored to normal and the causes of the abnormal exposure have been eliminated:

   (a) only those workers who are essential to the carrying out of repairs and other necessary work shall be permitted to work in the affected area;

   (b) the workers concerned shall be provided with protective clothing and individual respiratory protection equipment which they must wear; the exposure may not be permanent and shall be kept to the strict minimum of time necessary for each worker;

   (c) unprotected workers shall not be allowed to work in the affected area.

Article 8
Foreseeable exposure

1. For certain activities such as maintenance, in respect of which it is foreseeable that there is the potential for a significant increase in exposure of workers, and in respect of which all scope for further technical preventive measures for limiting workers' exposure has already been exhausted, the employer shall determine, after consultation of the workers and/or their representatives in the undertaking or establishment, without prejudice to the employer's responsibility, the measures necessary to reduce the duration of workers' exposure to the minimum possible and to ensure protection of workers while they are engaged in such activities.

Pursuant to the first sub-paragraph, the workers concerned shall be provided with protective clothing and individual respiratory protection equipment which they must wear as long as the abnormal exposure persists; that exposure may not be permanent and shall be kept to the strict minimum of time necessary for each worker.

2. Appropriate measures shall be taken to ensure that the areas in which the activities referred to in the first sub-paragraph of paragraph 1 take place are clearly demarcated and indicated or that unauthorized persons are prevented by other means from having access to such areas.

Article 9
Access to risk areas

Appropriate measures shall be taken by employers to ensure that access to areas in which the activities in respect of which the results of the assessment referred to in Article 3(2) reveal a risk to workers' safety or health take place are accessible solely to workers who, by reason of their work or duties, are required to enter them.

Article 10
Hygiene and individual protection

1. Employers shall be obliged, in the case of all activities for which there is a risk of contamination by carcinogens, to take appropriate measures to ensure that:

    (a) workers do not eat, drink or smoke in working areas where there is a risk of contamination by carcinogens;

    (b) workers are provided with appropriate protective clothing or other appropriate special clothing;

      separate storage places are provided for working or protective clothing and for street clothes;

    (c) workers are provided with appropriate and adequate washing and toilet facilities;

    (d) protective equipment is properly stored in a well-defined place;

      it is checked and cleaned if possible before, and in any case after, each use;

      defective equipment is repaired or replaced before further use.

2. Workers may not be charged for the cost of these measures.

Article 11
Information and training of workers

1. Appropriate measures shall be taken by the employer to ensure that workers and/or workers' representatives in the undertaking or establishment receive sufficient and appropriate training, on the basis of all available information, in particular in the form of information and instructions, concerning:

(a) potential risks to health, including the additional risks due to tobacco consumption;

(b) precautions to be taken to prevent exposure;

(c) hygiene requirements;

(d) wearing and use of protective equipment and clothing;

(e) steps to be taken by workers, including rescue workers, in the case of incidents and to prevent incidents.

The training shall be:

— adapted to take account of new or changed risk, and

— repeated periodically if necessary.

2. Employers shall inform workers of installations and related containers containing carcinogens, ensure that all containers, packages and installations containing carcinogens are labelled clearly and legibly, and display clearly visible warning and hazard signs.

Article 12
Information for workers

Appropriate measures shall be taken to ensure that:

(a) workers and/or any workers' representatives in the undertaking or establishment can check that this Directive is applied or can be involved in its application, in particular with regard to:

(i) the consequences for workers' safety and health of the selection, wearing and use of protective clothing and equipment, without prejudice to the employer's responsibility for determining the effectiveness of protective clothing and equipment;

(ii) the measures determined by the employer which are referred to in the first sub-paragraph of Article 8(1), without prejudice to the employer's responsibility for determining such measures;

(b) workers and/or any workers' representatives in the undertaking or establishment are informed as quickly as possible of abnormal exposures, including those referred to in Article 8, of the causes thereof and of the measures taken or to be taken to rectify the situation;

(c) the employer keeps an up-to-date list of the workers engaged in the activities in respect of which the results of the assessment referred to in Article 3(2) reveal a risk to workers' health or safety, indicating, if the information is available, the exposure to which they have been subjected;

(d) the doctor and/or the competent authority as well as all other persons who have responsibility for health and safety at work have access to the list referred to in sub-paragraph (c);

(e) each worker has access to the information on the list which relates to him personally;

(f)   workers and/or any workers' representatives in the undertaking or establishment have access to anonymous collective information.

Article 13
Consultation and participation of workers

Consultation and participation of workers and/or their representatives in connection with matters covered by this Directive, including the Annexes hereto, shall take place in accordance with Article 11 of Directive 89/391/EEC.

SECTION III
MISCELLANEOUS PROVISIONS

Article 14
Health surveillance

1.   The Member States shall establish, in accordance with national laws and/or practice, arrangements for carrying out relevant health surveillance of workers for whom the results of the assessment referred to in Article 3(2) reveal a risk to health or safety.

2.   The arrangements referred to in paragraph 1 shall be such that each worker shall be able to undergo, if appropriate, relevant health surveillance:

— prior to exposure,

— at regular intervals thereafter.

Those arrangements shall be such that it is directly possible to implement individual and occupational hygiene measures.

3.   If a worker is found to be suffering from an abnormality which is suspected to be the result of exposure to carcinogens, the doctor or authority responsible for the health surveillance of workers may require other workers who have been similarly exposed to undergo health surveillance.

In that event, a reassessment of the risk of exposure shall be carried out in accordance with Article 3(2).

4.   In cases where health surveillance is carried out, an individual medical record shall be kept and the doctor or authority responsible for health surveillance shall propose any protective or preventive measures to be taken in respect of any individual workers.

5.   Information and advice must be given to workers regarding any health surveillance which they may undergo following the end of exposure.

6.   In accordance with national laws and/or practice:

— workers shall have access to the results of the health surveillance which concern them, and

— the workers concerned or the employer may request a review of the results of the health surveillance.

7. Practical recommendations for the health surveillance of workers are given in Annex II.

8. All cases of cancer identified in accordance with national laws and/or practice as resulting from occupational exposure to a carcinogen shall be notified to the competent authority.

Article 15
Record-keeping

1. The list referred to in Article 12(c) and the medical record referred to in Article 14(4) shall be kept for at least 40 years following the end of exposure, in accordance with national laws and/or practice.

2. Those documents shall be made available to the responsible authority in cases where the undertaking ceases activity, in accordance with national laws and/or practice.

Article 16
Limit values

1. The Council shall, in accordance with the procedure laid down in Article 118a of the Treaty, set out limit values in Directives on the basis of the available information, including scientific and technical data, in respect of all those carcinogens for which this is possible, and, where necessary, other directly related provisions.

2. Limit values and other directly related provisions shall be set out in Annex III.

Article 17
Annexes

1. Annexes I and III may be amended in accordance only with the procedure laid down in Article 118a of the Treaty.

2. Purely technical adjustments to Annex II in the light of technical progress, changes in international regulations or specifications and new findings in the field of carcinogens shall be adopted in accordance with the procedure laid down in Article 17 of Directive 89/391/EEC.

Article 18
Use of data

The Commission shall have access to the use made by the competent national authorities of the information referred to in Article 14(8).

Article 19
Final provisions

1. Member States shall bring into force the laws, regulations and administrative provisions necessary to comply with this Directive not later than 31 December 1992.

Should Directives 67/548/EEC or 88/379/EEC be amended by amending Directives after notification of this Directive with respect to the substances and preparations referred to in Article 2(a) and (b), Member States shall bring into force the laws, regulations and administrative provisions necessary to introduce the amendments in question into the provisions referred to in the first sub-paragraph by the deadlines laid down for implementation of such amending Directives.

Member States shall forthwith inform the Commission that the provisions referred to in this paragraph have been brought into force.

2. Member States shall communicate to the Commission the provisions of national law already adopted or which they adopt in the future in the field governed by this Directive.

Article 20

This Directive is addressed to the Member States.

Done at Luxembourg, 28 June 1990.

ANNEX I
List of substances, preparations and processes
(Article 2(c))

1. Manufacture of auramine.

2. Work involving exposure to aromatic polycyclic hydrocarbons present in coal soots, tar, pitch, fumes or dust.

3. Work involving exposure to dusts, fumes and sprays produced during the roasting and electro-refining of cupro-nickel mattes.

4. Strong acid process in the manufacture of isopropyl alcohol.

ANNEX II
Practical recommendations for the health surveillance of workers
(Article 14(7))

1. The doctor and/or authority responsible for the health monitoring of workers exposed to carcinogens must be familiar with the exposure conditions or circumstances of each worker.

2. Health monitoring of workers must be carried out in accordance with the principles and practices of occupational medicine; it must include at least the following measures:

   — keeping records of a worker's medical and occupational history,

   — a personal interview,

   — where appropriate, biological monitoring, as well as detection of early and reversible effects.

   Further tests may be decided upon for each worker when he is the subject of health monitoring, in the light of the most recent knowledge available to occupational medicine.

ANNEX III
Limit values and other directly related provisions
(Article 16)

A. Limit values

   p.m.

B. Other directly related provisions

   p.m.

COUNCIL DIRECTIVE

of 26 November 1990

on the protection of workers from risks related to exposure to biological agents at work

(seventh individual Directive within the meaning of Article 16(1) of Directive 89/391/EEC)

[90/679/EEC]

THE COUNCIL OF THE EUROPEAN COMMUNITIES,

Having regard to the Treaty establishing the European Economic Community, and in particular Article 118a thereof,

Having regard to the proposal from the Commission, drawn up after consulting the Advisory Committee on Safety, Hygiene and Health Protection at Work,

In co-operation with the European Parliament,

Having regard to the Opinion of the Economic and Social Committee;

Whereas Article 118a of the Treaty provides that the Council shall adopt, by means of Directives, minimum requirements in order to encourage improvements, especially in the working environment, so as to guarantee better protection of the health and safety of workers;

Whereas that Article provides that such Directives shall avoid imposing administrative, financial and legal constraints in a way which would hold back the creation and development of small and medium-sized undertakings;

Whereas the Council Resolution of 27 February 1984 on a second action programme of the European Communities on safety and health at work provides for the development of protective measures for workers exposed to dangerous agents;

Whereas the communication from the Commission on its programme concerning safety, hygiene and health at work provides for the adoption of Directives to guarantee the safety and health of workers;

Whereas compliance with the minimum requirements designed to guarantee a better standard of safety and health as regards the protection of workers from the risks related to exposure to biological agents at work is essential to ensure the safety and health of workers;

Whereas this Directive is an individual Directive within the meaning of Article 16(1) of Council Directive 89/391/EEC of 12 June 1989 on the introduction of measures to encourage improvements in the safety and health of workers at work; whereas the provisions of that Directive are therefore fully applicable to the exposure of workers to biological agents, without prejudice to more stringent and/or specific provisions contained in the present Directive;

Whereas more precise knowledge of the risks involved in exposure to biological agents at work can be obtained through the keeping of records;

Whereas employers must keep abreast of new developments in technology with a view to improving the protection of workers' health and safety;

Whereas preventive measures should be taken for the protection of the health and safety of workers exposed to biological agents;

Whereas this Directive constitutes a practical aspect of the realization of the social dimension of the internal market;

Whereas, pursuant to Decision 74/325/EEC, the Advisory Committee on Safety, Hygiene and Health Protection at Work is consulted by the Commission on the drafting of proposals in this field,

HAS ADOPTED THIS DIRECTIVE:

SECTION I

GENERAL PROVISIONS

Article 1
Objective

1. This Directive, which is the seventh individual Directive within the meaning of Article 16(1) of Directive 89/391/EEC, has as its aim the protection of workers against risks to their health and safety, including the prevention of such risks, arising or likely to arise from exposure to biological agents at work.

    It lays down particular minimum provisions in this area.

2. Directive 89/391/EEC shall apply fully to the whole area referred to in paragraph 1, without prejudice to more stringent and/or specific provisions contained in this Directive.

3. This Directive shall apply without prejudice to the provisions of Council Directive 90/219/EEC of 23 April 1990 on the contained use of genetically modified micro-organisms and of Council Directive 90/220/EEC of 23 April 1990 on the deliberate release into the environment of generically modified organisms.

Article 2
Definitions

For the purpose of this Directive:

(a) 'biological agents' shall mean micro-organisms, including those which have been genetically modified, cell cultures and human endoparasites, which may be able to provoke any infection, allergy or toxicity;

(b) 'micro-organism' shall mean a microbiological entity, cellular or non-cellular, capable of replication or of transferring genetic material;

(c)   'cell culture' shall mean the in-vitro growth of cells derived from multi-cellular organisms;

(d)   'biological agents' shall be classified into four risk groups, according to their level of risk of infection:

1.   group 1 biological agent means one that is unlikely to cause human disease;

2.   group 2 biological agent means one that can cause human disease and might be a hazard to workers; it is unlikely to spread to the community; there is usually effective prophylaxis or treatment available;

3.   group 3 biological agent means one that can cause severe human disease and present a serious hazard to workers; it may present a risk of spreading to the community, but there is usually effective prophylaxis or treatment available;

4.   group 4 biological agent means one that causes severe human disease and is a serious hazard to workers; it may present a high risk of spreading to the community; there is usually no effective prophylaxis or treatment available.

Article 3
Scope — Determination and assessment of risks

1.   This Directive shall apply to activities in which workers are or are potentially exposed to biological agents as a result of their work.

2.   (a)   In the case of any activity likely to involve a risk of exposure to biological agents, the nature, degree and duration of workers' exposure must be determined in order to make it possible to assess any risk to the workers' health or safety and to lay down the measures to be taken.

(b)   In the case of activities involving exposure to several groups of biological agents, the risk shall be assessed on the basis of the danger presented by all hazardous biological agents present.

(c)   The assessment must be renewed regularly and in any event when any change occurs in the conditions which may affect workers' exposure to biological agents.

(d)   The employer must supply the competent authorities, at their request, with the information used for making the assessment.

3.   The assessment referred to in paragraph 2 shall be conducted on the basis of all available information including:

—   classification of biological agents which are or may be a hazard to human health, as referred to in Article 18;

—   recommendations from a competent authority which indicate that the biological agent should be controlled in order to protect workers' health when workers are or may be exposed to such a biological agent as a result of their work;

—   information on diseases which may be contracted as a result of the work of the workers;

— potential allergenic or toxigenic effects as a result of the work of the workers;

— knowledge of a disease from which a worker is found to be suffering and which has a direct connection with his

Article 4
Application of the various Articles in relation to assessment of risks

1. If the results of the assessment referred to in Article 3 show that the exposure and/or potential exposure is to a group 1 biological agent, with no identifiable health risk to workers, Articles 5 to 17 and Article 19 shall not apply.

   However, point 1 of Annex VI should be observed.

2. If the results of the assessment referred to in Article 3 show that the activity does not involve a deliberate intention to work with or use a biological agent but may result in the workers being exposed to a biological agent, as in the course of the activities for which an indicative list is given in Annex I, Articles 5, 7, 8, 10, 11, 12, 13 and 14 shall apply unless the results of the assessment referred to in Article 3 show them to be unnecessary.

SECTION II
EMPLOYERS' OBLIGATIONS

Article 5
Replacement

The employer shall avoid the use of a harmful biological agent if the nature of the activity so permits, by replacing it with a biological agent which, under its conditions of use, is not dangerous or is less dangerous to workers' health, as the case may be, in the present state of knowledge.

Article 6
Reduction of risks

1. Where the results of the assessment referred to in Article 3 reveal a risk to workers' health or safety, workers' exposure must be prevented.

2. Where this is not technically practicable, having regard to the activity and the risk assessment referred to in Article 3, the risk of exposure must be reduced to as low a level as necessary in order to protect adequately the health and safety of the workers concerned, in particular by the following measures which are to be applied in the light of the results of the assessment referred to in Article 3:

(a) keeping as low as possible the number of workers exposed or likely to be exposed;

(b) design of work processes and engineering control measures so as to avoid or minimize the release of biological agents into the place of work;

(c) collective protection measures and/or, where exposure cannot be avoided by other means, individual protection measures;

(d) hygiene measures compatible with the aim of the prevention or reduction of the accidental transfer or release of a biological agent from the workplace;

(e) use of the biohazard sign depicted in Annex II and other relevant warning signs;

(f) drawing up plans to deal with accidents involving biological agents;

(g) testing, where it is necessary and technically possible, for the presence, outside the primary physical confinement, of biological agents used at work;

(h) means for safe collection, storage and disposal of waste by workers, including the use of secure and identifiable containers, after suitable treatment where appropriate;

(i) arrangements for the safe handling and transport of biological agents within the workplace.

Article 7
Information for the competent authority

1. Where the results of the assessment referred to in Article 3 reveal a risk to workers' health or safety, employers shall, when requested, make available to the competent authority appropriate information on:

    — the results of the assessment;

    — the activities in which workers have been exposed or may have been exposed to biological agents;

    — the number of workers exposed;

    — the name and capabilities of the person responsible for safety and health at work;

    — the protective and preventive measures taken, including working procedures and methods;

    — an emergency plan for the protection of workers from exposure to a group 3 or a group 4 biological agent which might result from a loss of physical containment.

    2. Employers shall inform forthwith the competent authority of any accident or incident which may have resulted in the release of a biological agent and which could cause severe human infection and/or illness.

    3. The list referred to in Article 11 and the medical record referred to in Article 14 shall be made available to the competent authority in cases where the undertaking ceases activity, in accordance with national laws and/or practice.

Article 8
Hygiene and individual protection

1. Employers shall be obliged, in the case of all activities for which there is a risk to the health or safety of workers due to work with biological agents, to take appropriate measures to ensure that:

(a) workers do not eat or drink in working areas where there is a risk of contamination by biological agents;

(b) workers are provided with appropriate protective clothing or other appropriate special clothing;

(c) workers are provided with appropriate and adequate washing and toilet facilities, which may include eye washes and/or skin antiseptics;

(d) any necessary protective equipment is:

— properly stored in a well-defined place;

— checked and cleaned if possible before, and in any case after, each use;

— is repaired, where defective, or is replaced before further use;

(e) procedures are specified for taking, handling and processing samples of human or animal origin.

2. (a) Working clothes and protective equipment, including protective clothing referred to in paragraph 1, which may be contaminated by biological agents, must be removed on leaving the working area and, before taking the measures referred to in sub-paragraph (b), kept separately from other clothing.

(b) The employer must ensure that such clothing and protective equipment is decontaminated and cleaned or, if necessary, destroyed.

3. Workers may not be charged for the cost of the measures referred to in paragraphs 1 and 2.

Article 9
Information and training of workers

1. Appropriate measures shall be taken by the employer to ensure that workers and/ or any workers' representatives in the undertaking or establishment receive sufficient and appropriate training, on the basis of all available information, in particular in the form of information and instructions, concerning:

(a) potential risks to health;

(b) precautions to be taken to prevent exposure;

(c) hygiene requirements;

(d) wearing and use of protective equipment and clothing;

(e) steps to be taken by workers in the case of incidents and to prevent incidents.

2. The training shall be:

— given at the beginning of work involving contact with biological agents,

— adapted to take account of new or changed risks, and

— repeated periodically if necessary.

Article 10
Worker information in particular cases

1. Employers shall provide written instructions at the workplace and, if appropriate, display notices which shall, as a minimum, include the procedure to be followed in the case of:

    — a serious accident or incident involving the handling of a biological agent;

    — handling a group 4 biological agent.

2. Workers shall immediately report any accident or incident involving the handling of a biological agent to the person in charge or to the person responsible for safety and health at work.

3. Employers shall inform forthwith the workers and/or any workers' representatives of any accident or incident which may have resulted in the release of a biological agent and which could cause severe human infection and/or illness.

    In addition, employers shall inform the workers and/or any workers' representatives in the undertaking or establishment as quickly as possible when a serious accident or incident occurs, of the causes thereof, and of the measures taken or to be taken to rectify the situation.

4. Each worker shall have access to the information on the list referred to in Article 11 which relates to him personally.

5. Workers and/or any workers' representatives in the undertaking or establishment shall have access to anonymous collective information.

6. Employers shall provide workers and/or their representatives, at their request, with the information provided for in Article 7(1).

Article 11
List of exposed workers

1. Employers shall keep a list of workers exposed to group 3 and/or group 4 biological agents, indicating the type of work done and, whenever possible, the biological agent to which they have been exposed, as well as records of exposures, accidents and incidents, as appropriate.

2. The list referred to in paragraph 1 shall be kept for at least 10 years following the end of exposure, in accordance with national laws and/or practice.

    In the case of those exposures which may result in infections:

    — with biological agents known to be capable of establishing persistent or latent infections,

    — that, in the light of present knowledge, are undiagnosable until illness develops many years later,

    — that have particularly long incubation periods before illness develops,

    — that result in illnesses which recrudesce at times over a long period despite treatment, or

— that may have serious long-term sequelae,

the list shall be kept for an appropriately longer time up to 40 years following the last known exposure.

3. The doctor referred to in Article 14 and/or the competent authority for health and safety at work, and any other person responsible for health and safety at work, shall have access to the list referred to in paragraph 1.

Article 12
Consultation and participation of workers

Consultation and participation of workers and/or their representatives in connection with matters covered by this Directive including the Annexes shall take place in accordance with Article 11 of Directive 89/391/EEC.

Article 13
Notification to the competent authority

1. Prior notification shall be made to the competent authority of the use for the first time of:

    — group 2 biological agents,

    — group 3 biological agents,

    — group 4 biological agents.

    The notification shall be made at least 30 days before the commencement of the work.

    Subject to paragraph 2, prior notification shall also be made of the use for the first time of each subsequent group 4 biological agent and of any subsequent new group 3 biological agent where the employer himself provisionally classifies that biological agent.

2. Laboratories providing a diagnostic service in relation to group 4 biological agents shall be required only to make an initial notification of their intention.

3. Renotification must take place in any case where there are substantial changes of importance to safety or health at work to processes and/or procedures which render the notification out of date.

4. The notification referred to in this Article shall include:

    (a) the name and address of the undertaking and/or establishment;

    (b) the name and capabilities of the person responsible for safety and health at work;

    (c) the results of the assessment referred to in Article 3;

    (d) the species of the biological agent;

    (e) the protection and preventive measures that are envisaged.

SECTION III
MISCELLANEOUS PROVISIONS

Article 14
Health surveillance

1. The Member States shall establish, in accordance with national laws and practice, arrangements for carrying out relevant health surveillance of workers for whom the results of the assessment referred to in Article 3 reveal a risk to health or safety.

2. The arrangements referred to in paragraph 1 shall be such that each worker shall be able to undergo, if appropriate, relevant health surveillance:

    — prior to exposure,

    — at regular intervals thereafter.

    Those arrangements shall be such that it is directly possible to implement individual and occupational hygiene measures.

3. The assessment referred to in Article 3 should identify those workers for whom special protective measures may be required.

    When necessary, effective vaccines should be made available for those workers who are not already immune to the biological agent to which they are exposed or are likely to be exposed.

    If a worker is found to be suffering from an infection and/or illness which is suspected to be the result of exposure, the doctor or authority responsible for health surveillance of workers shall offer such surveillance to other workers who have been similarly exposed.

    In that event, a reassessment of the risk of exposure shall be carried out in accordance with Article 3.

4. In cases where health surveillance is carried out, an individual medical record shall be kept for at least 10 years following the end of exposure, in accordance with national laws and practice.

    In the special cases referred to in Article 11 (2) second sub-paragraph, an individual medical record shall be kept for an appropriately longer time up to 40 years following the last known exposure.

5. The doctor or authority responsible for health surveillance shall propose any protective or preventive measures to be taken in respect of any individual worker.

6. Information and advice must be given to workers regarding any health surveillance which they may undergo following the end of exposure.

7. In accordance with national laws and/or practice:

    — workers shall have access to the results of the health surveillance which concern them, and

    — the workers concerned or the employer may request a review of the results of the health surveillance.

8. Practical recommendations for the health surveillance of workers are given in Annex IV.

9. All cases of diseases or death identified in accordance with national laws and/or practice as resulting from occupational exposure to biological agents shall be notified to the competent authority.

Article 15
Health and veterinary care facilities other than diagnostic laboratories

1. For the purpose of the assessment referred to in Article 3, particular attention should be paid to:

    (a) uncertainties about the presence of biological agents in human patients or animals and the materials and specimens taken from them;

    (b) the hazard represented by biological agents known or suspected to be present in human patients or animals and materials and specimens taken from them;

    (c) the risks posed by the nature of the work.

2. Appropriate measures shall be taken in health and veterinary care facilities in order to protect the health and safety of the workers concerned.

    The measures to be taken shall include in particular:

    (a) specifying appropriate decontamination and disinfection procedures, and

    (b) implementing procedures enabling contaminated waste to be handled and disposed of without risk.

3. In isolation facilities where there are human patients or animals who are, or who are suspected of being, infected with group 3 or group 4 biological agents, containment measures shall be selected from those in Annex V column A in order to minimize the risk of infection.

Article 16
Special measures for industrial processes, laboratories and animal rooms

1. The following measures must be taken in laboratories, including diagnostic laboratories, and in rooms for laboratory animals which have been deliberately infected with group 2, 3 or 4 biological agents or which are or are suspected to be carriers of such agents:

    (a) Laboratories carrying out work which involves the handling of group 2, 3 or 4 biological agents for research, development, teaching or diagnostic purposes shall determine the containment measures in accordance with Annex V, in order to minimize the risk of infection .

    (b) Following the assessment referred to in Article 3, measures shall be determined in accordance with Annex V, after fixing the physical containment level required for the biological agents according to the degree of risk.

Activities involving the handling of a biological agent must be carried out:

— only in working areas corresponding to at least containment level 2, for a group 2 biological agent;

— only in working areas corresponding to at least containment level 3, for a group 3 biological agent;

— only in working areas corresponding to at least containment level 4, for a group 4 biological agent.

(c) Laboratories handling materials in respect of which there exist uncertainties about the presence of biological agents which may cause human disease but which do not have as their aim working with biological agents as such (i.e. cultivating or concentrating them) should adopt containment level 2 at least. Containment levels 3 or 4 must be used, when appropriate, where it is known or it is suspected that they are necessary, except where guidelines provided by the competent national authorities show that, in certain cases, a lower containment level is appropriate.

2. The following measures concerning industrial processes using group 2, 3 or 4 biological agents must be taken:

(a) The containment principles set out in the second sub-paragraph of paragraph 1(b) should also apply to industrial processes on the basis of the practical measures and appropriate procedures given in Annex VI.

(b) In accordance with the assessment of the risk linked to the use of group 2, 3 or 4 biological agents, the competent authorities may decide on appropriate measures which must be applied to the industrial use of such biological agents.

(c) For all activities covered by this Article where it has not been possible to carry out a conclusive assessment of a biological agent but concerning which it appears that the use envisaged might involve a serious health risk for workers, activities may only be carried out in workplaces where the containment level corresponds at least to level 3.

Article 17
Use of data

The Commission shall have access to the use made by the competent national authorities of the information referred to in Article 14(9).

Article 18
Classification of biological agents

1. In accordance with the procedure laid down in Article 118a of the Treaty, the Council shall adopt within six months of the date of implementation given in Article 20(1) a first list of group 2, group 3 and group 4 biological agents for Annex III.

2. Community classification shall be on the basis of the definitions in Article 2(d) points 2 to 4 (groups 2 to 4).

3. Pending Community classification Member States shall classify biological agents that are or may be a hazard to human health on the basis of the definition in Article 2(d) points 2 to 4 (groups 2 to 4).

4. If the biological agent to be assessed cannot be classified clearly in one of the groups defined in Article 2(d), it must be classified in the highest risk group among the alternatives.

Article 19

Annexes

Purely technical adjustments to the Annexes in the light of technical progress, changes in international regulations or specifications and new findings in the field of biological agents shall be adopted in accordance with the procedure laid down in Article 17 of Directive 89/391/EEC.

Article 20

Final provisions

1. Member States shall bring into force the laws, regulations and administrative provisions necessary to comply with this Directive not later than three years after the notification of this Directive. They shall forthwith inform the Commission thereof.

    However, in the case of the Portuguese Republic, the time limit referred to in the first sub-paragraph shall be five years.

2. Member States shall communicate to the Commission the provisions of national law already adopted or which they adopt in the field governed by this Directive.

Article 21

This Directive is addressed to the Member States.

Done at Brussels, 26 November 1990.

ANNEX I

INDICATIVE LIST OF ACTIVITIES
(Article 4 (2))

1.   Work in food production plants.

2.   Work in agriculture.

3.   Work activities where there is contact with animals and/or products of animal origin.

4.   Work in health care, including isolation and post mortem units.

5.   Work in clinical, veterinary and diagnostic laboratories, excluding diagnostic microbiological laboratories.

6.   Work in refuse disposal plants.

7.   Work in sewage purification installations.

ANNEX II

BIOHAZARD SIGN
(Article 6 (2) (e))

ANNEX III

COMMUNITY CLASSIFICATION
(Articles 18 and 2 (d))

For the record

ANNEX IV

PRACTICAL RECOMMENDATIONS FOR THE HEALTH SURVEILLANCE
OF WORKERS
(Article 14 (8))

The doctor and/or the authority responsible for the health surveillance of workers exposed to biological agents must be familiar with the exposure conditions or circumstances of each worker.

Health surveillance of workers must be carried out in accordance with the principles and practices of occupational medicine; it must include at least the following measures:

— keeping records of a worker's medical and occupational history;

— a personalized assessment of the workers' state of health;

— where appropriate, biological monitoring, as well as detection of early and reversible effects.

Further tests may be decided upon for each worker when he is the subject of health surveillance, in the light of the most recent knowledge available to occupational medicine.

COUNCIL DIRECTIVE

of 4 December 1990

on the operational protection of outside workers exposed to the risk of ionizing
radiation during their activities in controlled areas
[90/641/Euratom]

THE COUNCIL OF THE EUROPEAN COMMUNITIES.

Having regard to the Treaty establishing the European Atomic Energy Community, and in particular Articles 31 and 32 thereof,

Having regard to the proposal from the Commission, submitted following consultation with a group of persons appointed by the Scientific and Technical Committee from among scientific experts in the Member States, as laid down in Article 31 of the Treaty,

Having regard to the opinion of the European Parliament,

Having regard to the opinion of the Economic and Social Committee,

Whereas, Article 2(b) of the Treaty provides that the Community shall establish uniform safety standards to protect the health of workers and of the general public and ensure that they are applied in accordance with the procedures laid down in Chapter III of Title II of the Treaty;

Whereas, on 2 February 1959, the Council adopted Directives laying down the basic standards for the protection of the health of workers and of the general public against the dangers arising from ionizing radiations, as amended by Directives 80/836/Euratom and 84/467/Euratom;

Whereas Title VI of Directive 80/836/Euratom lays down the fundamental principles governing operational protection of exposed workers;

Whereas Article 40(1) of that Directive provides that each Member State shall take all necessary measures to ensure the effective protection of exposed workers;

Whereas Article 20 and 23 of that Directive establish a classification of areas of work and categories of exposed workers according to the level of exposure;

Whereas the workers performing activities in a controlled area within the meaning of the said Articles 20 and 23 can belong to the personnel of the operator or be outside workers;

Whereas Article 3 of Directive 50/836/Euratom concerning the activities referred to in Article 1 of that Directive provides that they should be reported or subject to prior authorization in cases decided upon by each Member State;

Whereas outside workers are liable to be exposed to ionizing radiation in several controlled areas in succession in one and the same Member State or in different Member States; whereas these specific working conditions require an appropriate radiological monitoring system;

Whereas any radiological monitoring system for outside workers must provide protection equivalent to that offered the operator's established workers, by means of common provisions;

Whereas, pending the introduction of a uniform Community-wide system, account should also be taken of the radiological monitoring systems for outside workers which may exist in the Member States;

Whereas, to optimize the protection of outside workers, it is necessary to define clearly the obligations of outside undertakings and operators, without prejudice to the contribution that outside workers themselves have to make to their own protection;

Whereas the system for the radiological protection of outside workers also applies as far as practicable to the case of a self-employed worker with the status of outside undertaking,

HAS ADOPTED THIS DIRECTIVE:

TITLE I

Purpose and definitions

Article 1

The purpose of this Directive is to supplement Directive 80/836/Euratom thereby optimizing at Community level operational protection arrangements for outside workers performing activities in controlled areas.

Article 2

For the purposes of this Directive:

— 'controlled area' means any area subject to special rules for the purposes of protection against ionizing radiation and to which access is controlled, as specified in Article 20 of Directive 80/836/Euratom;

— 'operator' means any natural or legal person who under national law, is responsible for a controlled area in which an activity required to be reported under Article 3 of Directive 80/836/Euratom is carried on;

— 'outside undertaking' means any natural or legal person, other than the operator, including members of his staff, performing an activity of any sort in a controlled area;

— 'outside worker' means any worker of category A, as defined in Article 23 of Directive 80/836/Euratom, performing activities of any sort in a controlled area, whether employed temporarily or permanently by an outside undertaking, including trainees, apprentices and students within the meaning of Article 10 of that Directive, or whether he provides services as a self-employed worker;

— 'radiological monitoring system' means measures to apply the arrangement set out in Directive 80/836/Euratom, and in particular in Title VI thereof, during the activities of outside workers.

— 'activities carried out by a worker' means any service or services provided by an outside worker in a controlled area for which an operator is responsible.

TITLE II

Obligations of Member States' competent authorities

Article 3

Each Member State shall make the performance of the activities referred to in Article 2 of Directive 80/836/Euratom by outside undertakings subject to reporting or prior authorization as laid down in accordance with Title II of the aforementioned Directive, in particular Article 3 thereof.

Article 4

1. Each Member State shall ensure that the radiological monitoring system affords outside workers equivalent protection to that for workers employed on a permanent basis by the operator.

2. Pending the establishment, at Community level, of a uniform system for the radiological protection of outside workers, such as a computer network, recourse shall be had:

    (a) on a transitional basis, in accordance with the common provisions set out in Annex I, to

        — a centralized national network, or

        — the issuing of an individual radiological monitoring document to every outside worker, in which case the common provisions of Annex II shall also apply;

    (b) in the case of cross-frontier outside workers, and until the date of establishment of a system within the meaning of paragraph 2, to the individual document referred to in (a).

TITLE III

Obligations of outside undertakings and operators

Article 5

Outside undertakings shall, either directly or through contractual agreements with the operators, ensure the radiological protection of their workers in accordance with the relevant provisions of Titles III to VI of Directive 80/836/Euratom, and in particular:

(a) ensure compliance with the general principles and the limitation of doses referred to in Articles 6 to 11 thereof;

(b) provide the information and training in the field of radiation protection referred to in Article 24 thereof;

(c) guarantee that their workers are subject to assessment of exposure and medical surveillance under the conditions laid down in Articles 26 and 28 to 38 thereof;

(d) ensure that the radiological data of the individual exposure monitoring of each of their workers within the meaning of Annex I, part II to this Directive are kept up to date in the networks and individual documents referred to in Article 4(2).

Article 6

1. The operator of a controlled area in which outside workers perform activities shall be responsible, either directly or through contractual agreements, for the operational aspects of their radiological protection which are directly related to the nature of the controlled area and of the activities.

2. In particular, for each outside worker performing activities in a controlled area, the operator must:

   (a) check that the worker concerned has been passed as medically fit for the activities to be assigned to him;

   (b) ensure that, in addition to the basic training in radiation protection referred to in Article 5(1)(b), he has received specific training in connection with the characteristics of both the controlled area and the activities;

   (c) ensure that he has been issued with the necessary personal protective equipment;

   (d) also ensure that he receives individual exposure monitoring appropriate to the nature of the activities, and any operational dosimetric monitoring that may be necessary;

   (e) ensure compliance with the general principles and limitation of doses referred to in Articles 6 to 11 of Directive 80/836/Euratom;

   (f) ensure or take all appropriate steps to ensure that after every activity the radiological data of individual exposure monitoring of each outside worker within the meaning of Annex I, Part III, are recorded.

TITLE IV

Obligations of outside workers

Article 7

Every outside worker shall be obliged to make his own contribution as far as practicable towards the protection that the radiological monitoring system referred to in Article 1 is intended to afford him.

TITLE V
Final provisions

Article 8

1. Member States shall bring into force not later than 31 December 1993, the laws, regulations and administrative provisions necessary to comply with this Directive. They shall forthwith inform the Commission thereof.

2. When Member States adopt the measures referred to in paragraph 1, they shall contain a reference to this Directive or shall be accompanied by such reference on the occasion of their official publication. The methods of making such a reference shall be laid down by the Member States.

3. Member States shall communicate to the Commission the main provisions of domestic law which they adopt in the field governed by this Directive.

Article 9

This Directive is addressed to the Member States.

Done at Brussels, 4 December 1990.

ANNEX I

PROVISIONS COMMON TO THE NETWORKS AND INDIVIDUAL DOCUMENTS REFERRED TO IN ARTICLE 4(2)

PART I

1. Any radiological monitoring system of the Member States for outside workers must comprise the following three sections:

   — particulars concerning the outside workers' identity;

   — particulars to be supplied before the start of any activity;

   — particulars to be supplied after the end of any activity.

2. The competent authorities of the Member States shall take the measures necessary to prevent any forgery or misuse of, or illegal tampering with, the radiological monitoring system.

3. Data on the outside worker's identity must also include the worker's sex and date of birth.

PART II

Before the start of any activity, the data to be supplied via the radiological monitoring system to the operator or his approved medical practitioner by the outside undertaking or an authority empowered to that end must be as follows:

   — the name and address of the outside undertaking;

   — the medical classification of the outside worker in accordance with Article 35 of Directive 80/836/Euratom;

   — the date of the last periodic health review;

   — the results of the outside worker's individual exposure monitoring.

PART III

The data which the operator must record or have recorded by the authority empowered to that end in the radiological monitoring system after the end of any activity must be as follows:

   — the period covered by the activity;

   — an estimate of any effective dose received by the outside worker;

   — in the event of non-uniform exposure, an estimate of the dose-equivalent in the different parts of the body;

   — in the event of internal contamination, an estimate of the activity taken in or the committed dose.

ANNEX II

PROVISIONS ADDITIONAL TO THOSE OF ANNEX I CONCERNING THE INDIVIDUAL RADIOLOGICAL MONITORING DOCUMENT

1. The individual radiological monitoring document issued by the Member States' competent authorities for outside workers shall be a non-transferable document.

2. Pursuant to Annex I, Part I (2), individual documents shall be issued by the Member States' competent authorities, which shall give each individual document an identification number.

COMMISSION DIRECTIVE

of 29 May 1991

on establishing indicative limit values by implementing Council Directive
80/1107/EEC on the protection of workers from the risks related to exposure to
chemical, physical and biological agents at work
[91/322/EEC]

THE COMMISSION OF THE EUROPEAN COMMUNITIES,

Having regard to the Treaty establishing the European Economic Community,

Having regard to Council Directive 80/1107/EEC of 27 November 1980 on the
protection of workers from the risks related to exposure to chemical, physical and
biological agents at work, as last amended by Directive 88/642/EEC, and in
particular the first sub-paragraph of Article 8(4) thereof,

Having regard to the opinion of the Advisory Committee on Safety, Hygiene and
Health Protection at Work,

Whereas the third sub-paragraph of Article 8(4) of Directive 80/1107/EEC states
that indicative limit values shall reflect expert evaluations based on scientific data;

Whereas the aim of fixing these values is the harmonization of conditions in this
area while maintaining the improvements made;

Whereas the Directive constitutes a practical step towards the achievement of the
social dimension of the internal market;

Whereas occupational exposure limit values should be regarded as an important
part of the overall approach to ensuring the protection of the health of workers at
the workplace;

Whereas an initial list of occupational exposure limit values can be established for
agents for which similar values exist in the Member States, giving priority to agents
which are found at places of work and are likely to have an effect on the health of
workers; whereas this list can be based on existing scientific data as far as the effects
on health are concerned, although for certain agents these data are very limited;

Whereas in addition it may be necessary to establish occupational exposure limit
values for shorter periods taking into account the effects arising from short term
exposure;

Whereas a reference method covering, inter alia, assessment of exposure and
measuring strategy for occupational exposure limit values is contained in Directive
80/1107/EEC;

Whereas, in view of the importance of obtaining reliable measurements of
exposure in relation to occupational exposure limit values, it may be necessary in
the future to establish appropriate reference methods;

Whereas occupational exposure limit values need to be kept under review and will need to be revised if new scientific data indicate that they are no longer valid;

Whereas, for some agents it will be necessary in the future to consider all absorption pathways, including the possibility of penetration through the skin, in order to ensure the best possible level of protection;

Whereas the measures laid down in this Directive are in conformity with the opinion of the Committee set up pursuant to Article 9 of Directive 80/1107/EEC,

HAS ADOPTED THIS DIRECTIVE:

Article 1

Indicative limit values, of which Member States shall take account, inter alia, when establishing the limit values referred to in Article 4(4)(b) of Directive 80/1107/EEC are listed in the Annex.

Article 2

1.   Member States shall bring into force the provisions necessary to comply with this Directive by 31 December 1993. They shall immediately inform the Commission thereof.

   When Member States adopt these provisions, these shall contain a reference to this Directive or shall be accompanied by such reference at the time of their official publication. The procedure for such reference shall be adopted by Member States.

2.   Member States shall communicate to the Commission the provisions of national law which they adopt in the field governed by this Directive.

Article 3

This Directive is addressed to the Member States.

Done at Brussels, 29 May 1991

ANNEX

INDICATIVE LIMIT VALUES FOR OCCUPATIONAL EXPOSURE

Einecs (1)	CAS (2)	Name of agent	Limit values (3)	
			mg/m³ (4)	ppm (5)
2 001 933	54-11-5	Nicotine (6)	0.5	–
2 005 791	64-18-6	Formic acid	9	5
2 005 807	64-19-7	Acetic acid	25	10
2 006 596	67-56-1	Methanol	260	200
2 008 352	75-05-8	Acetonitrile	70	40
2 018 659	88-89-1	Picric acid (6)	0.1	–
2 020 495	91-20-3	Naphtalene	50	10
2 027 160	98-95-3	Nitrobenzene	5	1
2 035 852	108-46-3	Resorcinol (6)	45	10
2 037 163	109-89-7	Diethylamine	30	10
2 038 099	110-86-1	Pyridine (6)	15	5
2 046 969	124-38-9	Carbon dioxide	9 000	5 000
2 056 343	144-62-7	Oxalic acid (6)	1	–
2 069 923	420-04-2	Cyanamide (6)	2	–
2 151 373	1305-62-0	Calcium dihydroxide (6)	5	–
2 152 361	1314-56-3	Disphosphorus pentaoxide (6)	1	–
2 152 424	1314-80-3	Disphosphorus pentasulphide (6)	1	–
2 152 932	1319-77-3	Cresols (all isomers) (6)	22	5
2 311 161	7440-06-4	Platinum (metallic) (6)	1	–
2 314 843	7580-67-8	Lithium hydride (6)	0.025	–
2 317 781	7726-95-6	Bromine (6)	0.7	0.1
2 330 603	10026-13-8	Phosphorus pentachloride (6)	1	–
2 332 710	10102-43-9	Nitrogen monoxide	30	25
	8003-34-7	Pyrethrum	5	–
		Barium (soluble compounds as Ba) (6)	0.5	–
		Silver (soluble compounds as Ag) (6)	0.01	–
		Tin (inorganic compounds as Sn) (6)	2	–

(1) Einecs: *European Inventory of Existing Chemical Substances.*
(2) CAS: *Chemical Abstract Service Number.*
(3) Measured or calculated in relation to a reference period of eight hours.
(4) Mg/m³ = milligrams per cubic metre of air at 20°C and 101.3 KPa (760 mm mercury pressure).
(5) Ppm = parts per million by volume in air (ml/m³).
(6) Existing scientific data on health effects appear to be particularly limited.

of 25 June 1991

amending Directive 83/477/EEC on the protection of workers from the risks
related to exposure to asbestos at work

(second individual Directive within the meaning of Article 8 of Directive 80/1107/EEC)

[91/382/EEC]

THE COUNCIL OF THE EUROPEAN COMMUNITIES,

Having regard to the Treaty establishing the European Economic Community,
and in particular Article 118a thereof,

Having regard to the proposal from the Commission drawn up following consultation with the Advisory Committee on Safety, Hygiene and Health Protection at
Work,

In co-operation with the European Parliament,

Having regard to the opinion of the Economic and Social Committee,

Whereas Article 118a of the Treaty provides that the Council shall adopt, by means
of directives, minimum requirements for encouraging improvements, especially
in the working environment, to ensure a better level of protection of the safety and
health of workers;

Whereas, under the terms of that Article, such directives are to avoid imposing administrative, financial and legal constraints in a way which would hold back the
creation and development of small and medium-size undertakings.

Whereas the communication from the Commission on its programme concerning
safety, hygiene and health at work provides for the adoption of directives designed
to guarantee the safety and health of workers;

Whereas the Council, in its resolution of 21 December 1987 on safety, hygiene and
health at work, took note of the Commission's intention of submitting to the
Council in the near future minimum requirements at Community level concerning protection against the risks resulting from dangerous substances, including
carcinogenic substances; whereas it considered that in this connection the principle of substitution using a recognized non-dangerous or less dangerous substance should be taken as a basis;

Whereas asbestos is a particularly hazardous agent which can cause serious illness
and which is found in various forms in a large number of circumstances at work;

Whereas, in view of the progress made in scientific knowledge and technology and
in the light of experience gained in applying Council Directive 83/447/EEC of 19
September 1983 on the protection of workers from the risks related to exposure
to asbestos at work (second individual Directive within the meaning of Article 8 of
Directive 80/1107/EEC), the protection of workers should be improved and the
action levels and limit values laid down in Directive 83/477/EEC should be
reduced;

Whereas the prohibition of the application of asbestos by means of the spraying process is not sufficient to prevent asbestos fibres being released into the atmosphere; whereas other working procedures that involve the use of certain materials containing asbestos must also be prohibited;

Whereas a decision cannot yet be taken establishing a single method for measurement of asbestos-in-air concentrations at Community level;

Whereas this Directive should be reviewed by 31 December 1995, taking account, in particular, of progress made in scientific knowledge and technology and of experience gained in applying this Directive;

Whereas Decision 74/325/EEC, as last amended by the 1985 Act of Accession, provides that the Advisory Committee on Safety, Hygiene and Health Protection at Work is to be consulted by the Commission for the purpose of drafting proposals in this field,

HAS ADOPTED THIS DIRECTIVE:

Directive 83/477/EEC is hereby amended as follows:

1.   Article 3(3) shall be replaced by the following:

'3.   If the assessment referred to in paragraph 2 shows that the concentration of asbestos fibres in the air at the place of work in the absence of any personal protective equipment is, at the option of the Member States, at a level as measured or calculated:

(a)   for chrysotile

— lower than 0,20 fibres per cm^3 in relation to an eight-hour reference period, and/or

— lower than a cumulative dose of 12,00 fibre-days per cm^3 over a three-month period;

(b)   for all other forms of asbestos either alone or in mixtures, including mixtures containing chrysotile:

— lower than 0,10 fibres per cm^3 in relation to an eight-hour reference period, and/or

— lower than a cumulative dose of 6,00 fibre-days per cm^3 over a three-month period,

Articles 4, 7, 13, 14(2), 15 and 16 shall not apply.'

Article 5 shall be replaced by the following:

'Article 5

The application of asbestos by means of the spraying process and working procedures that involve using low-density (less than $1g/cm^3$) insulating or sound-proofing materials which contain asbestos shall be prohibited.'

In point (1) of Article 7, the third paragraph shall be replaced by the following:

'In accordance with Article 118a of the Treaty and taking account in particular of progress made in scientific knowledge and technology and of experience gained in applying this Directive, the Council shall review the provisions of the first sentence of the first paragraph by 31 December 1995, with a view to establishing a single method for measurement of asbestos-in-air concentrations at Community level;'.

Article 8 shall be replaced by the following:

'Article 8

The following limit values shall be applied:

(a) concentration of chrysotile fibres in the air at the place of work:

0,60 fibres per cm^3 measured or calculated in relation to an eight-hour reference period;

(b) concentration in the air at the place of work of all other forms of asbestos fibres, either alone or in mixtures, including mixtures containing chrysotile:

0,30 fibres per cm^3 measured or calculated in relation to an eight-hour reference period.'

Article 9 shall be replaced by the following:

'Article 9

1. Without prejudice to the third paragraph of point 1 of Article 7, in accordance with Article 118a of the Treaty and taking account in particular of progress made in scientific knowledge and technology and of experience gained in applying this Directive, the Council shall review the provisions of this Directive by 31 December 1992.

2. The amendments required to adapt the Annexes to this Directive to take account of technical progress shall be made in accordance with the procedure described in Articles 9 and 10 of Council Directive 80/1107/EEC of 27 November 1980 on the protection of workers from the risks related to exposure to chemical, physical and biological agents at work.

6. Article 12 is hereby amended as follows:

(a) the following sub-paragraph shall be added to paragraph 2:

'At the request of the competent authorities, the plan shall include information on the following:

— the nature and probable duration of the work,

— the place where the work is carried out,

— the methods applied where the work involves the handling of asbestos or of materials containing asbestos,

— the characteristics of the equipment used for:

— protection and decontamination of those carrying out the work,

— protection of other persons present on or near the worksite.'.

(b) the following paragraph shall be added:

'3. At the request of the competent authorities, the plan referred to in paragraph 1 must be notified to them before the start of the projected work.'.

Article 2

1. Member States shall bring into force the laws, regulations and administrative provisions necessary to comply with this Directive not later than 1 January 1993.

They shall forthwith inform the Commission thereof.

When Member States adopt these measures, they shall contain a reference to this Directive or shall be accompanied by such reference on the occasion of their official publication. The methods of making such a reference shall be laid down by the Member States.

The date 1 January 1993 shall, however, be replaced by 1 January 1996 in the case of asbestos-mining activities.

However, as regards the Hellenic Republic:

— the date referred to in the first sub-paragraph shall be 1 January 1996,

— the date referred to in the fourth sub-paragraph shall be 1 January 1999.

2. Member States shall communicate to the Commission the provisions of national law which they adopt in the field governed by this Directive.

Article 3

This Directive is addressed to the Member States.

Done at Luxembourg, 25 June 1991.

COUNCIL DIRECTIVE

of 25 June 1991

supplementing the measures to encourage improvements in the safety and
health at work of workers with a fixed-duration employment relationship or a
temporary employment relationship
[91/383/EEC]

THE COUNCIL OF THE EUROPEAN COMMUNITIES,

Having regard to the Treaty establishing the European Economic Community,
and in particular Article 118a thereof,

Having regard to the proposal from the Commission,

In co-operation with the European Parliament,

Having regard to the opinion of the Economic and Social Committee,

Whereas Article 118a of the Treaty provides that the Council shall adopt, by means
of Directives, minimum requirements for encouraging improvements, especially
in the working environment, to guarantee a better level of protection of the safety
and health of workers;

Whereas, pursuant to the said Article, Directives must avoid imposing administrative, financial and legal constraints which would hold back the creation and development of small and medium-sized undertakings;

Whereas recourse to forms of employment such as fixed-duration employment
and temporary employment has increased considerably;

Whereas research has shown that in general workers with a fixed-duration employment relationship or temporary employment relationship are, in certain sectors,
more exposed to the risk of accidents at work and occupational diseases than other
workers;

Whereas these additional risks in certain sectors are in part linked to certain
particular modes of integrating new workers into the undertaking; whereas these
risks can be reduced through adequate provision of information and training from
the beginning of employment;

Whereas the Directives on health and safety at work, notably Council Directive 89/
391/EEC of 12 June 1989 on the introduction of measures to encourage improvements in the safety and health of workers at work, contain provisions intended to
improve the safety and health of workers in general;

Whereas the specific situation of workers with a fixed-duration employment
relationship or a temporary employment relationship and the special nature of the
risks they face in certain sectors calls for special additional rules, particularly as
regards the provision of information, the training and the medical surveillance of
the workers concerned;

Whereas this Directive constitutes a practical step within the framework of the
attainment of the social dimension of the internal market,

HAS ADOPTED THIS DIRECTIVE:

SECTION I
SCOPE AND OBJECT

Article 1
Scope

This Directive shall apply to:

1. employment relationships governed by a fixed-duration contract of employment concluded directly between the employer and the worker, where the end of the contract is established by objective conditions such as: reaching a specific date, completing a specific task or the occurrence of a specific event;

2. temporary employment relationships between a temporary employment business which is the employer and the worker, where the latter is assigned to work for and under the control of an undertaking and/or establishment making use of his services.

Article 2
Object

1. The purpose of this Directive is to ensure that workers with an employment relationship as referred to in Article 1 are afforded, as regards safety and health at work, the same level of protection as that of other workers in the user undertaking and/or establishment.

2. The existence of an employment relationship as referred to in Article 1 shall not justify different treatment with respect to working conditions in as much as the protection of safety and health at work are involved, especially as regards access to personal protective equipment.

3. Directive 89/391/EEC and the individual Directives within the meaning of Article 16(1) thereof shall apply in full to workers with an employment relationship as referred to in Article 1, without prejudice to more binding and/or more specific provisions set out in this Directive.

SECTION II
GENERAL PROVISIONS

Article 3
Provision of information to workers

Without prejudice to Article 10 of Directive 89/391/EEC, Member States shall take the necessary steps to ensure that:

1. before a worker with an employment relationship as referred to in Article 1 takes up any activity, he is informed by the undertaking and/or establishment making use of his services of the risks which he faces;

2. such information:

    — covers, in particular, any special occupational qualifications or skills or special medical surveillance required, as defined in national legislation, and

    — states clearly any increased specific risks, as defined in national legislation, that the job may entail.

Article 4
Workers' training

Without prejudice to Article 12 of Directive 89/391/EEC, Member States shall take the necessary measures to ensure that, in the cases referred to in Article 3, each worker receives sufficient training appropriate to the particular characteristics of the job, account being taken of his qualifications and experience.

Article 5
Use of workers' services and medical surveillance of workers

1. Member States shall have the option of prohibiting workers with an employment relationship as referred to in Article 1 from being used for certain work as defined in national legislation, which would be particularly dangerous to their safety or health, and in particular for certain work which requires special medical surveillance, as defined in national legislation.

2. Where Member States do not avail themselves of the option referred to in paragraph 1, they shall, without prejudice to Article 14 of Directive 89/391/EEC, take the necessary measures to ensure that workers with an employment relationship as referred to in Article 1 who are used for work which requires special medical surveillance, as defined in national legislation, are provided with appropriate special medical surveillance.

3. It shall be open to Member States to provide that the appropriate special medical surveillance referred to in paragraph 2 shall extend beyond the end of the employment relationship of the worker concerned.

Article 6
Protection and prevention services

Member States shall take the necessary measures to ensure that workers, services or persons designated, in accordance with Article 7 of Directive 89/391/EEC, to carry out activities related to protection from and prevention of occupational risks are informed of the assignment of workers with an employment relationship as referred to in Article 1, to the extent necessary for the workers, services or persons designated to be able to carry out adequately their protection and prevention activities for all the workers in the undertaking and/or establishment.

SECTION III
SPECIAL PROVISIONS

Article 7
Temporary employment relationships: information

Without prejudice to Article 3, Member States shall take the necessary steps to ensure that:

1. before workers with an employment relationship as referred to in Article 1(2) are supplied, a user undertaking and/or establishment shall specify to the temporary employment business, *inter alia*, the occupational qualifications required and the specific features of the job to be filled;

2.   the temporary employment business shall bring all these facts to the attention of the workers concerned.

Member States may provide that the details to be given by the user undertaking and/or establishment to the temporary employment business in accordance with point 1 of the first sub-paragraph shall appear in a contract of assignment.

Article 8
Temporary employment relationships: responsibility

Member States shall take the necessary steps to ensure that:

1.   without prejudice to the responsibility of the temporary employment business as laid down in national legislation, the user undertaking and/or establishment is/ are responsible, for the duration of the assignment, for the conditions governing performance of the work;

2.   for the application of point 1, the conditions governing the performance of the work shall be limited to those connected with safety, hygiene and health at work.

SECTION IV
MISCELLANEOUS PROVISIONS

Article 9
More favourable provisions

This Directive shall be without prejudice to existing or future national or Community provisions which are more favourable to the safety and health protection of workers with an employment relationship as referred to in Article 1.

Article 10
Final provisions

1.   Member States shall bring into force the laws, regulations and administrative provisions necessary to comply with this Directive by 31 December 1992 at the latest. They shall forthwith inform the Commission thereof.

When Member States adopt these measures, the latter shall contain a reference to this Directive or shall be accompanied by such reference on the occasion of their official publication. The methods of making such a reference shall be laid down by the Member States.

2.   Member States shall forward to the Commission the texts of the provisions of national law which they have already adopted or adopt in the field covered by this Directive.

3.   Member States shall report to the Commission every five years on the practical implementation of this Directive, setting out the points of view of workers and employers.

The Commission shall bring the report to the attention of the European Parliament, the Council, the Economic and Social Committee and the Advisory Committee on Safety, Hygiene and Health Protection at Work.

4. The Commission shall submit to the European Parliament, the Council and the Economic and Social Committee a regular report on the implementation of this Directive, due account being taken of paragraphs 1, 2 and 3.

Article 11

This Directive is addressed to the Member States.

Done at Luxembourg, 25 June 1991